ital Electronics

171 910

Digital Electronics

D. C. Green
MTech, CEng, MIEE

5th Edition

NORWICH CITY COLLEGE LIBRARY			
Stock No.	171910		
Cla	621·3815 GRE		
Cat.		Proc.	

Addison Wesley Longman Limited
Edinburgh Gate, Harlow
Essex CM20 2JE
England

and Associated Companies throughout the World

© D. C. Green 1982, 1986
© Addison Wesley Longman Limited 1988, 1993, 1999

The right of D. C. Green to be identified
as author of this Work has been asserted by him in
accordance with the Copyright, Design and
Patents Act 1988.

All rights reserved; no part of this publication may be
reproduced, stored in any retrieval system, or transmitted
in any form or by any means, electronic, mechanical,
photocopying, recording, or otherwise without either the prior
written permission of the Publishers or a licence permitting
restricted copying in the United Kingdom issued by the
Copyright Licensing Agency Ltd., 90 Tottenham Court Road,
London W1P 9HE.

First published 1982 under the title *Digital Techniques and Systems* by Pitman Publishing Limited

Second edition 1986
Third edition 1988
Fourth edition 1993
Fifth edition 1999

ISBN 0 582 317363

British Library Cataloguing-in-Publication Data
A catalogue record for this book is
available from the British Library.

Set by 35 in 10/12 pt Times
Printed in Malaysia, PP

Contents

Preface

This book provides a comprehensive coverage of the basic techniques and circuits employed in modern digital electronics.

Chapter 1 gives an introduction to both digital signals and to some of the many varied applications of digital technology. This chapter also introduces the reader to the practical exercises that are included in the majority of the following chapters. Chapter 2 is concerned with binary and hexadecimal numbers and arithmetic. Chapters 3 through to 8 consider the circuits that are employed in combinational logic designs, such as gates, adders, multiplexers and decoders. Throughout these chapters the emphasis has been placed on the 74LS logic family because devices in this family are readily available, are cheap, and, most importantly for someone learning about digital electronics, are easy to use. Chapters 9, 10 and 11 consider sequential logic circuits, such as flip-flops, counters and shift registers, while chapters 12 and 13 cover memories and programmable logic devices, respectively. Visual displays are the concern of chapter 14, and both analogue-to-digital and digital-to-analogue converters are the subject of chapter 15.

The final chapter, 16, in the book gives a number of exercises that can be carried out using the electronic design software package Electronics Workbench. Each exercise in this chapter corresponds to a practical exercise in an earlier chapter and it has been given the same number, e.g. exercise EWB 10.1 corresponds to practical exercise 10.1. A reader with access to a PC and Electronics Workbench can download the Electronics Workbench files from Addison Wesley Longman's website at ftp://ftp.awl.co.uk/pub/awl-he/engineering/green/dig_electron. The circuit to be investigated will then appear in the circuit window. In most cases the same IC(s) is/are employed in both the practical exercise and the Electronics Workbench exercise; when this is not the case it is because the IC in question is not included in the Electronics Workbench library.

Throughout the book, digital electronic devices are shown using either one, or both, of their IEC symbols and their pinouts. The IEC symbols are the standard method of representing devices and they are increasingly employed. Any student of digital electronics needs to become well acquainted with their use.

A large number of worked examples are given throughout the text to illustrate various points. Most chapters include one, or more, practical exercises; these exercises require only the use of components and equipment that should be available in any laboratory where such practical work is carried out. Hopefully, the practical exercises will provide some familiarity with the handling of digital ICs and other components, and with the use of instruments such as the CRO. Each chapter concludes with a number of exercises and the majority of them are provided with a worked solution that will be found at the back of the book.

D.C.G.

Preface

1 Digital signals and systems

After reading this chapter you should be able to:

(a) Name some of the applications of modern digital electronics.
(b) State some of the advantages of using digital technology.
(c) Describe the parameters of digital waveforms.
(d) Convert between frequency and period for a periodic clock waveform.
(e) Sketch the timing waveform for any digital waveform.
(f) Understand the requirement for two-state devices in digital electronics.

The sounds produced by the human voice and by musical instruments vary continuously in both amplitude and frequency and they are said to be *analogue signals*. In a communication system sound waves are applied to a telephone transmitter, or to a microphone, to be converted into an analogue electrical signal whose waveform is (assuming zero distortion) a replica of the incident sound waveform. At the other end of the system the analogue electrical signal must be applied to a telephone receiver, or to a loudspeaker, for audible sounds to be produced. Most naturally occurring physical quantities are analogue in their nature in that their value, be it pressure, velocity, weight, or some other quantity, vary continuously with time. Analogue signals include sinusoidal, exponential, and triangular waveforms. A circuit that preserves the waveform of an analogue signal as it is processed is said to be an analogue (or linear) circuit. Examples of linear circuits that are to be found in the home include a radio receiver and the television receiver (although, ever increasingly, more and more of their internal functions are being carried out using digital techniques). Other well-known examples of analogue devices are the speedometer and the petrol gauge in the dashboard of a car, both of which give a continuously varying indication.

The amplitude of a *digital signal* does *not* vary continuously; instead it may only take up any one of a number n of defined values. The amplitude of the digital signal will suddenly change from one value to another and it will never have an undefined value. The number n of defined values could, for example, be 10 to give a decimal digital system. However, decimal digital electronic systems are not employed in practice, since it is difficult to design electronic circuitry that could accurately, and consistently, differentiate between the 10 voltage levels that would have to be employed. A *binary digital system* has $n = 2$ and so it may only have either one of two values; it may either be HIGH or it may be LOW. The use of binary digital signals allows relatively simple *two-state devices* to be employed which can only ever be in either one of two possible states, ON or OFF.

Examples of two-state devices are: an LED which is either glowing visibly or is dim; a buzzer which is either producing an audible sound or is not; an electrical switch which either completes or breaks a circuit; or a semiconductor diode that is either fully conducting (turned ON) or is non-conducting (turned OFF).

The advantages to be gained from the use of digital techniques instead of analogue techniques arise largely from the use of just the two voltage levels. Digital circuitry operates by switching transistors ON and OFF and does not need to produce or to detect precise values of voltage and/or current at particular points in an equipment or system. Because of this it is easier and cheaper to mass-produce digital circuitry than it is analogue circuitry. Also, the binary nature of the signals makes it much easier to consistently obtain a required operating performance from a large number of circuits. Digital circuits are more reliable than analogue circuits because faults will not often occur through variations in performance caused by changing values of components, misaligned coils, and so on. Again, the effects of noise and interference are very much reduced in a digital system since the digital pulses can always be regenerated and made like new whenever their waveshape is becoming distorted to the point where errors are likely. This is not possible in an analogue system where the effects of unwanted noise and interference signals is to degrade the signal permanently.

Two-state devices

Ideal two-state device

The ideal two-state device will have the following characteristics: (a) be turned ON or OFF by binary 1 and 0 logic levels, (b) change from one state to the other instantaneously, and (c) have infinite resistance when OFF and zero resistance when ON.

A manual switch has the disadvantages of slow speed of operation, large physical size and contact bounce, but it will satisfy requirement (c). Contact bounce is the creation of sporadic and irregular voltage pulses as the switch contact(s) is/are made or broken. As the switch is operated its contact(s) may make and break several times, with each make and break producing a voltage pulse. The voltage pulse may appear as several unwanted pulses which may well cause erroneous operation of a digital circuit. To prevent this from happening a *debouncing circuit* is often employed (p. 207).

Transistor switches

Bipolar and field effect transistors can be switched very rapidly by an applied voltage. A logic 1 voltage will turn the transistor ON, when a large collector, or drain, current will flow to cause the transistor to *saturate*. A logic 0 voltage applied to the transistor will turn it OFF and then zero current flows through the device. High-speed switching is necessary to reduce the power dissipated in the transistor as it changes state; when it is ON or OFF the power dissipated is very small since either the current through, or the voltage across, the device is very small. However, neither type of transistor is able to satisfy requirement (c) fully.

The signal applied to a transistor that is employed as a two-state device either turns the transistor full ON or it turns the transistor OFF; when the transistor is ON its output voltage is LOW, at the saturation value $V_{CE(SAT)}$, and when the transistor is OFF its output voltage is HIGH, at the power supply voltage V_{CC}. Most digital circuitry employs *positive logic* in which a HIGH voltage represents logic 1 and a LOW voltage represents logic 0. *Negative logic* in which a HIGH voltage represents logic 0 and a LOW voltage represents logic 1 is used in digital communications and a few digital circuits.

Digital signals

A binary digital signal voltage may be either *unipolar* or *bipolar*. A unipolar signal uses only one polarity voltage, either positive or negative, and it switches between this voltage and (nominally) zero voltage. A bipolar signal switches between a positive voltage and a negative voltage.

Figures 1.1(a) and (b) shows two examples of unipolar binary digital voltages and Fig. 1.2 shows an example of a bipolar binary digital voltage.

Fig. 1.1 *Unipolar digital voltage*

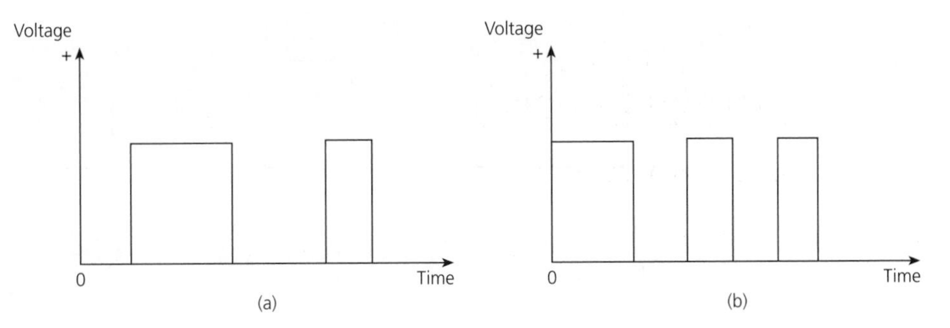

Fig. 1.2 *Bipolar digital voltage*

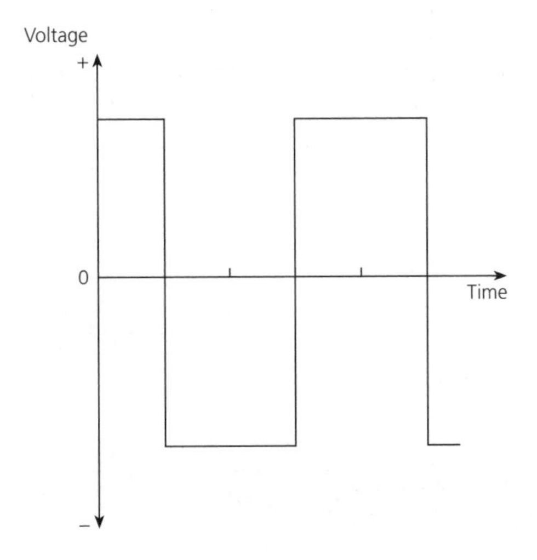

A digital signal consists of a number of logic 1s and 0s that may represent numbers, letters, symbols or control signals. Each 1 or 0 is a binary digit or *bit* and a combination of eight bits, which make a *byte*, may be used to represent one character in the *American Standard Code for Information Interchange* (ASCII) system. Three examples of the ASCII code are shown in Fig. 1.3. A digital circuit responds to the logic level of each bit and not the actual voltage which is allowed to vary between limits that are discussed in chapter 5.

Many digital circuits are required to be timed precisely and this timing is provided by a *clock* waveform. Figure 1.4 shows a clock waveform; the waveform repeats itself at regular time intervals, and so it is said to be *periodic*. The *period T* of the clock waveform is the length of time from the leading edge of one pulse to the leading edge of the next pulse. The *clock frequency* is equal to the reciprocal of the clock period

$$f = 1/T \tag{1.1}$$

Fig. 1.3 *ASCII code examples*

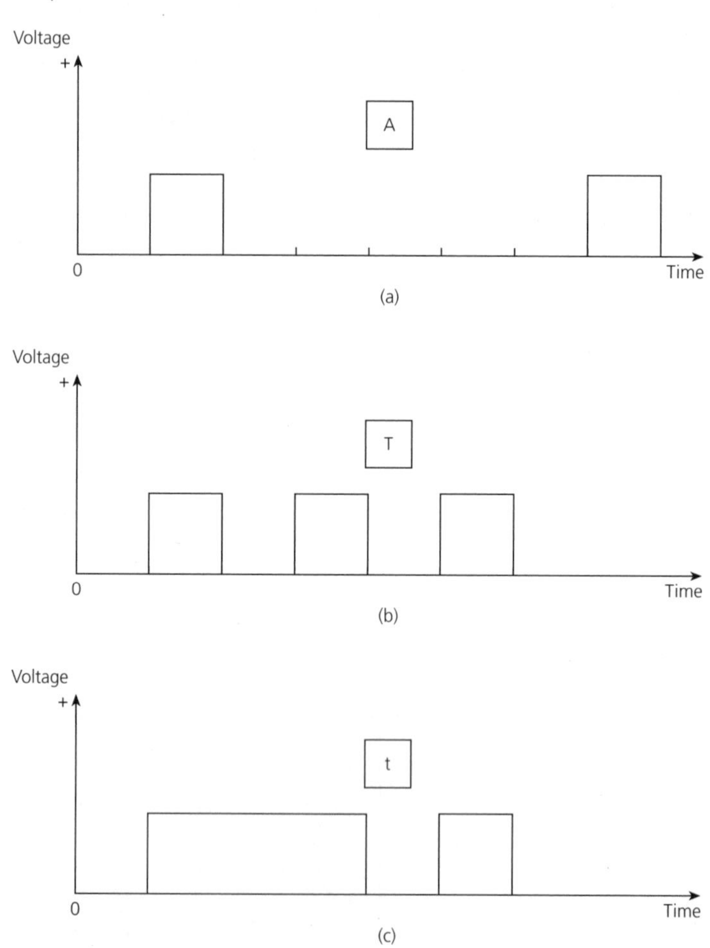

EXAMPLE 1.1

Calculate the frequency of the clock waveform whose period is 1 μs.

Solution

$f = 1/T = 1/(1 \times 10^{-6}) = 1$ MHz (*Ans.*)

Duty cycle and mark–space ratio

Referring to Fig. 1.4, the *duty cycle* of a rectangular waveform is the ratio t_1/T. The *mark–space ratio* is the ratio t_1/t_2. For a square waveform, $t_1 = t_2$ and then the duty cycle = 0.5 and the mark–space ratio = 1.

Rise-time and fall-time

A practical digital waveform cannot have its leading edges and trailing edges rise or fall instantaneously. Instead a practical pulse waveform will be of the form shown in Fig. 1.5. The rise-time t_r of the waveform is the time it takes for the voltage to increase from 10 to 90% of its peak value. Conversely, the fall-time t_f is the time in which the voltage falls from 90 to 10% of its peak value. Typical values for digital signals are a few nanoseconds.

The *pulse width* is the time for which the amplitude of the pulse is greater than 50% of its peak value.

Fig. 1.4 *Duty cycle and mark–space ratio*

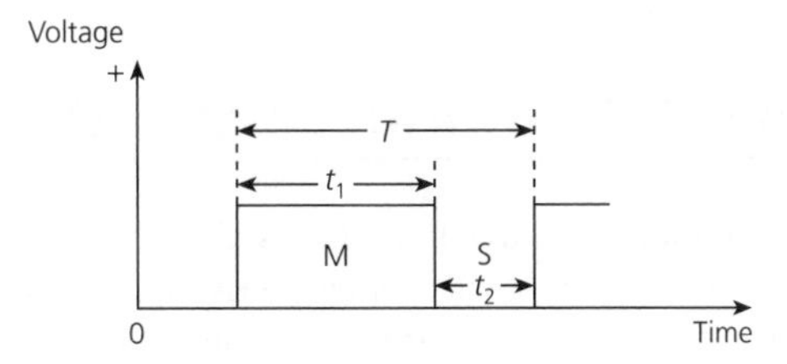

Fig. 1.5 *Rise-time and fall-time*

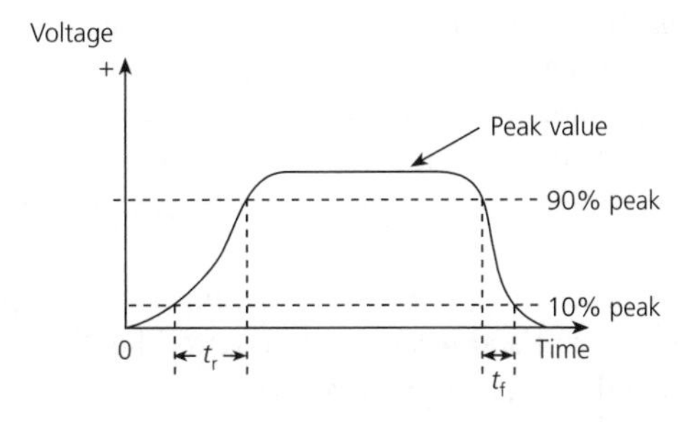

Fig. 1.6 *Transmission of data word 00100110: (a) serial and (b) parallel*

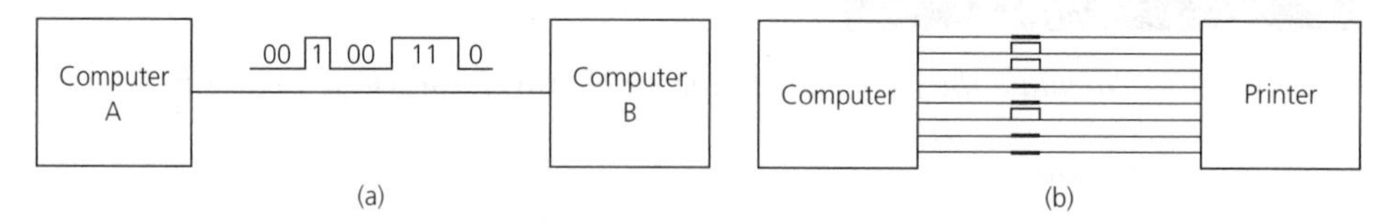

Serial and parallel transmission of data

Data may be transmitted from one point to another using either serial or parallel transmission. Serial transmission employs a single conductor and earth return, or a pair of conductors, as the data path. The bits making up a digital word are transmitted sequentially. Parallel transmission of data uses a separate conductor for each bit in a digital word. This means that an 8-bit word needs eight conductors for its transmission.

Serial transmission is relatively slow but it is much cheaper. Parallel transmission is used for the movement of data within a computer and usually to and from a nearby peripheral, such as a printer. The groups of conductors employed for this purpose are known as a *bus*, and are either 16 or 32 bits wide, and in some larger computers perhaps 64 bits wide. The difference between serial and parallel transmission is shown in Figs 1.6(a) and (b).

Digital media

Digital media is the integration of different formats of data including text, images, graphics, video and audio. It provides a means of presenting and communicating information in an easily understood format.

The digital computer and the microprocessor

A digital computer is a machine that is able to input data and then process it in accordance with the *program* that it is currently running, and then output the result(s) of the operation either to memory or to a peripheral device such as a printer. A program is a set of instructions that tells the computer what it is to do.

The digital computer is an integral part of the day-to-day operation of many firms and organizations, ranging from Government departments, commercial concerns such as banks and insurance companies, to industrial firms in all branches of engineering and science. Computers are employed for the calculation of wages and salaries, taxes, pensions, bills and accounts; for the storage of medical, scientific and engineering data; and for the rapid booking of aircraft seats, theatre tickets and foreign holidays. Computers are also used to carry out complex scientific and engineering calculations, to control engineering processes in factories, to control the operation of telephone exchanges and military equipment and weapons, to control the distribution networks for gas, water and electricity, and for many other purposes.

Fig. 1.7 *Digital computer*

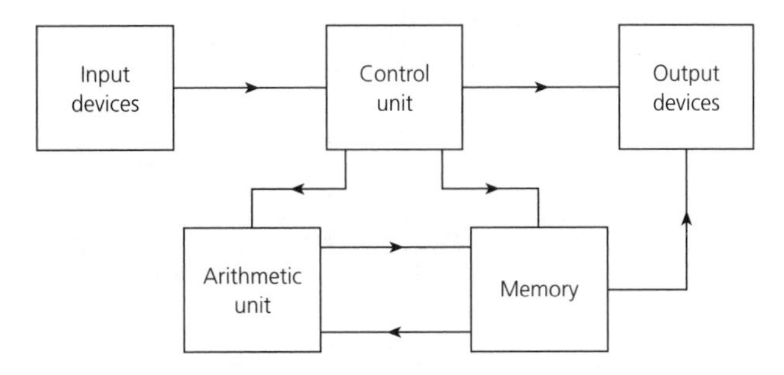

Perhaps the most commonly met example is the cash dispenser that is mounted in the wall of so many bank and building society branches. The dispenser is linked by a telephone line to a main frame computer; when a cash request is made the computer will check the account of the card holder and if it contains sufficient funds will send back a signal to the cash dispenser authorizing it to pay out the sum requested.

A block diagram of a digital computer is shown in Fig. 1.7. Some of the memory is provided internally to the computer in ICs known as RAM and ROM, or on a hard disc, while the remainder is provided externally to the computer in a floppy disc or some kind of magnetic storage.

The instructions contained in the program are taken sequentially (one after the other) from the memory under the direction of the control unit. Each instruction causes the arithmetic unit to perform arithmetic and logic operations on the data, also taken from the memory. The results of the calculations can be stored in the memory or they can be held temporarily in a part of the arithmetic unit known as an accumulator. When a calculation has been completed, the control unit will transfer the results either to a memory device or to an output device, which will (probably) produce the results in printed form. Alternatively, a VDU may be employed to give a visual display of the results, or the results may be transmitted over a *data link* to a distant point where they are needed. The input devices used to feed information into a computer are usually a keyboard, a disc drive, or a modem.

Microprocessors

A microprocessor is an integrated circuit that is able to control the operation of a wide variety of equipment. A microprocessor can be built into the equipment whose operation it is to control. Figure 1.8 shows the basic block diagram of a microprocessor system; the microprocessor chip contains various registers, an arithmetic unit and control circuitry. The memory and the input/output interface circuits are also provided by integrated circuits.

As an example, microprocessor control of modern radio receivers and systems is increasingly used in the latest equipment. A microprocessor can be programmed to control the tuning, the gain and the selectivity of a receiver as the receiving conditions alter. Remote HF radio stations can be distantly controlled by means of a microprocessor; the functions controlled being the selection of the frequency to be transmitted

Fig. 1.8 *Microprocessor system*

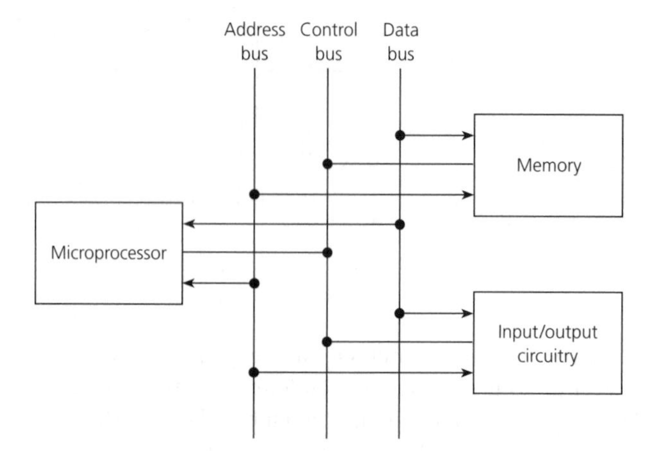

and the frequency to be received as determined by a pre-set schedule. Also, the performance of the station can be monitored and any faults or degradation of service detected and recorded.

Other examples are: (a) an automatic washing machine which goes through its wash cycle under the control of an in-built microprocessor, and (b) a gas-fired central heating system.

The vast majority of data terminals employ the *International Alphabet 5* code (IA5); this is a 7-bit code that is commonly known as the *ASCII* code (short for American Standard Code for Information Interchange) (see Appendix A). Sometimes it is necessary to avoid having two, or more, bits changing value between successive numbers. Then, the *Gray code* is employed (see Appendix B).

The Internet

The *Internet* is a network of linked computers which are located at different points all over the world that provides easy communication between persons and organizations no matter where they are located. Schools, colleges, universities, offices, factories and individual persons can gain access to information and very fast bi-directional communication.

Both the telephone and FAX are limited in their applications because they both deal with only one kind of information. When a computer is connected to the Internet it is able to do everything a telephone and a FAX system can do, plus exchanging any of the following: computer software, documents, drawings, photographs, sound signals, text information and video clips. The Internet is able to make copies of any information available to many people at the same time and/or change the information in any way required.

The Internet has three main parts: (a) Electronic mail, (b) Newsgroup, and (c) the World Wide Web. These three parts may be used separately or all together as required.

Electronic mail

Electronic mail (e-mail) is the electronic equivalent of the ordinary mail that delivers letters to the home. An e-mail user with a message to send types the message into their computer, specifies the destination address, and then transmits the message into the Internet. Within a few seconds (usually) the message will arrive at its destination. Unlike a telephone call, an e-mail message does not require immediate attention but it can wait in the computer until such time as the recipient looks at it. This aspect of e-mail is somewhat similar to FAX but there is no need for the message to be printed out and, if it is, the print quality is generally much better.

E-mail can be *off-line* or *on-line*. With off-line operation a message is typed before a connection to the Internet is established. All messages can then be transmitted together when a connection is set up. At the same time any messages that are waiting will be down-loaded into the computer and can be read. With on-line operation e-mail messages are typed and transmitted while the computer is connected to the Internet and any incoming messages are down-loaded immediately. Off-line operation is cheaper than on-line.

The address of an e-mail customer is of the form: name@where.co.uk (assuming that the address is in the UK).

Newsgroups

Newsgroups – also known as *Usenet* – offer places, or discussion areas, in which people can exchange views and information and are hence a sort of public electronic bulletin board. When a message is sent to a discussion area anyone can read that message. If anyone replies to the message the reply will also be readable by everyone who accesses that area. Newsgroups are used for entertainment, for debate, and for providing answers to queries over a very wide range of topics.

World Wide Web

The World Wide Web (WWW) is a sort of electronic magazine that contains contributions from a very large number of sources. Each location on the WWW, known as a *web site*, contains links (highlighted words or phrases) that can be clicked with a mouse to lead to further related topics. These, in turn, contain links to even more topics and so on. Each web site has a unique address; a typical address is http://www.name.co.uk.

The WWW is open to everyone, and anyone can provide information, i.e. create their own web site. Typical information to be found at web sites includes: details of holidays and flights available from travel agents/firms, stock market prices, summaries of research reports and weather forecasts.

Internet access

The basic block diagram of the Internet is shown in Fig. 1.9. The large computer system consists of powerful mini and main-frame computers that are permanently linked together by very high-speed data links. The large computer sites in the UK are fully interconnected and also permanently linked to the Internet in the USA by submarine fibre optic cable. The computers operate for 24 hours a day and 365 days a year. To

Fig. 1.9 *The Internet*

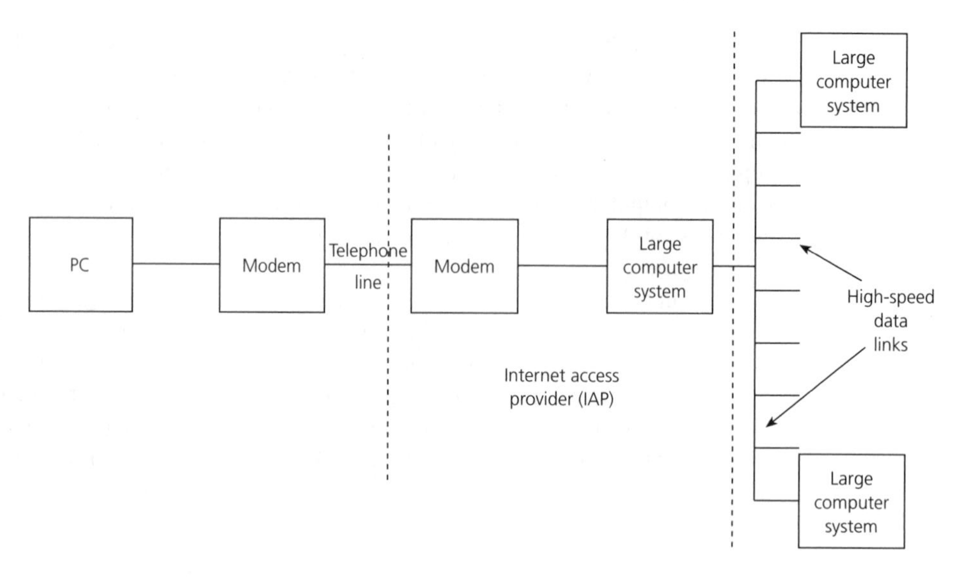

access the Internet a user requires a modern telephone socket connected to the *public switched telephone network* (PSTN), a computer, a modem and suitable software. It is also necessary to subscribe to one of the *Internet Access Providers* (IAPs). In the UK, BT offers several services that provide access to the Internet. These services include: (a) BT Internet, which offers dial-up access to customers, (b) BTNet Internet, which connects business customer's private networks to the Internet, and (c) Dial IP, which provides dial-up access to host computer sites using Internet technology.

An *on-line service* (OLS), e.g. Compuserve and AOL, provides other services as well as Internet access. An OLS is perhaps more like an electronic magazine that offers Internet access as an extra. The block diagram of an OLS is the same as Fig. 1.9 except that OLS replaces IAP. An OLS system includes magazine articles, e-mail and facilities for inter-connecting customers, as well as providing Internet access. It is easier to use an OLS than a IAP because an OLS provides an on-screen index that lists all the facilities available, but it is more expensive to use than an IAP.

Digital equipment and systems

Hand-calculators are nowadays in common use and provide another example of the use of digital circuitry. All calculators are able to carry out the basic mathematical procedures, while many are provided with several more advanced mathematical facilities. Some models are programmable. The circuitry contained within a calculator is complex and these devices have only become practical since the advent of integrated circuits.

Many cash registers and weighing scales used in the shops are electronic and these provide a readout in digital form of the total money to be paid and, once the money offered has been entered by the operator, of the change to be given, as well as a printout of the purchases.

Fig. 1.10 *Indication of measured quantity by (a) analogue and (b) digital method*

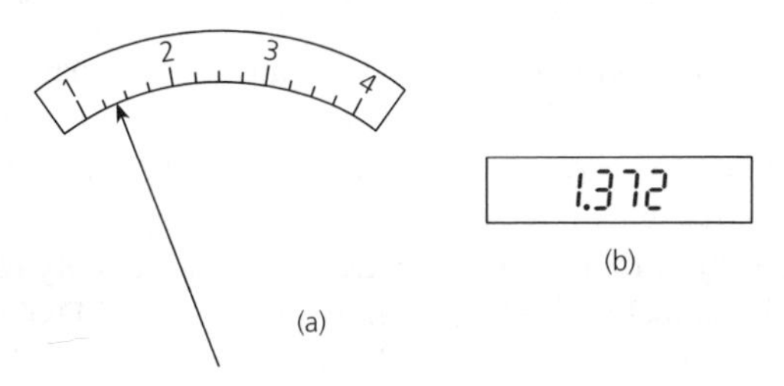

Electronic point-of-sale terminals are used by supermarkets and department stores. The operator has to identify the item being purchased and enter its details and price, often by scanning its bar code, when its description and its price are displayed upon a screen. The terminal will total the costs of the items purchased and print out a detailed receipt. The terminal is usually linked to a computer which maintains a record of the items sold and compares it with the remaining stock levels and automatically orders new supplies when necessary.

Electronic fund-transfer-at-point-of-sale (EFTPoS) enables a customer in a shop to pay for goods using a credit or debit card. The shop assistant merely has to swipe the plastic card through a card reader and this reads off the customer's name and the card number before it prints out a voucher which the customer is required to sign. Details of the transaction are transmitted over a telephone line to the card company's computer centre.

In engineering, digital readouts of data are often more convenient and accurate than analogue readings. Digital voltmeters and frequency meters or counters are particularly suited to measurement applications where a large number of repetitive readings are to be made. The advantage of digital meters is most noticeable when relatively unskilled personnel are employed to carry out the tests. With a digital instrument the operator can read at a glance the value of the displayed parameter, but very often an analogue reading requires care if reasonable accuracy is to be obtained. This point is illustrated by Fig. 1.10. When the pointer (Fig. 1.10a) is in between two scale markings, some doubt exists as to the value indicated. No such doubt is present with the digital instrument; its indicated value (Fig. 1.10b) is easy to read. Similarly, digital thermometers and weighing scales are in common use and are easier to read than the older analogue versions.

Data-loggers are digital circuits that convert the analogue output from a transducer (a resistance strain gauge, a thermocouple, or a potentiometer, for example) into a digital form so that the measured parameter can be recorded on a magnetic tape.

Another application of computers and digital techniques that is gaining in importance in the modern world is in the field of transport. The movement of vehicles in a large transport system can be controlled and monitored by a computer. Railway companies, for example, have introduced a computerized system for the optimized control of their freight traffic. Each truck has its movements continuously monitored and the computer works out and augments the best way of moving the trucks around the network in order to carry the maximum amount of freight in the most economic manner. In many large cities, computers control the traffic lights that direct the flow of traffic across road junctions. The computer continuously monitors the number of cars passing and waiting to pass the various junctions and varies the frequency of the traffic light operations to optimize the flow of traffic.

Compact discs and digital audio tape

The traditional methods of recording music on to magnetic tape use a recording head to convert the analogue sound signal directly into the corresponding analogue magnetic signal. Later, the music is re-created by passing the magnetic tape over a play head to convert the magnetic signal into sound. The recording process and the later re-play of the music are both subject to noise interference and better quality recordings can be obtained by the use of digital techniques. For both compact disc (CDs) and digital audio tape (DAT) the electrical signal is first converted into digital form and the digitized information is stored on the CD or DAT. When the CD is later inserted into a CD player the digital signal is first converted back to analogue form before the music signal is sent to the loudspeaker.

Telephone transmission systems

Speech signals can be transmitted over purely analogue circuits, but digital transmission using *pulse code modulation* (PCM) is employed for the core network in the UK. At present the access network is still predominantly analogue, but the eventual aim is for this also to be digital. ISDN is an integrated telephone network in which all signal transmission, signalling, and exchange switching is achieved using digital techniques. One advantage of digital techniques is that they allow all kinds of signals, such as speech, music, television, telegraphy and data, to be transmitted over the same circuit. The distribution of VHF sound broadcast signals and the audio signals of television broadcasts from studios to transmitters is carried out digitally using PCM. Recently, the introduction of digital terrestrial television has been announced.

Teletext services are transmitted by both the BBC and the IBA – using the names CEEFAX and ORACLE, respectively – to provide information to the home. The data are transmitted digitally using some of the lines in each field of the television signal which are not modulated by the video signal. In the television receiver a digital decoder is provided to recover and display the incoming information on the television screen.

Private mobile radio systems operating in the VHF and the UHF bands can also be controlled by a computer to ensure optimum performance as the mobiles move around the service area. Such systems are used by organizations that employ a large number of mobiles, such as BT in the UK. A public mobile radio system, integrated with the telephone network and known as cellular radio, also depends upon computer control. The older systems employ analogue techniques but the newer global system for mobile communications (GSM) employs digital circuitry throughout.

Practical exercises

Most of the chapters in this book include a number of practical exercises. The components required to carry out these practical exercises are as follows:
Resistors: eight 270 Ω, two 1 kΩ, three 5.1 kΩ, two 7.5 kΩ, one 10 kΩ, two 20 kΩ, one 39 kΩ and one 82 kΩ.
Capacitors: one each of 220 pF, 0.01 μF, 0.1 μF, and 1 μF.

Diodes: two 1N914 signal diodes and five TIL 209 LEDs.
Integrated digital circuits: two 74LS00 quad 2-input NAND gate, one 74LS02 quad 2-input NOR gate, one 74LS08 quad 2-input AND gate, two 74LS27 triple 3-input NOR gate, one 74LS32 quad 2-input OR gate, one 74LS47 LED decoder/driver, one 74LS74 dual D flip-flop, one 74LS75 quad D latch, two 74LS76 dual J-K flip-flop, one 74LS90 decade counter, two 74LS93 binary counters, one 74LS95 shift register, one 74LS112A dual J-K flip-flop, one 74LS139 dual 2-to-4 decoder/demultiplexer, one 74LS153 dual 4-to-1 multiplexed, one 74LS161A 4-bit synchronous counter, one 74LS193 4-bit asynchronous up/down counter, one 74LS194 universal shift register, and one 74LS283 full-adder.
Other integrated circuits: one 555 timer, one 741 op-amp, one DAC 0808 digital-to-analogue converter, plus one 7-segment LED display. 74HC devicer can be used instead of 74LS devices except for the 74LS47, the 74LS75, the 74LS76, the 74LS90, the 74LS93, the 74LS95, the 74LS194 and the 74LS283. These devices do not exist in the 74HC logic family.
A *breadboard* and possibly a number of switches.

The equipment that is needed to perform the practical exercises is: a dual-beam CRO, a multimeter, power supplies (for ±5 V, ±15 V and +10 V), and a pulse generator. The pulse generator could be a commercial instrument, or a single circuit could be built permanently and used for several exercises. A suitable circuit would use the 555 timer IC and is given in practical exercise 1.1 on p. 14. In all of the exercises it may prove desirable for the power supply to the breadboard to be decoupled by a 10 or 33 μF electrolytic capacitor.

The digital integrated circuits chosen for the practical exercises are all members of the TTL 74LS series logic family. The 74LS devices are used because they are cheap, readily available, and are not prone to damage by static electricity, as are CMOS devices. This choice makes it much easier to build circuits since care has to be taken when fitting CMOS ICs. A further point is that the 4000 CMOS family is little used today and the newer advanced CMOS families (AC/ACT/AHC/AHCT) do not (as yet) include anywhere as many devices as 74LS TTL. Where an advanced CMOS equivalent to a quoted 74LS device exists it can usually replace the quoted IC, but its pinout will, of course, have to be checked.

The circuits given in the practical exercises have been drawn in three different ways: (a) using the pinout of the IC(s), (b) using the logic symbol of the IC(s), or (c) drawing the IC pins in the most convenient order to simplify the circuit. This has been done since the circuits found in the literature are drawn using either method (a) or (c); while the use of logic symbols is, as yet, in its infancy it will, in all probability, become more common in the future. In any case, some familiarity with modern logic symbols is desirable.

The circuits given in the practical exercises are best made up on a breadboard of some kind, preferably one that consists of a large number of holes, or sockets, on a 0.1 inch grid (typically 550 on a 6 inch board). The sockets are joined together in groups of about five. Component leads and wires can be pushed into the holes where they are held firmly (hopefully!) in place by spring contacts. Fig. 1.11 shows a breadboard with a few components inserted into a circuit; groups of sockets are indicated by the thin lines. Some types have continuous contact rows at both the top and bottom for use as + and – power supply lines. The use of a breadboard allows a circuit to be (relatively) rapidly built and makes it easy to alter connections if errors are made and/or to change component values. Each exercise will provide some experience of the assembly of components on a board and of the likelihood of an error.

Fig. 1.11 *(a) Breadboard and (b) resistor and IC inserted on breadboard*

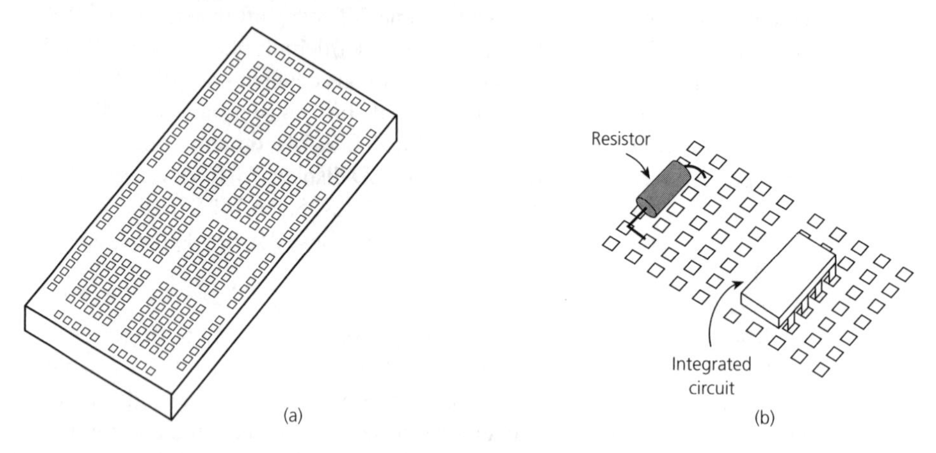

PRACTICAL EXERCISE 1.1

To build a low-frequency pulse generator using a 555 timer IC connected as an astable multivibrator.

Components and equipment: one 555 timer IC, two 7.5 kΩ resistors, one 0.1 μF capacitor, one 0.01 μF capacitor, two 1N914 signal diodes (or equivalent). Breadboard. Dual-beam CRO. Power supply.

The pinout of the LM555C timer IC is shown in Fig. 1.12(a). The basic astable multivibrator circuit is shown in Fig. 1.12(b). The frequency of oscillation is given by

$$f = 1/[0.69C_1 (R_1 + 2R_2)] \text{ Hz} \tag{1.2}$$

The duty cycle of the output waveform is determined by the ratio $(R_1 + R_2)/(R_1 + 2R_2)$. This means that the duty cycle must always be greater than 0.5 and so the output waveform can never be square. A square waveform can be approached by making R_2 several times larger than R_1.

Procedure:

(a) Build the circuit in permanent form on stripboard so that it can be used for later practical exercises.

(b) Connect the output of the circuit to the CRO and observe the displayed waveform. Measure its frequency and account for any discrepancy between the measured and calculated values.

(c) Measure the duty cycle of the output waveform and state its value. Calculate the mark–space ratio.

(d) Measure the rise-time of the leading edge of a pulse and the fall-time of a trailing edge.

Fig. 1.12 *(a) LM555C timer pinout and (b) astable multivibrator*

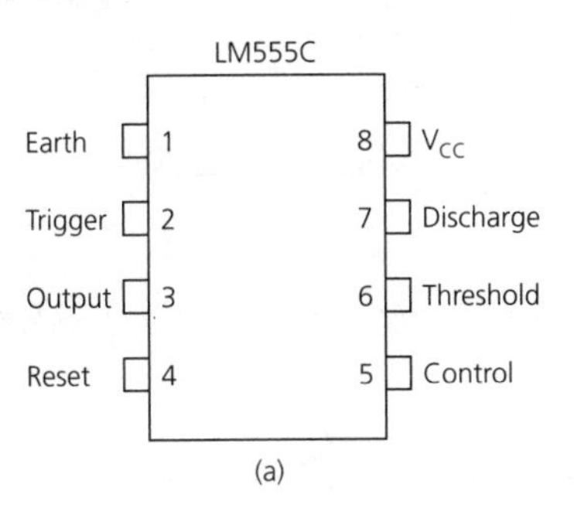

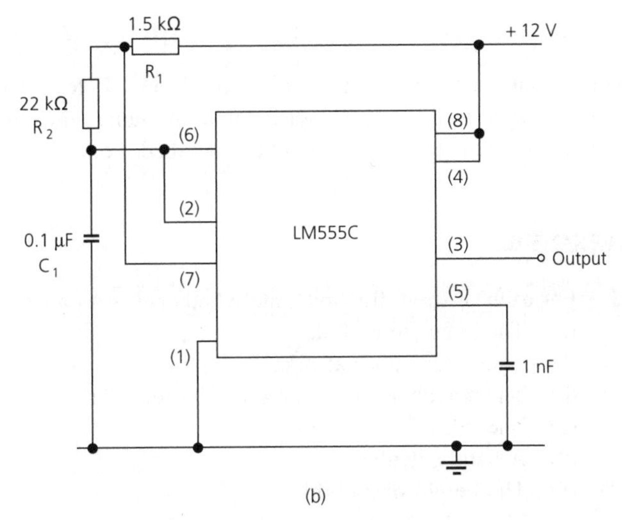

Light emitting diodes (LEDs)

In many of the circuits used for practical exercises the logical state(s) of the output(s) is/are indicated by one, or more, LEDs. These devices are discussed in chapter 14 but, simply, an LED will glow visibly when turned ON and a current of some 10–20 mA flows. Each LED needs a current-limiting resistor connected in series with it, otherwise its life may be short.

Switches

In most of the practical exercises various inputs are required to be connected to either logic 1 or logic 0 voltage level. In the figures this requirement is drawn in the manner shown in Fig. 1.13. The connection to either logic 1 or logic 0 can be made using a flying lead which is permanently connected to the input and moved to either 1 or 0 when required. Alternatively, a switch can be employed. There are several types to choose from; keyboard, push-to-break, push-to-make, rotary, slide and toggle. Any of these

Fig. 1.13

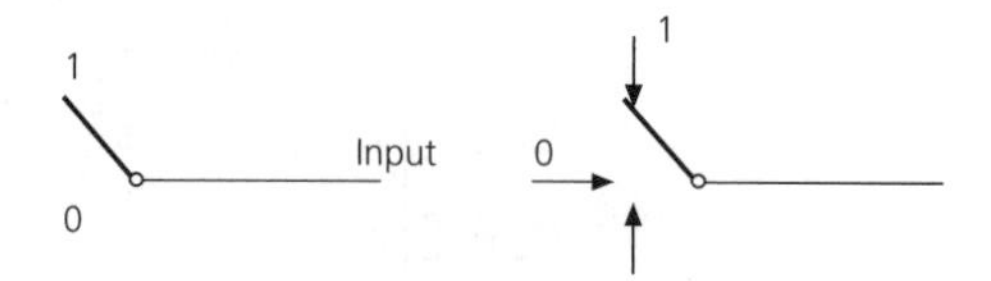

Fig. 1.14 *Four types of switch: (a) SPST, (b) SPDT, (c) SPST and (d) DPST*

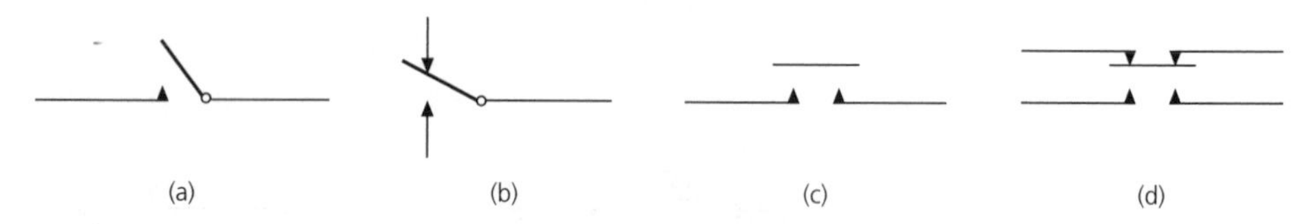

switches will have a number of poles and throws. Poles are the number of separate circuits the switch is able to make or break at the same time, and throws are the number of positions into which each pole may be switched. Four examples are shown in Fig. 1.14.

EXERCISES

1.1 For each of the following state whether it is an analogue or a digital quantity
- (a) The pages in a book.
- (b) The temperature at night.
- (c) The waveband switch on a radio receiver.
- (d) The ink left in a fountain pen.
- (e) A hand calculator.
- (f) The height of a child.
- (g) The indicator on a garage petrol pump.

1.2 A digital signal voltage varies between the values +9 V and + 0.2 V. Is this
- (a) A unipolar or a bipolar signal?
- (b) If the +9 V voltage represents logic 1, is this an example of positive or negative logic?

1.3 Draw a bipolar binary digital voltage that varies between ±12 V with a clock frequency of 10 MHz. Determine the periodic time of the waveform.

1.4 Discuss the reasons why a digital computer employs digital electronic circuitry and not analogue circuits. Give a reason why binary digital circuitry is employed and not decimal digital.

1.5 Which of the following can be classified as being analogue quantities?
- (a) The speed of a car.
- (b) The number of seconds in a minute.
- (c) Binary numbers.
- (d) A digital watch.
- (e) An electric light switch.
- (f) A dimmer switch.

1.6 Draw a positive unipolar digital waveform which has 8 bits alternately 0 and 1 if logic 1 = 4.5 V and logic 0 = 0.2 V and the bit width is 1 μs.

1.7 Draw a clock waveform that has a periodic time of 240 ns. What is its frequency?

1.8 A repetitive waveform is at +5 V for 1 μs and at 0 V for 2 μs. Calculate
(a) Its duty cycle.
(b) Its mark–space ratio.

1.9 Draw and label a digital signal that has a duty cycle of 60%, a pulse width of 200 ns, a rise-time of 20 ns, and a fall-time of 30 ns.

1.10 A digital transmission system transmits bits at (a) 4800 bits/s, (b) 140 Mbits/s, and (c) 565 Mbits/s. Calculate the bit width (or bit time) in each case.

2 Binary and hexadecimal arithmetic

After reading this chapter you should be able to:

(a) Identify numbering systems other than decimal.
(b) Determine the weighting factor for each digit position in each of the decimal, binary, octal and hexadecimal numbering systems.
(c) Convert a number given in one numbering system into its equivalent form in any other system.
(d) Perform arithmetic calculations using binary and hexadecimal numbers.
(e) Manipulate 1s complement and 2s complement numbers.
(f) Describe the format of BCD numbers.

The numbering system used in every-day life is the *decimal* (*denary*) system in which numbers have a base of 10. Every number consists of the sum of 0–9 units, 0–9 tens, 0–9 hundreds, 0–9 thousands, and so on. A number such as 2573 means $(2 \times 1000) + (5 \times 100) + (7 \times 10) + (3 \times 1)$, or $2 \times 10^3 + 5 \times 10^2 + 7 \times 10^1 + 3 \times 10^0$. Since only the digits 0 through to 9 are used in the decimal system, the number 10 requires two digits, i.e. 10, indicating $1 \times 10^1 + 0 \times 10^0$.

It is not essential for the base, or *radix*, of a numbering system to be 10; instead the base may be any chosen number. If the base is 2, the *binary system* is obtained; if the base is 8 the *octal system* is obtained; if the base is 16 the *hexadecimal system* results. To indicate the base, if there is any chance that the base might be mistaken, a subscript is used, e.g. 10_2 means a binary number, 10_{10} means a decimal number and 10_8 is an octal number. Usually, a hexadecimal number is indicated by the letter H.

The binary numbering system is employed in digital electronics, in digital computers and microprocessors since it requires only two digits, 1 and 0. This allows the use of electronic devices with just two states, either HIGH or LOW (ON or OFF), as discussed in chapter 1. Although binary numbers are advantageous for digital circuitry and systems they are not easy for people to follow. For example, the number 1624_{10} is easy to read but its equivalent in binary, 11001011000, is not. For this reason binary numbers are usually converted into any one of their decimal, octal or hexadecimal equivalents before being read out of the system.

Table 2.1 *Powers of 2*

2^7	2^6	2^5	2^4	2^3	2^2	2^1	2^0
128	64	32	16	8	4	2	1

Table 2.2

11	0	0	0	1	0	1	1
43	0	1	0	1	0	1	1
63	0	1	1	1	1	1	1
111	1	1	0	1	1	1	1

Binary numbers

In digital electronic systems, the active devices employed are operated as switches and have two stable states, ON and OFF. For this reason, the binary numbering system is used, in which only two digits, 1 and 0, are allowable. Larger numbers are obtained by utilizing the various powers of 2. The *least-significant* digit of a binary number represents a multiple (0 or 1) of 1 and is (normally) written at the right-hand side of the number. The next digit to the left indicates either zero or one 2, and so on, as shown in Table 2.1.

The highest-value digit in a binary number is known as the *most-significant digit* (MSD), and the lowest-value number is called the *least-significant digit* (LSD).

The value of each power of 2 is given in the table and any desired number can be obtained by the correct choice of 0s and 1s. Thus the number 21, for example, is equal to 16 + 4 + 1 and it is therefore given by 00010101 in the 8-unit binary code, or by 10101 if only 5 bits are used.

Some other binary equivalents of denary numbers are given in Table 2.2, a 7-unit code being assumed.

Base or radix conversion

Decimal to binary

There are two main ways in which a number can be converted from decimal into binary.

(a) Consider the decimal number 100. From Table 2.1, the highest power of 2 in this number is $2^6 = 64$. Therefore there is a 1 in the 2^6 column. Subtract 64 from 100 to get 36; this is higher than $2^5 = 32$ and hence there is a 1 in the 2^5 column also. Now subtract 32 from 36 to get 4; this means that there are no $2^4 = 16$ or $2^3 = 8$

components in the binary number and hence a 0 goes in each of those columns. There should then be a 1 in the $2^2 = 4$ column, to make the binary number equal to 36, followed by 0s in both of the last two columns. Thus the binary equivalent of 100_{10} is 1100100.

(b) To convert a decimal integer number into its binary equivalent, the decimal number should be repeatedly divided by 2, and each time the remainder, which will be either 0 or 1, should be noted. Eventually the number will be reduced to 1, at which stage further division will not give an integer number, and so the quotient 1 is considered to be a remainder of 1. The required binary number is then obtained by writing down the remainders in reverse order because the procedure produces the LSD first and the MSD last.

EXAMPLE 2.1

Convert 38 into binary.

Solution

Method (a)

2^5	2^4	2^3	2^2	2^1	2^0
1	0	0	1	1	0 (*Ans.*)

Method (b)

Number	38	19	9	4	2	1
Remainder	0	1	1	0	0	1

Therefore, 38 = 100110 (*Ans.*)

EXAMPLE 2.2

Convert 277 into binary.

Solution

Method (a)

2^8	2^7	2^6	2^5	2^4	2^3	2^2	2^1	2^0	
1	0	0	0	1	0	1	0	1	(*Ans.*)

Method (b)

Number	277	138	69	34	17	8	4	2	1
Remainder	1	0	1	0	1	0	0	0	1

Therefore, 277 = 100010101 (*Ans.*)

Decimal fractions to binary fractions

The binary number system is not limited to whole numbers. To the right of the binary point come the fractional numbers listed in Table 2.3.

The table can be used to convert a decimal fraction into the equivalent binary fraction or a repeated multiplication method may be employed.

Table 2.3 ***Fractional powers of 2***

2^{-1}	2^{-2}	2^{-3}	2^{-4}	2^{-5}	2^{-6}	2^{-7}
0.5	0.25	0.125	0.0625	0.03125	0.015625	0.0078125

(a) Consider the decimal fraction 0.59375. Compare it with the most-significant column in the table; 0.59375 is larger than 0.5 so put a 1 into the first place after the binary point and then subtract 0.5. This leaves 0.09375. The next column in the table represents $2^{-2} = 0.25$, which is too big, so put a 0 in this position. Similarly, 2^{-3} is also too large and another 0 is necessary. The next column is $2^{-4} = 0.0625$ and, hence, a 1 goes into the next place. The remainder is now $0.09375 - 0.0625 = 0.03125$. This goes, with zero remainder, into the 2^{-5} column. Thus, the equivalent binary fraction is 0.10011. [Note: 10011 is 19/32 = 0.59375.]

(b) The alternative method is to multiply the decimal fraction repeatedly by 2 and each time note the integer number obtained. The required binary fraction is then obtained by reading the integers from left to right.

EXAMPLE 2.3

Convert 0.125 into binary.

Solution

Method (a)

2^{-1}	2^{-2}	2^{-3}	2^{-4}
0	0	1	0 (*Ans.*)

Method (b)

Fraction	0.125	0.25	0.5	1.0
Integers		0	0	1

Therefore, 0.125 = 0.001 (*Ans.*)

EXAMPLE 2.4

Convert 0.426 to binary.

Solution

Method (a)

2^{-1}	2^{-2}	2^{-3}	2^{-4}	2^{-5}	2^{-6}
0	1	1	0	1	1 (*Ans.*)

Method (b)

Fraction	0.426	0.852	1.704	1.408	0.8156	1.632	1.264
Integers		0	1	1	0	1	1

Therefore, 0.426 = 0.011011 etc. (*Ans.*)

The answer to example (2.4) requires many binary places to give a precise result, but usually only a few places are calculated depending on the required accuracy.

When the denominator of a decimal fraction is a power of 2 the conversion from decimal to binary becomes easier since the LSD of the binary fraction is known immediately. For example, consider 5/8; the numerator in binary is 101 and the LSD must appear in the $1/8 = 2^{-3}$ column. Therefore, 5/8 = 0.101.

EXAMPLE 2.5

Convert into binary (a) 9/16, (b) 22 7/16, and (c) 30.6875.

Solution

(a) 9 = 1001, hence, 9/16 = 0.1001 (*Ans.*)
(b) 7 = 0111, hence, 22 7/16 = 10110.0111 (*Ans.*)
(c) 11110.1011 (*Ans.*)

Binary fraction to decimal fraction

The conversion of binary into their denary equivalents is best achieved using Table 2.3 on p. 21.

EXAMPLE 2.6

Convert 0.01101 into decimal.

Solution

From Table 2.3, 0.0101 = 0.25 + 0.0625 = 0.3125 (*Ans.*)

EXAMPLE 2.7

Convert 10110.101 into denary.

Solution

Using Tables 2.2 and 2.3, 10110.101
$= 1 \times 16 + 0 \times 8 + 1 \times 4 + 1 \times 2 + 0 \times 1 + 1 \times 0.5 + 0 \times 0.25 + 1 \times 0.125$
= 22.625 (*Ans.*)

Table 2.4 $A + B$

A	*B*	*Sum*	*Carry*
0	0	0	0
1	0	1	0
0	1	1	0
1	1	0	1

Table 2.5

Carry-in	*B*	*A*	*Sum*	*Carry-out*
0	0	0	0	0
0	0	1	1	0
0	1	0	1	0
0	1	1	0	1
1	0	0	1	0
1	0	1	0	1
1	1	0	0	1
1	1	1	1	1

Binary arithmetic

The processes of binary addition, subtraction, multiplication and division are essentially the same as in arithmetic but are, of course, restricted to the use of the two digits 1 and 0.

Binary addition

The rules for the addition of binary numbers are given in Table 2.4.

When two 1s are added together their sum is 10 and so the sum in that order of unit is 0 with a carry of 1.

For the LSD there are only the two digits A and B to consider. The second-significant digit, and all the following digits, may have a third digit to consider, which is the carry digit, which may come from the previous column. If a carry is produced it must be added to the next more-significant column as a carry-in. This gives the eight possible combinations shown in Table 2.5.

Regardless of which input is at 1, the sum of two 0s and a 1 is 1 without a carry, the sum of two 1s and a 0 is 0 with a carry, and the sum of three inputs at 1 is 1 with a carry.

EXAMPLE 2.8

Add the following decimal numbers using binary arithmetic (a) 2 + 1; (b) 3 + 1; (c) 3 + 4.

Solution

(a)
$$\begin{array}{r} 10 \\ +01 \\ \hline 11 \end{array} = 3\ (\textit{Ans.})$$

(b)
$$\begin{array}{r} 11 \\ +01 \\ \hline 101 \end{array} = 4\ (\textit{Ans.})$$

(c)
$$\begin{array}{r} 11 \\ +100 \\ \hline 111 \end{array} = 7\ (\textit{Ans.})$$

EXAMPLE 2.9

Add the binary numbers 10111 and 01101.

Solution

$$\begin{array}{rl} 10111 & = 23 \\ +01101 & = 13 \\ \hline 100100 & = 36\ (\textit{Ans.}) \end{array}$$

EXAMPLE 2.10

Convert 33 3/8 and 22 5/8 into binary. Add the two binary numbers and convert the results into decimal form.

Solution

$$\begin{array}{lr} 33\ 3/8 = & 100001.011 \\ 22\ 5/8 = & +\ 10110.101 \\ \hline & 111000.000 \end{array} = 32 + 16 + 8 = 56\ (\textit{Ans.})$$

The addition process can be applied to as many numbers as necessary as shown by the next example.

EXAMPLE 2.11

Add 1011.011, 101.101 and 110.1.

Solution

$$\begin{array}{rl} 1011.011 & = 11.375 \\ 101.101 & = 5.625 \\ +\ 110.100 & = 6.5 \\ \hline 10111.100 & = 23.5 \quad (\textit{Ans.}) \end{array}$$

Binary subtraction

Binary subtraction can be carried out in either one of two different ways. The easiest method on paper is similar to that used for the subtraction of one decimal number from another. When binary subtraction is to be carried out electronically, however, it is easier if *complements* are employed. The most-significant digit (MSD) of a binary number can be used as a *sign bit* that indicates whether the number is negative or positive. If the sign bit is 0 the number is positive and if the sign bit is 1 the number is negative. For example, $00110111 \equiv +55_{10}$ and $11010101 \equiv -85_{10}$. The sign and magnitude form of a binary number has the disadvantage that separate processes will be required for addition and subtraction. It is desirable to be able to carry out subtraction using an addition process and this means that either *1s complement* or *2s complement* representation of signed numbers is employed. When the MSD is interpreted as a sign bit then when it is 0 the remaining bits give the true magnitude of the number, e.g. 011010010 = +210. Conversely, if the sign bit is 1 the remaining bits are in either 1s or 2s complement form.

The use of 1s or 2s complement arithmetic depends upon the digital circuitry that is employed with the majority of modern equipment preferring the 2s complement.

A subtraction problem can be treated as the addition of two signed numbers, i.e. $X - Y = X + (-Y)$. The use of signed numbers gives negative differences when $X < Y$ and a negative sum when both numbers are negative. When the sum of two 1s or 2s complement numbers is negative the sign bit of the sum is 1; then the magnitude of the sum is in complement form. The true magnitude of a number appearing in its complement form is obtained by finding the complement of that complemented number.

Direct subtraction

The rules for performing binary subtraction are given in Table 2.6.

When a number is to be subtracted from a smaller number (always 0 – 1), a 1 must be borrowed from the next column to the left. This 1 is a power of 2 higher and, hence, the difference, or remainder, becomes 2 – 1 = 1.

When a borrow is needed the A^0 column must borrow from the A^1 column. When this occurs, e.g. $A - B = 2 - 1 = 10 - 01$, A^0 increases by 2 and A^1 becomes 0. In the A^0 column 2 – 1 = 1 and in the A^1 column 0 – 0 = 0. Hence, the difference is 01 or 1_{10}.

Table 2.6 *A – B*

A	*B*	*Difference*	*Borrow*
0	0	0	0
1	0	1	0
0	1	1	1
1	1	0	0

Sometimes it is necessary to borrow from two, or more, columns to the left. Consider the simple sum 4 – 1

EXAMPLE 2.12

Subtract 10101 from 11011.

Solution

$$\begin{array}{r} 11011 = 27 \\ \underline{-10101} = 21 \\ 00110 = \ \ 6 \end{array}$$

$$\begin{array}{r} 100 = 4 \\ -001 = 1 \end{array}$$

Here it is necessary to borrow 1 from the 2^2 column. Its value is 2 in the 2^1 column; 1 remains in that column and the other 1 is taken to the 2^0 column where its value is 2. Then in the 2^0 column $2 - 1 = 1$, and in the 2^1 column $1 - 0 = 1$. The 2^2 column now contains 0. Subtracting now gives $011 = 3$.

EXAMPLE 2.13

Using binary arithmetic subtract 3 from (a) 8, and (b) 16.

Solution

(a) $$\begin{array}{r} 1000 = 8 \\ -0011 = 3 \end{array}$$

The 1 in the 2^0 column cannot be taken from the 0, so a borrow from a higher column is necessary. The only 1 available is in the 2^3 column. Using this, it appears as 2 ($\times$ 4) in the 2^2 column, here 1 remains and 1 is moved to the 2^1 column. Here again 1 remains and 1 is moved to the 2^0 column where its value is 2. In the 2^0 column $2 - 1 = 1$, in the 2^1 column $1 - 1 = 0$, and in the 2^2 column $1 - 0 = 1$. Hence, the answer is $0101 = 5$ (*Ans.*)

(b) $$\begin{array}{r} 10000 = 16 \\ \underline{-00011} = \ \ 3 \\ 01101 = 13 \end{array}$$ (*Ans.*)

Ones complement

The 1s complement of a binary number is obtained by complementing every digit in the number, i.e. by changing each 1 to 0 and each 0 to 1. The number to be subtracted is 1s complemented whether or not it is the smaller number of the two and then the two numbers are added together. If there is a carry in the sum that is beyond the most significant digit of the two numbers, that carry must be added to the least-significant digit. The carry bit is often called the *end-around carry* (EAC) and it can occur only when the remainder is a positive number. The absence of the EAC means that the remainder is negative and is in its complemented form. Overflow will occur if the sum of the two numbers being 'added' exceeds the available range of numbers and then an incorrect result will be obtained.

(a) 1s complement: 1 11010010 is negative. The 1s complement is 1 00101101 = −45.
(b) 2s complement: 1 11010010 is negative. The 2s complement is 1 00101101 − 1 = 1 00101100 = −44.

EXAMPLE 2.14

Express −52 in (a) 1s complement, and (b) 2s complement form. Use 8 bits.

Solution

(a) 52 = 00110100. The 1s complement is 1 11001011 (*Ans.*)
(b) 2s complement = 1 11001011 + 1 = 1 11001100 (*Ans.*)

EXAMPLE 2.15

Use the 1s complement method to determine (a) 15 – 10, (b) 10 – 15, (c) 7/8 – 5/8, and (d) 3/4 – 13/16.

Solution

(a) 15 = 1111 and 10 = 1010. The 1s complement of 1010 is 0101, hence,

```
  1111
+0101
 10100
```

There is a carry of 1 which must be added to the sum to give 0101 = 5 (*Ans.*)

(b) The 1s complement of 15 is 0000, hence,

```
  1010
+0000
  1010
```

There is no carry past the MSD and this means that the result is negative and in complement form. Complementing, gives 0101 = –5 (*Ans.*)

(c) 7/8 = 0.111 and 5/8 = 0.101. Therefore,

```
  0.111
+0.010
  1.001
```

Since the MSD is 1 the result is positive and adding the 1 to the LSD gives 0.010 = 0.25 (*Ans.*)

(d) 3/4 = 0.110 and 13/16 = 0.1101. Therefore,

```
 0.1100
+0.0010
 0.1110
```

The MSD is 0 which means that the result is negative. Taking the complement gives 0.001 = −1/16 (*Ans.*)

EXAMPLE 2.16

Subtract 10.625 from 8.75.

Solution

8.75 = 1000.110 and 10.625 = 1010.101. The 1s complement of 10.625 is 0101.010 and, hence,

```
 1000.110
+0101.010
 1110.000
```

There is no 1 carried over, so the result is negative. Complementing gives 0001.111 = −1.875 (*Ans.*)

Twos complement

One disadvantage of 1s complement arithmetic is that it has two values for zero, i.e. + 0 = 00000000 and −0 = 11111111. Also, binary subtraction is more easily carried out electronically if 2s complement arithmetic is employed instead of 1s complement arithmetic. The 2s complement of a binary number is equal to the 1s complement of the number plus 1. It can also be found by either (a) subtracting the number from the next power of 2 that is higher than the number, or (b) complementing only the bits that are to the left of the least-significant 1 bit. The range of an 8-bit number is from $+127_{10}$ to -128_{10}; if this range is exceeded incorrect results will be obtained.

EXAMPLE 2.17

Find the 2s complement of 101101.

Solution

Method (a)
Ones complement = 010010. Twos complement = 010010 + 1 = 010011 (*Ans.*)

Method (b)

```
 1000000
− 101101
  010011 (Ans.)
```

Method (c)
Complementing all bits to the left of the least-significant 1 gives 010011 (*Ans.*)

If an end-around carry is produced it is an indication that the result of the subtraction is positive, and no carry indicates a negative result.

EXAMPLE 2.18

Use (a) 1s complement, and (b) 2s complement arithmetic to calculate (i) 22 – 7, and (ii) 7 – 22.

Solution

(a) (i) 22 = 10110 and 7 = 00111. The 1s complement of 7 is 11000. Hence,

$$\begin{array}{r} 10110 \\ +11000 \\ \hline 101110 \end{array}$$

There is a carry of 1, adding this to the sum gives 01111 = +15 (*Ans.*)

(ii) The 2s complement is 11001 and, hence,

$$\begin{array}{r} 110110 \\ +\ 11001 \\ \hline 101111 \end{array}$$

There is a carry, so the result is positive and 01111 = +15 (*Ans.*)

(b) (i) The 1s complement of 22 is 01001 and, hence,

$$\begin{array}{r} 00111 \\ +01001 \\ \hline 10000 \end{array}$$

There is no carry, so the result is negative. Complementing gives 01111 = –15 (*Ans.*)

(ii) The 2s complement is 01010 and, hence,

$$\begin{array}{r} 00111 \\ +01010 \\ \hline 10001 \end{array}$$

There is no carry, so the result is negative. Complementing gives 0110 and adding 1 gives 01111 = -15_{10} (*Ans.*)

EXAMPLE 2.19

Subtract 10101 from 11011 using the 2s complement method.

Solution

$$\begin{array}{r} 11011 = 27 \\ -10101 = 21 \end{array} = \begin{array}{r} 11011 \\ +01011 \\ \hline 100110 \end{array} \begin{array}{l} \\ \text{2s complement} \\ = +6 \end{array}$$

There is a carry, so the result is positive (*Ans.*)

EXAMPLE 2.20

Subtract 11011 from 10101 using the 2s complement method.

Solution

$$\begin{array}{r} 10101 = 21 \\ -11011 = 27 \end{array} = \begin{array}{r} 10101 \\ \underline{+00101} \\ 011010 \end{array} \text{ 2s complement}$$

There is no carry so, the result is negative. The 1s complement is 00101 and adding 1 gives 00110 = -6_{10} (*Ans.*)

EXAMPLE 2.21

Subtract 11101 from 01011 using the 2s complement method.

Solution

$$\begin{array}{r} 01011 = 11 \\ -11101 = 29 \end{array} = \begin{array}{r} 01011 \\ \underline{+00011} \\ 001110 \end{array} \text{ 2s complement}$$

There is no carry, so the result is negative. The 1s complement is 01110 − 1 = 01101 and complementing gives 10010 = −18 (*Ans.*)

EXAMPLE 2.22

Subtract 11101 from 01011 using the 1s complement method.

Solution

$$\begin{array}{r} 01011 = 11 \\ -11101 = 29 \end{array} = \begin{array}{r} 01011 \\ \underline{+00010} \\ 001101 \end{array} \text{ 1s complement}$$

There is no carry, so the result is negative. The 1s complement is 010010 = -18_{10} (*Ans.*)

EXAMPLE 2.23

Subtract 01011 from 11101 using the 1s complement method.

Solution

$$\begin{array}{r} 11101 = 29 \\ -01011 = 11 \end{array} = \begin{array}{r} 11101 \\ \underline{+10100} \\ 110001 \end{array} \text{ 1s complement}$$

The positive sign digit (1) must now be shifted around to the right-hand side of the number and added to the digit already there. Hence, the difference is 10010 = 18. (*Ans.*)

EXAMPLE 2.24

Add $+112_{10}$ and $+65_{10}$ using 2s complement arithmetic.

Solution

$$\begin{array}{rl} 128 = & 01110000 \\ 65 = & \underline{+01000001} \\ & 10110001 = -49_{10}! \end{array}$$

The error has occurred because the correct answer of 177_{10} is larger than the maximum range number 127_{10} (*Ans.*)

Binary multiplication

The product of two binary numbers is 1 only if both the digits are 1, otherwise it is 0. This means that binary multiplication is just a matter of shifting and adding. As a simple example, consider $A \times B$, where $A = 5$ and $B = 3$. The binary value of 5 = 0101, and 3 = 0011. First, multiply the 2^0 column of number A by B. Next, multiply the 2^1 bit of A by B and shift the result one place to the left before writing it down. Repeat this for each other bit in number A. Finally, take the sum of the *partial products* to obtain the result.

EXAMPLE 2.25

Multiply 11011 by 10101.

Solution

$$\begin{array}{rl} 11011 & = 27 \\ \underline{\times 10101} & = 21 \\ 11011 & \\ 00000 & \\ 11011 & \text{partial products} \\ 00000 & \\ \underline{11011} & \\ 1000110111 & = 1 + 2 + 4 + 16 + 32 + 512 \\ & = 567 \ (\textit{Ans.}) \end{array}$$

Binary multiplication can also be carried out with fractional numbers. The partial products are written without binary points and the number of places in the fractional part of the product is made equal to the sum of the number of places in the fractional parts of the two numbers.

EXAMPLE 2.26

Multiply 3/8 by 1/4.

Solution

3/8 = 0.011 and 1/4 = 0.010. Therefore,

```
   0.011
 ×0.010    (6 binary places)
0.000011 = 0.09375 = 3/32 (Ans.)
```

EXAMPLE 2.27

Multiply 4.75 by 3.625 using binary arithmetic.

Solution

4.75 = 100.11 and 3.625 = 11.101. Hence,

```
      100.110
     ×011.101
       100110
      000000
     100110
    100110
   100110
10001.001110 = 17.21875 = 17 7/32 (Ans.)
```

Binary division

Binary division is carried out in a similar manner to division using decimal numbers. Suppose the divisor has three digits; then if the first three digits of the dividend are equal to, or larger than, the divisor put a 1 in the quotient and subtract the divisor from the first three digits. Next bring down the next bit of the dividend to be the LSD of the remainder. If the remainder is larger than the divisor put another 1 into the quotient, if it is smaller insert a 0. If a 1 is inserted, subtract the divisor from the remainder and bring down the next bit of the dividend and repeat the procedure. At each step in the procedure the divisor is either smaller than the bit group value or it is larger. Whenever it is smaller the quotient bit is 1, and whenever it is larger the quotient bit is 0.

EXAMPLE 2.28

Divide 50 by 5 using binary division.

Solution

50 = 110010 and 5 = 101.

```
                    101     (quotient)
(divisor)   101)110010     (dividend)
                101
                00101
                  101
                  000
```

The quotient = 101 = 5 (*Ans.*)

Since 101 is smaller than the first three bits of the dividend (110) the first quotient bit is 1. Taking 101 from 110 gives 001 and bringing down the next (fourth bit) 0 gives 10. This is smaller than the divisor (101) so that the next bit in the quotient is 0. Then bringing down the final 1 from the dividend gives 101 which equals the divisor, so the last quotient bit is 1. In this example there is no remainder.

EXAMPLE 2.29

Divide 11001 by 100.

Solution

```
       110.01
100)11001
    100
     100
     100
      001
      000
       010
       000
        100
        100
        000
```

Hence, result = 110.01 = 6.25 (*Ans.*)

Octal numbers

The *octal numbering system* is sometimes employed as a more convenient way of grouping binary numbers to make them easier to read. The base, or radix, of the octal system is eight and, hence, the digits 0 through to 7 are used. Numbers 8 and 9 never appear

Table 2.7

8^4	8^3	8^2	8^1	8^0
4096	512	64	8	1

since $8 = 10_8$ and $9 = 11_8$. Electronically, it is easier, and hence cheaper, to convert from binary to octal than it is to convert from binary to decimal. Thus, with the octal number system, Table 2.7 shows that $17_{10} = 21_8$ and $100_{10} = 1 \times 64 + 4 \times 8 + 4 = 144_8$.

It is also possible to determine the equivalent octal number of a given decimal number by dividing successively by 8. The remainders obtained give the octal digits, starting with the least significant. Thus, (a) 17/8 = 2 remainder 1, 2/8 = 0 remainder 2, so that $17_{10} = 21_8$, and (b) 100/8 = 12 remainder 4, 12/8 = 1 remainder 4, 1/8 = 0 remainder 1. Hence, $100_{10} = 144_8$.

EXAMPLE 2.30

Convert 2004_{10} into octal.

Solution

$$\begin{array}{r} 250 \\ 8\overline{)2004} \\ \underline{2000} \\ 4 \end{array} \quad \text{(least-significant digit)}$$

$$\begin{array}{r} 31 \\ 8\overline{)250} \\ \underline{248} \\ 2 \end{array} \quad \text{(next-significant digit)}$$

$$\begin{array}{r} 3 \\ 8\overline{)31} \\ \underline{24} \\ 7 \end{array} \quad \text{(next digit)}$$

$$\begin{array}{r} 0 \\ 8\overline{)3} \end{array} \quad \text{(most-significant digit)}$$

Therefore, $2000_{10} = 3724_8$ (*Ans.*)

[To check: from Table 2.3, $3724_8 = 3 \times 512 + 7 \times 64 + 2 \times 8 + 4 = 2004_{10}$.]

Binary-to-octal conversion

Converting a binary number into octal is simply a matter of dividing the binary number into groups of three starting from the least-significant bit. Each group of bits is then given an octal number according to the value of the group. The value of the binary number 101101101 is difficult to see immediately but the equivalent octal number of 101 101 101, i.e. 555_8, is much easier.

EXAMPLE 2.31

(a) Convert 11101101101110 into octal. (b) Convert 7325_8 into binary.

Solution

(a) 011 101 101 110 = 3556_8 (*Ans.*)
(b) 7325_8 = 111 011 010 101 (*Ans.*)

Binary-coded decimal numbers

In the *binary-coded decimal* (BCD) system each of the 10 decimal digits 0 through to 9 is represented by a 4-bit binary code. Each decimal digit is individually converted into its binary equivalent. Since a 4-bit code allows the use of numbers up to 15 (1111) the BCD system includes some redundant states. Binary-coded decimal is widely employed in digital electronics because it is relatively easy to code and/or decode. It is always used for visual numeric displays such as digital clocks and voltmeters.

EXAMPLE 2.32

(a) Write down the binary number 101111001 in BCD. (b) Write down the BCD number 2543 in binary.

Solution

(a) 1 0111 1001 = 179 (*Ans.*)
(b) 2543 = 10 0101 0100 0011 (*Ans.*)

Hexadecimal numbers

Microprocessors use address and data buses that carry bits in groups of 4, or multiples of 4 such as 8, 16 and 32. Because of this it is generally more useful to employ the 4-bit *hexadecimal numbering system* which has a base of 16, rather than the octal system. Since groups of 4 bits are employed the hexadecimal system provides values from 0 through to 15 (see Table 2.8). Numbers higher than 9 are indicated by the first letters in the alphabet: A = 10, B = 11, C = 12, D = 13, E = 14, and F = 15.

A hexadecimal number is usually indicated by a following H. Thus, 16_{10} = 10H, 17_{10} = 11H, 27_{10} = 1BH, and 257_{10} = 101H and so on.

Table 2.8

16^3	16^2	16^1	16^0
4096	256	16	1

Binary-to-hexadecimal conversion

Converting a binary number to hexadecimal requires the binary number to be divided into 4-bit groups, starting with the LSB, and indicating the value of each group by a hexadecimal digit. For example, the binary number 110 1101 1100 0111 is equal to 6DC7H.

Hexadecimal-to-binary conversion

To convert a hexadecimal number to the equivalent binary number each hexadecimal digit is changed into the equivalent 4-bit binary group. Thus, 3FA = 0011 1111 1010.

Hexadecimal-to-decimal conversion

A hexadecimal number is converted into decimal by the use of Table 2.8. From the table, 7A3H is equal to $7 \times 256 + 10 \times 16 + 3 \times 1 = 1955_{10}$, and 3F4H is equal to $3 \times 256 + 15 \times 16 + 4 = 1012_{10}$.

Decimal-to-hexadecimal conversion

Although the conversion from decimal to hexadecimal can be carried out by successively dividing the decimal number by 16, this, in itself, is not too simple. It is probably better to use Table 2.3 again. Consider the decimal number 750: 750_{10} requires $2 \times 256 = 512$ and this leaves $750 - 512 = 238$. In turn, 238 needs $14 \times 16 = 224$ which leaves $238 - 224 = 14$. Therefore, 750_{10} = 2EEH.

EXAMPLE 2.33

Convert 498_{10} to hexadecimal (a) using Table 2.8, and (b) by dividing successively by 16.

Solution

(a) $498 = 1 \times 256 + 15 \times 16 + 2$ = 1F2H (*Ans.*)

(b) 498/16 = 31 remainder = 2 (least-significant digit)
31/16 = 1 remainder 15,
1/16 = 0 remainder = 1 (most-significant digit).

Therefore, 498_{10} = 1F2H (*Ans.*)

Decimal fraction-to-hexadecimal fraction conversion

To convert a decimal fraction into the corresponding hexadecimal fraction multiply the decimal fraction repeatedly by 16. The whole number which results from each product is the wanted hexadecimal digit beginning with the most-significant digit.

EXAMPLE 2.34

Convert 0.12_{10} to hexadecimal.

Solution

$0.12 \times 16 = 1.92$ hex digit $= 1$
$0.92 \times 16 = \text{E}.72$ $= \text{E}$
$0.72 \times 16 = \text{B}.52$ $= \text{B}$
$0.52 \times 16 = 8.32$ $= 8$
Therefore, $0.12_{10} = 0.1\text{EB8H}$ (*Ans.*)
[To check: $0.1\text{EB85H} = 1 \times 16^{-1} + 14 \times 16^{-2} + 11 \times 16^{-3} + 8 \times 16^{-4} = 0.0625 + 0.0547 + 2.685 \times 10^{-3} + 1.22 \times 10^{-4} = 0.12^{110}$.]

Hexadecimal arithmetic

Addition

The sum of two hexadecimal digits is the same as their equivalent decimal sum provided the sum is less than 16. If the sum is greater than 16, then 16 must be subtracted from the sum to obtain the decimal sum, and a carry of 1 is produced. For example, 3H + 5H = 8H (3 + 5 = 8), AH + 3H = DH (10 + 3 = 13), AH + 9H = 13H (10 + 9 = 16 + 3), and CH + 9H = 15H (12 + 9 = 16 + 5).

EXAMPLE 2.35

Add 2096_{10} to 0110110101101110 using hexadecimal arithmetic.

Solution

$2096 = 8 \times 256 + 3 \times 16 + 0 = 830\text{H}$
0110 1101 011 1110 = 6D6EH
Adding, 6D6E + 830 = 7S9EH (*Ans.*)

Subtraction

The subtraction of one hexadecimal number from another is similar to decimal subtraction except that whenever a 1 is borrowed from a column to the left its value is 16 and not 10.

EXAMPLE 2.36

Solve (a) 34H – 0DH, and (b) B7H – 4DH.

Solution

(a)

```
  34H
 -0DH
  27H (Ans.)
```

[D cannot be subtracted from 4 so borrow 1 from the 3 in the 16^1 column. The 3 becomes 2 and the carried 1 becomes 16 in the 16^0 column. Now the sum is $(16 + 4) - 13 = 7$. In the 16^1 column the sum is $2 - 0 = 2$.]

(b)

```
  B7H
 -4DH
  6AH (Ans.)
```

[D cannot be taken from 7 so 1 is borrowed from the 16^1 column making B become A. In the 16^0 column $(16 + 7) - 13 = 10 = A$. In the 16^1 column $A - 4 = 6$.]

Electronically, hexadecimal subtraction is best carried out using a complement method.

EXAMPLE 2.37

Subtract 33H from B8H.

Solution

```
  B8H
 - 33H
  85H (Ans.)
```

Using the complement method: take each digit away from 15 to obtain the 15s complement of 33. Thus,

```
  B8H
 +CCH
 184H
```

Add the carry to the right-hand pair of digits to get 85H, as before (*Ans.*)

A carry of 1 indicates a positive answer.

EXAMPLE 2.38

Subtract B8H from 33H.

Solution

The 15s complement of B8 is 47H. Hence,

$$\begin{array}{r} 33H \\ 47H \\ \hline +7AH \end{array}$$

There is no carry, so the result is negative (*Ans.*)

Complementing again gives 85H (*Ans.*)

[To check: 33H = 48 + 3 = 51 and B8H = 11 × 16 + 8 = 184. 51 − 184 = −133.]

EXERCISES

2.1 Add
- (a) 11101110, 1010101, and 1110011.
- (b) 11100101, 11010001, and 10110111.

2.2 Subtract
- (a) $73_{10} - 46_{10}$.
- (b) $33_{10} - 17_{10}$.
- (c) $56_{10} - 75_{10}$.

using
- (i) Direct binary subtraction.
- (ii) Ones complement.
- (iii) Twos complement arithmetic.

2.3 Use 2s complement arithmetic to perform
- (a) $127_{10} - 46_{10}$.
- (b) $127_{10} - 128_{10}$.
- (c) $1_{10} - 3_{10}$.

2.4 Use 1s complement arithmetic to subtract
- (a) 46_{10} from 127_{10}.
- (b) $-37_{10} - 91_{10}$.

2.5 Use 2s complement arithmetic to solve $58_{10} - 33_{10} + 14_{10} + (-45_{10})$.

2.6 (a) Convert from decimal to octal
- (i) 212.
- (ii) 399.
- (iii) 42.
- (iv) 1000.

(b) Convert from octal to decimal
(i) 212.
(ii) 377.
(iii) 42.
(iv) 26.

2.7 (a) Convert the following hexadecimal numbers into decimal
(i) AFH.
(ii) 2BCH.
(iii) 77H.
(iv) 1EAH.
(b) Convert from decimal to hexadecimal
(i) 100.
(ii) 328.
(iii) 1000.
(iv) 7300.

2.8 Add
(a) (i) 10110111 + 11100011.
(ii) 11010101 + 1111101.
(iii) 1111 + 1101.
(iv) 10110110 + 11011.
(b) (i) 47H + AAH.
(ii) BBH + 172H.
(iii) 40DH + 27H.
(iv) FFH + EAH.

2.9 Subtract
(a) (i) 101101 – 111.
(ii) 1001 – 11.
(iii) 10001101 – 100011.
(iv) 110111 – 101.
(b) (i) AAH – 74H.
(ii) 172H – BBH.
(iii) 40DH – 27H.
(iv) FFH – EAH.

2.10 Write down the 2s complement of each of the following negative decimal numbers
(a) -7_{10}.
(b) -77_{10}.
(c) -126_{10}.
(d) -200_{10}.

2.11 Determine the hexadecimal equivalent of
(a) -16_{10}.
(b) -66_{10}.
(c) -166_{10}.
(d) -200_{10}.

2.12 (a) Determine the highest power of 2 in each of the following decimal numbers: 1024, 27, 1444, and 511.
(b) Convert each of the numbers into binary.

2.13 (a) Convert each of the following decimal fraction numbers into binary 27.46865, 5 21/64, 4 3/8.
(b) Convert from binary into decimal each of the following 101101.1, 1110111.011, 111.0101, 10001.1101.

2.14 Convert each of the following numbers to their 2s complement form
(a) 101101.
(b) 0.101101.
(c) 11.010.
(d) 1110111.
(e) 0.10101.

2.15 For each of the following work out the product in binary code. Then convert both the problem and the answer to decimal form and confirm your answers
(a) 1101 × 10101.
(b) 10101 × 1010.
(c) 11010 × 1011.
(d) 11110 × 1100.
(e) 1011 × 10.01.
(f) 10.011 × 1.01.

2.16 Solve each of the following divisions and check your answer by converting both problem and answer to decimal
(a) 1111/11.
(b) 10101/101.
(c) 10110111/1100.
(d) 110111/10110.
(e) 110.11/0.11.
(f) 111.01/110.11.

2.17 (a) Convert the following binary numbers to (a) BCD and (b) octal
(i) 11001.
(ii) 1010110.
(iii) 1010010011.
(iv) 110001.00101.
(b) Convert the following BCD numbers to octal
(i) 75.
(ii) 34.
(iii) 973.
(iv) 1035.

2.18 (a) Convert the hexadecimal numbers ABH and 45H to
(i) Binary.
(ii) Decimal.
(iii) Octal

(b) Convert the hexadecimal numbers DEH, 5FEH and 123H to
   (i) Decimal.
   (ii) Binary.

2.19 Calculate $X - Y$ for each of the following cases
(a) $X = 111.01$ and $Y = 100.1$.
(b) $X = 100001$ and $Y = 110$.
(c) $X = 1001.01$ and $Y = 11.1$.

2.20 For each number in exercise 2.19 determine the product XY.

2.21 (a) Determine the smallest and the largest hexadecimal numbers that can be used in a 16-bit digital system.
(b) Calculate the number of different analogue values that may be represented.
(c) A 12-bit system stores the binary number 1000 1011 0011. Convert this to (i) decimal and (ii) hexadecimal.

2.22 Write down (a) the 1s complement and (b) the 2s complement of each of the following
$+23_{10}$.
-23_{10}.
-56_{10}.
$+56_{10}$.
$+90_{10}$.
-90_{10}.

2.23 (a) Convert from binary to decimal
   (i) 1101101.
   (ii) 0101011.
   (iii) 1011111.
   (iv) 1100000.
(b) Convert from decimal into binary
   (i) 109.
   (ii) 43.
   (iii) 95.
   (iv) 96.

2.24 (a) Convert from binary into octal
   (i) 1101101.
   (ii) 0101022.
   (iii) 10111111.
   (iv) 11000000.
(b) Convert from octal into binary
   (i) 42.
   (ii) 24.
   (iii) 77.
   (iv) 1572.

2.25 (a) Convert from binary into hexadecimal
- (i) 110101011011.
- (ii) 101100111000.
- (iii) 101101100110111.

(b) Convert from hexadecimal to binary
- (i) 4AH.
- (ii) BDH.
- (iii) 142H.
- (iv) BC2H.

2.26 (a) Convert from decimal into octal
- (i) 212.
- (ii) 399.
- (iii) 42.
- (iv) 1000.

(b) Convert from octal into decimal
- (i) 218_8.
- (ii) 366_8.
- (iii) 42_8.
- (iv) 26_8.

2.27 Evaluate using the 2s complement method
- (a) 11011101 – 1101.
- (b) 10110111 – 100001.
- (c) 1000000000000 – 1010101.
- (d) 11011000 – 101111.

2.28 Convert
- (a) 1011.101 to decimal.
- (b) 33.3_{10} to binary.
- (c) 25.4H to binary.

2.29 Convert to binary
- (a) 106.5031.
- (b) 6.3.
- (c) 10.8125.
- (d) 50.65.

3 Logic gates

> After reading this chapter you should be able to:
>
> (a) Understand the basic logic functions.
> (b) Recognize the symbol for, and the logic diagram of, each type of gate.
> (c) Describe the operation of each type of gate.
> (d) Construct the truth table of each type of gate.
> (e) Draw timing diagrams for gates.
> (f) Know the Boolean equations that describe the operation of each type of gate.

A logic gate is a circuit that performs a logic function on a number of input binary signals. The logic gate is the basic building block from which many different kinds of logic circuit can be constructed. The signals at the input and output terminals of a gate are either at a HIGH voltage level or at a LOW voltage level. A gate will produce one output level whenever certain combinations of input levels are present, and the other output level whenever any other combination is present at the inputs. Logic gates are readily available in integrated circuit form and the various logic families in common use will be discussed and their characteristics compared in chapter 7. In this chapter the emphasis will be on the various types of gates and the ways in which they can be interconnected to perform different logical functions. The types of gate to be considered are the AND, NOT, OR, NAND, NOR, exclusive-OR, and the exclusive-NOR or coincidence gate. The IEC (International Electrotechnical Commission) symbols for each of these gates are given in Fig. 3.1. The American gate symbols are also given. Positive logic is assumed throughout this chapter; that is, logic 1 is represented by the more-positive or HIGH voltage, and logic 0 by the less-positive voltage or LOW.

Logic functions are defined by a *truth table*. This is a table which lists all the possible combinations of the input variables applied to a gate. Each variable can only be at either the logic 1 or the logic 0 voltage level. The output of a logic gate is generally labelled as F. The combinations are usually listed in binary counting order, i.e. 000, 001, 010, etc. In addition, a logic, or *Boolean* equation, can be employed to represent a logic function. The interconnection of logic gates to obtain a specified circuit is known as *logic design*.

A lamp may represent logic 1 when it is turned ON and glows visibly, and logic 0 when it is turned OFF. A manual switch may be regarded as being at logic 1 when it is closed and at logic 0 when it is open.

Fig. 3.1 *Gate symbols (⊵ denotes active-LOW output; ⊴ denotes active-LOW input)*

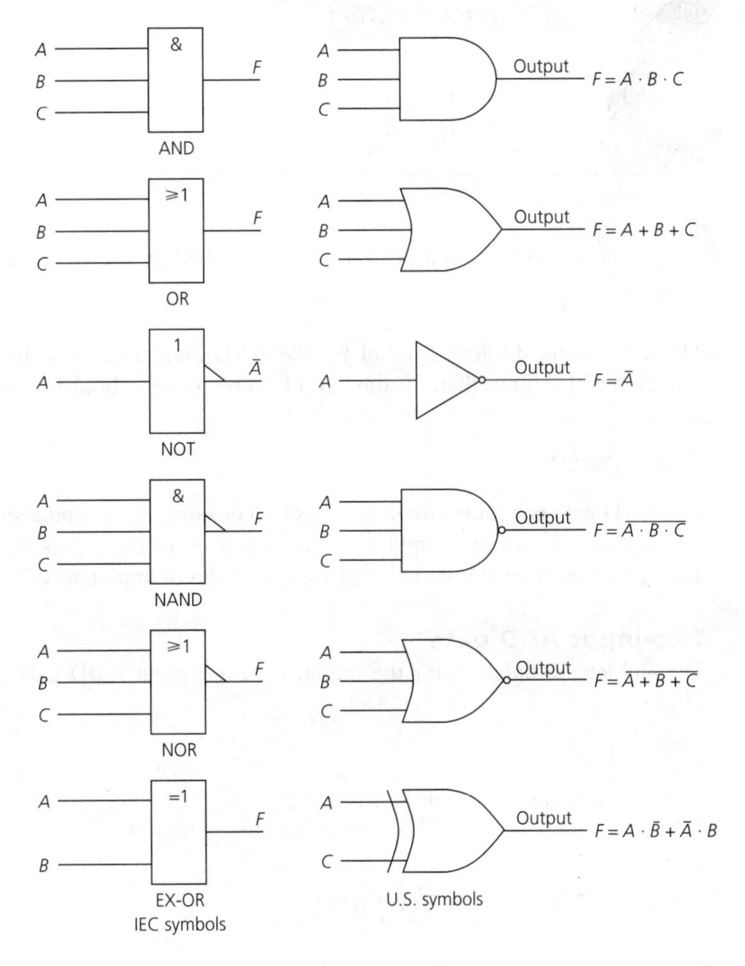

Fig. 3.2 *The AND logic function*

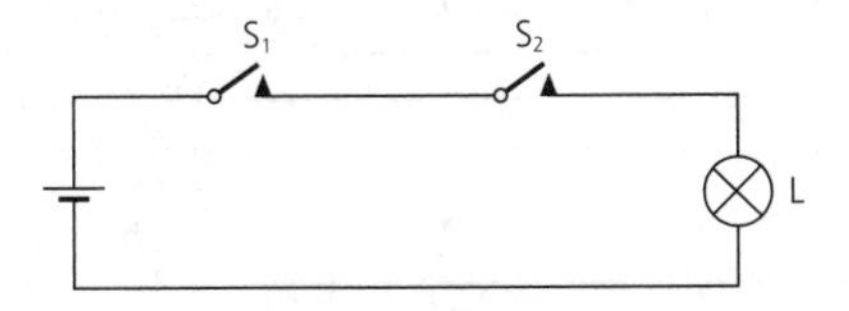

AND logic function

Figure 3.2 shows a lamp connected in series with two switches S_1 and S_2 and a d.c. voltage supply. For a current to flow in the circuit and the lamp to light both switches must be closed. The operation of the circuit can be described by its *truth table* which is given in Table 3.1.

Table 3.1 *AND logic function*

Switch S_1	0	1	0	1
Switch S_2	0	0	1	1
Lamp L	0	0	0	1

The action of the circuit can also be described by the Boolean equation:

$$L = S_1 \cdot S_2 \quad (3.1)$$

The dot · is the Boolean symbol for the AND logical function but it is often omitted. Equation (3.1) shows that, for the lamp L to be ON or 1, both S_1 *and* S_2 must be ON or 1.

AND gate

The AND gate is a logic circuit that has two or more inputs and a single output terminal. The logical state of the output is 1 only when *all* of the inputs are also at logical 1. If any one or more of the inputs is at logical 0, the output state will also be at 0.

Two-input AND gate

The Boolean expression for the output F of a 2-input AND gate is given by equation (3.2)

$$F = A \cdot B \quad (3.2)$$

The truth table is given in Table 3.2.

Table 3.2 *Two-input AND gate*

A	0	1	0	1
B	0	0	1	1
F	0	0	0	1

EXAMPLE 3.1

Determine the output F of a 2-input AND gate when (a) $A = B = 0$; (b) $A = 1$ and $B = 0$. (c) $A = B = 1$.

Solution

From Table 3.2, (a) $F = 0.0 = 0$, (b) $F = 1.0 = 0$, and (c) $F = 1.1 = 1$ (*Ans.*)

Table 3.3 *Three-input AND gate*

A	0	1	0	0	1	1	0	1
B	0	0	1	0	1	0	1	1
C	0	0	0	1	0	1	1	1
F	0	0	0	0	0	0	0	1

Three-input AND gate

The output F of an AND gate with three inputs A, B, and C is

$$F = A \cdot B \cdot C \tag{3.3}$$

The truth table of a 3-input AND gate is given in Table 3.3 and shows that the output is at logical 1 only when A AND B AND C are at 1. The number of combinations of the input variables is equal to 2^n, where n is the number of variables. Thus, the truth table of the 2-input AND gate would require $2^2 = 4$, the 3-input AND gate $2^3 = 8$, and the 4-input gate would require $2^4 = 16$ columns, and so on.

EXAMPLE 3.2

Figure 3.3(a) shows a digital circuit constructed using AND gates. Write down the truth table of the circuit. Simplify the circuit.

Fig. 3.3

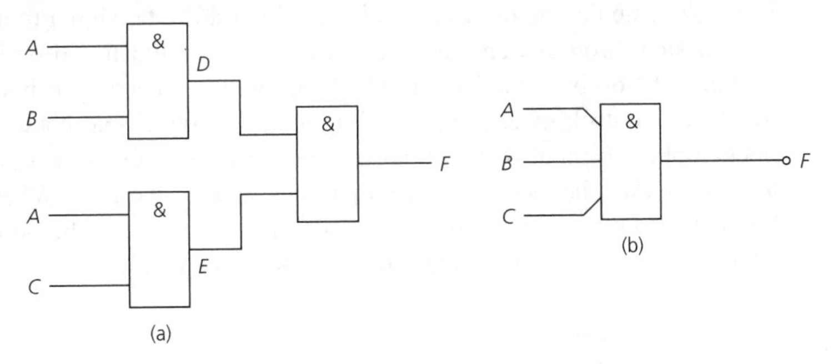

Solution

The truth table is given in Table 3.4. The final column of this truth table is the same as the bottom row of Table 3.3. Hence, the required logic function could be produced by a single 3-input AND gate as shown in Fig. 3.3(b). This is the first indication that it is often possible to simplify a logic circuit. Groups of ANDed terms are often known as product terms (*Ans.*)

Table 3.4

A	B	$D = A \cdot B$	C	$E = A \cdot C$	$F = DE = ABC$
0	0	0	0	0	0
1	0	0	0	0	0
0	1	0	0	0	0
1	1	1	0	0	0
0	0	0	1	0	0
1	0	0	1	1	0
0	1	0	1	0	0
1	1	1	1	1	1

Fig. 3.4 *Enabling a signal*

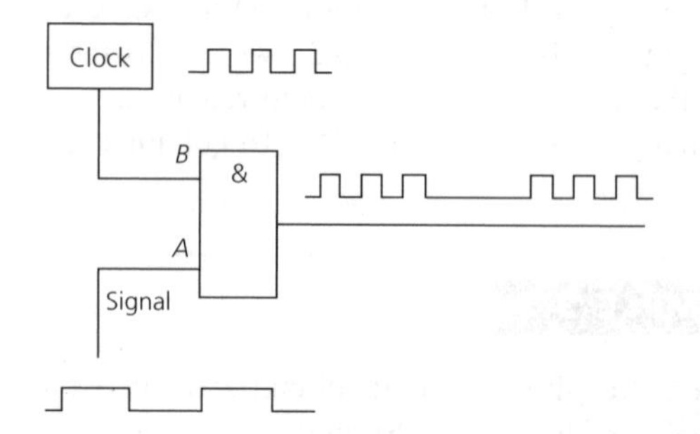

Enabling and inhibiting a signal

The AND gate can be used to enable or inhibit a digital signal to allow, or prevent, its transmission through a circuit. The basic concept is illustrated by Fig. 3.4.

Since the output of a 2-input AND gate will be at 1 only if both its inputs A and B are at 1, a control, or enable, signal applied to input A can control the passage of the clock applied to input B. When the enable signal is at the logic 0, or LOW, level it will stop, or *inhibit*, the clock at B from passing through the gate. When input A is at logic 1, it will allow, or *enable*, the clock applied to B to pass to the output. When the clock waveform is inhibited the output of the AND gate is LOW.

Practical gates

Gates are manufactured in different *logic families* and these are discussed in chapter 7. The easiest logic family to use is probably the TTL LS family (p. 163) and devices in this family are used as examples throughout this book. LS devices are not (usually) employed for new designs when a 74HC alternative is probably available. Devices in the other 74 CMOS logic families (p. 168) may also be available and these provide an alternative that can replace the LS device in each example and practical exercise; however, all CMOS devices require careful handling if they are not to be damaged and, therefore, introductory practical exercises are probably best carried out using LS devices. Logic gates are available in packages that contain 1, 2, 3 or 4 identical gates.

Fig. 3.5 *A 74LS08 quad 2-input AND gate: (a) pinout and (b) logic symbol*

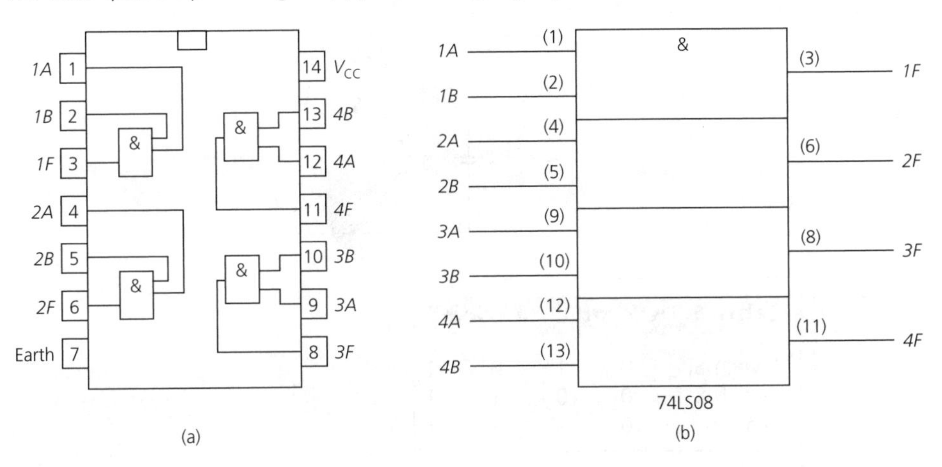

The pin connections (often known as the *pinout*) of the 74LS08 quad 2-input AND gate are shown in Fig. 3.5(a) and its IEC symbol is shown in Fig. 3.5(b). Also available are the 74LS11 triple 3-input AND gate and the 74LS21 dual 4-input AND gate. Some gates have on open-collector output and this is indicated on the logic symbol by ◊. The 74HC08 has the same pinout and logic symbol.

EXAMPLE 3.3

Draw the output waveform of a 3-input AND gate with the three input waveforms shown in Fig. 3.6(a), (b) and (c).

Fig. 3.6

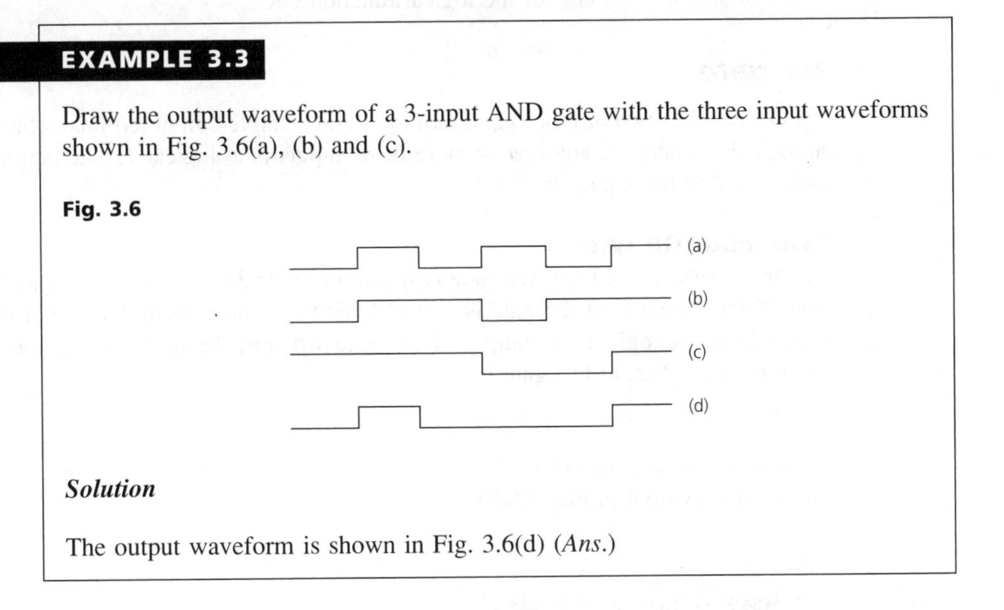

Solution

The output waveform is shown in Fig. 3.6(d) (*Ans.*)

OR logic function

Current will flow in the circuit shown in Fig. 3.7 and light the lamp if either switch S_1 OR switch S_2 OR both switches are closed or ON. Only if both switches are open (0)

Fig. 3.7 *The OR logic function*

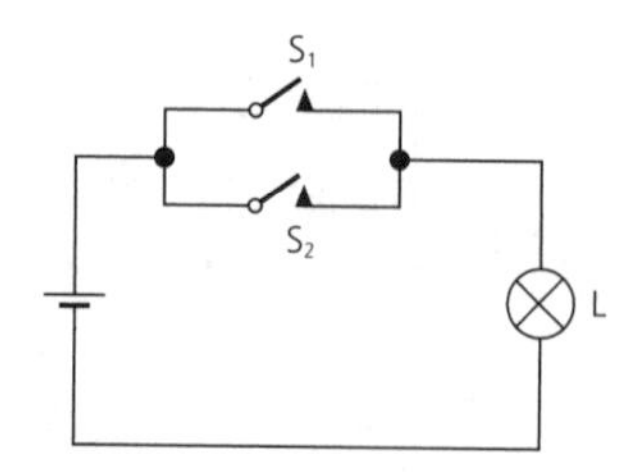

Table 3.5 ***OR logic function***

Switch S_1	0	1	0	1
Switch S_2	0	0	1	1
Lamp L	0	1	1	1

will the lamp be unlit (0). Table 3.5 gives the truth table for the circuit. The Boolean equation describing the action of the circuit is

$$L = S_1 + S_2 \tag{3.4}$$

The + sign is the symbol for the logical function OR.

OR gate

An OR gate has two or more input terminals and a single output terminal which will be at logical 1 whenever any one or more of its inputs is at logical 1. The output is LOW only when all the inputs are LOW.

Two-input OR gate

The truth table of a 2-input OR gate is shown in Table 3.6. When inputs A and B are at logic 0 the output F of the gate is also at logic 0. If either input A or input B or both inputs is/are at logic 1 the output of the gate will also be at 1. Hence, the Boolean equation for a 2-input OR gate is

$$F = A + B \tag{3.5}$$

The pin connections of the 74LS32 quad 2-input OR gate are shown in Fig. 3.8(a) and its IEC symbol in Fig. 3.8(b).

Table 3.6 ***2-input OR gate***

A	0	1	0	1
B	0	0	1	1
F	0	1	1	1

Fig. 3.8 *A 74LS32 quad 2-input OR gate: (a) pinout and (b) logic symbol*

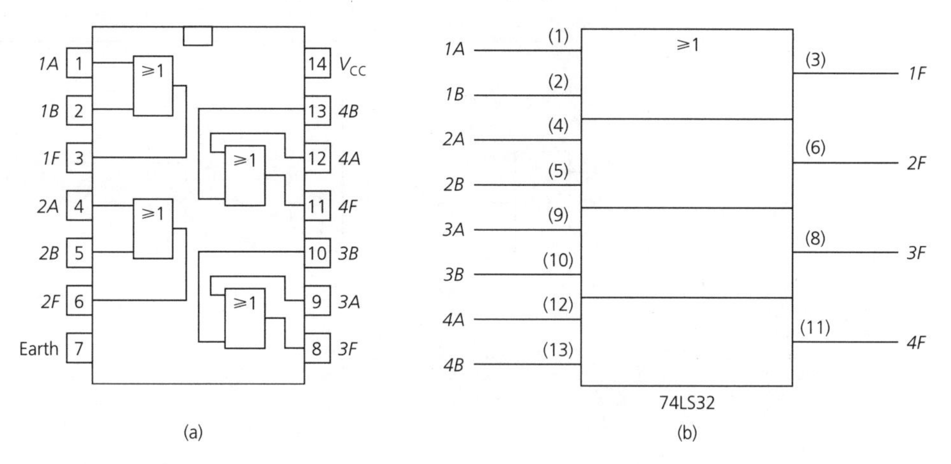

Table 3.7 ***Three-input OR gate***

A	0	1	0	0	1	1	0	1
B	0	0	1	0	1	0	1	1
C	0	0	0	1	0	1	1	1
F	0	1	1	1	1	1	1	1

Three-input OR gate

The Boolean expression for the output of a 3-input OR gate is given by equation (3.6)

$$F = A + B + C \tag{3.6}$$

The truth table of a 3-input OR gate is given in Table 3.7. Groups of ORed terms are often known as *sum terms*.

EXAMPLE 3.4

Show how the 74LS32 quad 2-input OR gate IC can be connected to produce a 4-input OR gate.

Solution

$A + B + C + D = (A + B) + (C + D)$ and this means that three of the IC's gates must be connected to give the required 4-input OR gate. Figure 3.9(a) shows the logic diagram, and Fig. 3.9(b) the required practical connections (*Ans.*)

Fig. 3.9

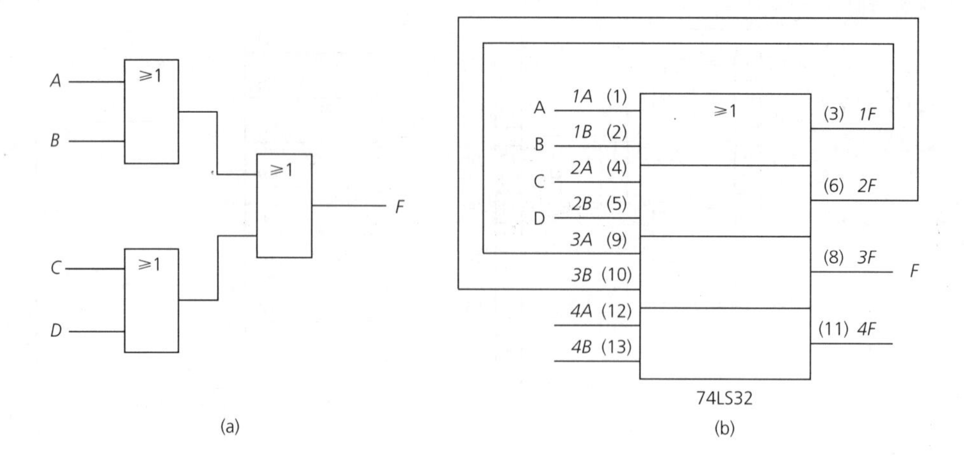

Fig. 3.10 *OR gate used to enable/inhibit a signal*

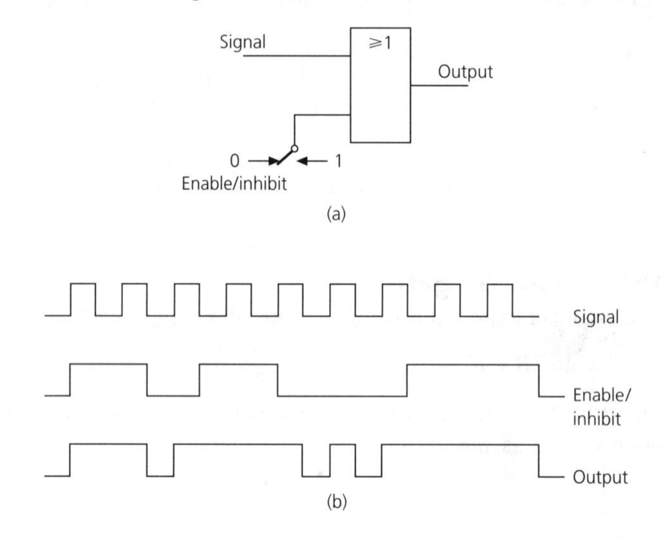

Enabling and inhibiting a digital signal

The OR gate can also be employed to enable or inhibit a digital waveform (see Fig. 3.10(a)). It differs from the AND gate version of Fig. 3.4 in that the enable signal is made HIGH to inhibit and LOW to enable. When inhibited the output of the circuit is HIGH. This is shown by the waveforms in Fig. 3.10(b).

EXAMPLE 3.5

Write down the truth table of the circuit given in Fig. 3.11 and, hence, show that the output F can be obtained in a simpler manner.

Fig. 3.11

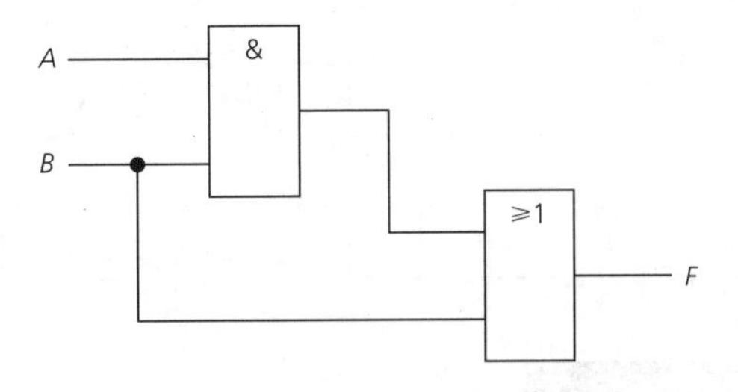

Solution

Table 3.8 shows the truth table of the circuit.

The output F of the circuit is always the same as the input B, and this means that no gates are required since input B may be connected directly to output F (*Ans.*)

Table 3.8

A	0	1	0	1
B	0	0	1	1
AB	0	0	0	1
$F = AB + B$	0	1	1	1

EXAMPLE 3.6

The rectangular waveforms shown in Fig. 3.12 are applied to the inputs of (a) a 2-input AND gate and (b) a 2-input OR gate. Draw the output waveform of each gate. Assume positive logic.

Fig. 3.12

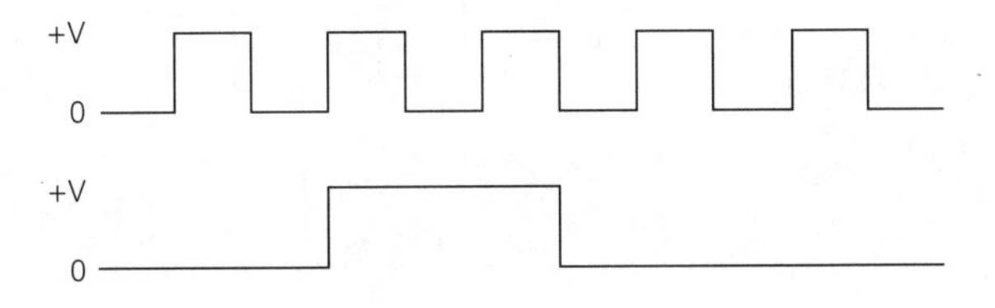

Solution

(a) The output of the AND gate will be 1 only when both of its input waveforms are 1. Figure 3.13(a) shows the output waveform (*Ans.*)
(b) The output of the OR gate will be 1 when either or both of its input waveforms are 1. Hence, the output waveform will be as given in Fig. 3.13(b) (*Ans.*)

Fig. 3.13

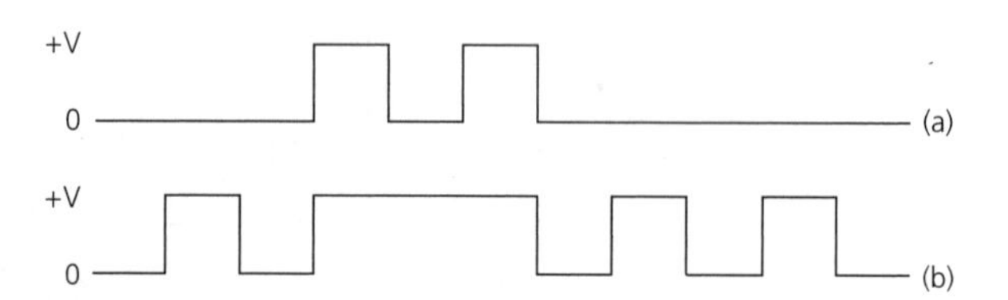

EXAMPLE 3.7

Determine the logic circuit whose truth table is given in Table 3.9. Draw the circuit.

Table 3.9

A	*B*	*C*	*F*
0	0	0	0
1	0	0	0
0	1	0	0
1	1	0	1
0	0	1	1
1	0	1	1
0	1	1	1
1	1	1	1

Solution

The output *F* is at logic 1 either if *A* AND *B* are 1, OR if *C* is 1, OR if *A* AND *B* AND *C* are 1. This means that inputs *A* and *B* are connected to a 2-input AND gate whose output is applied, along with input *C*, to a 2-input OR gate. The circuit is shown in Fig. 3.14 (*Ans.*)

Fig. 3.14

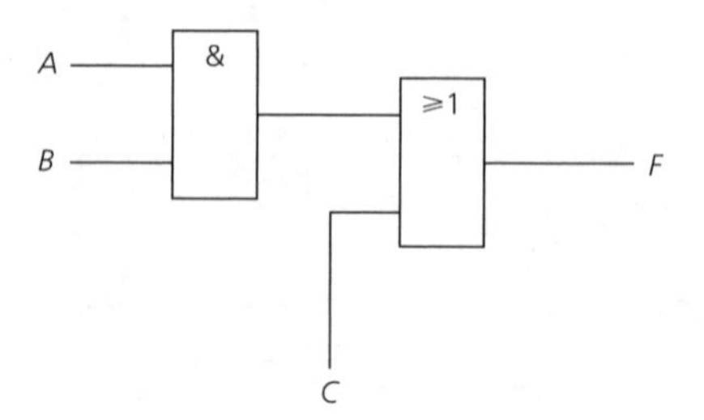

EXAMPLE 3.8

(a) Write down the Boolean equations that describe the logic circuits given in Figs 3.15(a) and (b). (b) Write down the truth table for each circuit and compare them. Comment on the results.

Fig. 3.15

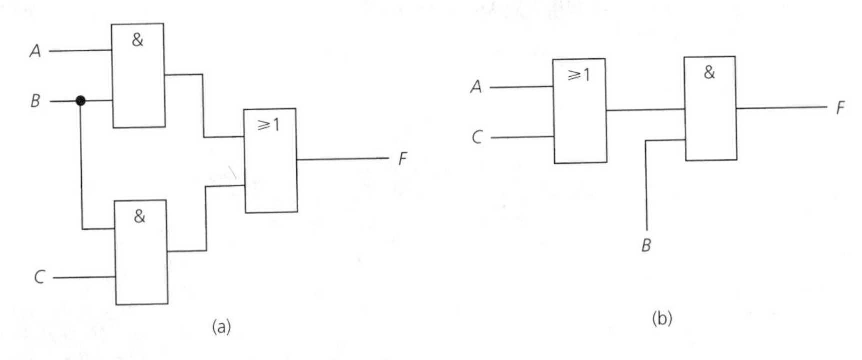

Table 3.10

A	0	1	0	0	1	1	0	1
B	0	0	1	0	1	0	1	1
AB	0	0	0	0	1	0	0	1
C	0	0	0	1	0	1	1	1
BC	0	0	0	0	0	0	1	1
F	0	0	0	0	1	0	1	1

Table 3.11

A	0	1	0	0	1	1	0	1
B	0	0	1	0	1	0	1	1
C	0	0	0	1	0	1	1	1
$A + C$	0	1	0	1	1	1	1	1
F	0	0	0	0	1	0	1	1

Solution

(a) For circuit (a) $F = A \cdot B + B \cdot C$ and for circuit (b) $F = (A + C) \cdot B$ (*Ans.*)

(b) The truth table for circuit (a) is given in Table 3.10 and the truth table for circuit (b) is shown in Table 3.11 (*Ans.*)

Comparing the rows in the tables that give the outputs F of the two circuits it can be seen that they are identical. This shows that the same logical function can often be produced by different combinations of gates. In this example, circuit (b) requires only two gates to perform the desired function as opposed to the three gates needed by circuit (a), and so circuit (b) would probably be chosen.

PRACTICAL EXERCISE 3.1

To investigate the operation of the 74LS08 AND gate and the 74LS32 OR gate, or the 74HC08 and the 74HC32.

Components and equipment: one 74LS08 (or 74HC08) quad 2-input AND gate IC, one 74LS32 (or 74HC32) quad 2-input OR gate IC. One LED. One 270 Ω resistor. Power supply. Breadboard.

Procedure:

Fig. 3.16

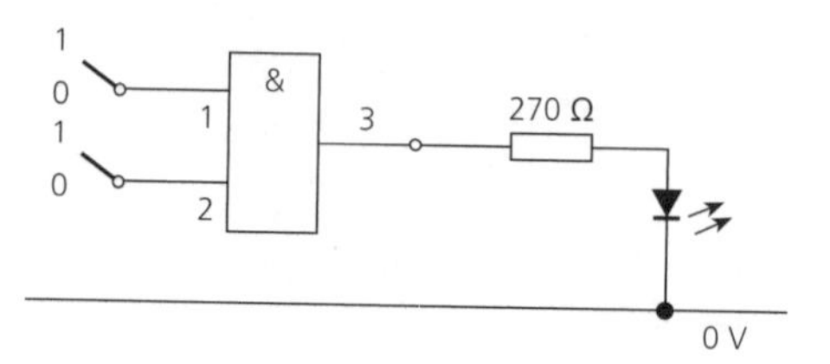

(a) Connect up the circuit shown in Fig. 3.16. Apply a 5 V d.c. power supply between pins 7 and 14, with the positive supply terminal on pin 14. The resistor is required to limit the current that flows in the LED when it is turned ON. When inhibited, the output of the circuit is LOW. [The operation of LEDs and the calculation of the value of the required series resistor are considered in chapter 14 (p. 312).] The LED will glow visibly whenever the output of the gate (pin 3) goes HIGH, i.e. is at logic 1.

(b) (i) Connect pin 1 to the 5 V supply and pin 2 to 0 V and note whether or not the LED glows. (ii) Swop over the connections to pins 1 and 2 and again note the logical state of the LED. (iii) Connect both inputs first to 5 V and then to 0 V and each time note the logical state of the LED. Draw up a truth table of the results (LED glowing = 1, LED dim = 0) and compare it with Table 3.2.

(c) Leave both input pins disconnected (floating) and observe the LED. Determine the logic state of the output of each gate. Comment on the result.

(d) Now remove the AND gate IC from the circuit and replace it with the OR gate IC. [The pin connections are the same.] Repeat procedures (b) and (c).

NOT logic function

The NOT logical function is performed by the circuit shown in Fig. 3.17. Before the switch S_1 is operated, current flows in the circuit and the lamp is lit, or ON. Operation of the switch stops current flowing in the circuit and turns the lamp OFF. The truth table for this circuit is given in Table 3.12.

The Boolean equation for the circuit is

$$L = \overline{S} \tag{3.7}$$

The bar over a symbol means 'NOT that symbol'. In the equation, it means 'NOT S'.

Fig. 3.17 *The NOT function*

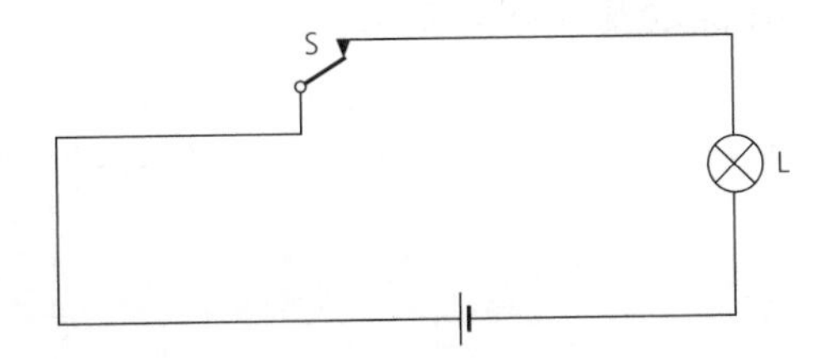

Table 3.12 NOT function

S	0	1
L	1	0

Table 3.13 NOT gate

A	0	1
F	1	0

Fig. 3.18 *A 74LS04 hex inverter: (a) pinout and (b) logic symbol*

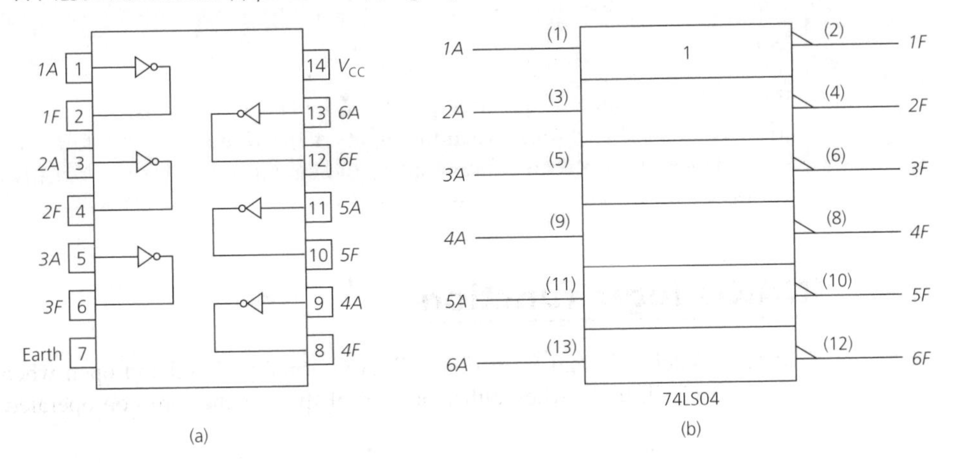

NOT gate

A NOT gate has a single input and a single output and it produces an output signal whose logic level is the inverse of that of the input signal. If the input is at logic 1 the output will be at logic 0 and vice versa. A NOT gate is used to *complement*, or invert a digital signal. The truth table of a NOT gate is given in Table 3.13.

The NOT gate is often known as an inverter. The hex inverter consists of six inverters in one IC package, e.g. the 74LS04 or 74HC04 whose pin connections are given in Fig. 3.18(a) and logic symbol in Fig. 3.18(b).

EXAMPLE 3.9

(a) Write down the truth table of the circuit shown in Fig. 3.19.
(b) Write down the Boolean equation representing the circuit.

Fig. 3.19

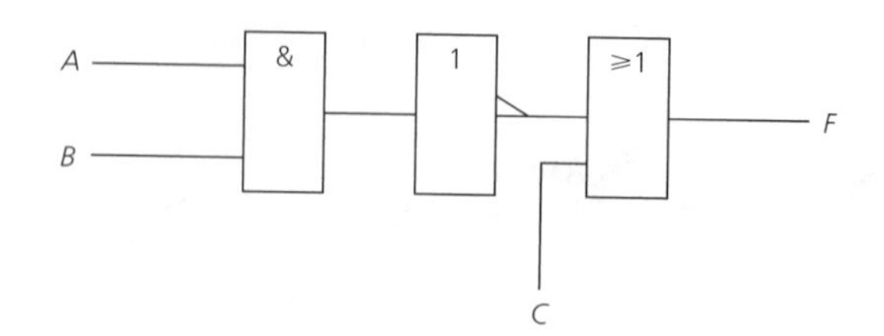

Table 3.14

A	0	1	0	1	0	1	0	1
B	0	0	1	1	0	0	1	1
$A \cdot B$	0	0	0	1	0	0	0	1
$\overline{A \cdot B}$	1	1	1	0	1	1	1	0
C	0	0	0	0	1	1	1	1
F	1	1	1	0	1	1	1	1

Solution

(a) The truth table of the circuit is given in Table 3.14 (*Ans.*)
(b) The output F is HIGH when the inputs A and B are both LOW OR if C is HIGH. Hence, the Boolean expression for the circuit is $F = \overline{A \cdot B} + C$ (*Ans.*)

NAND logic function

The two switches S_1 and S_2 in Fig. 3.20 are normally closed and open when operated. The lamp L will be lit when either or both of the switches are non-operated. If switch

Fig. 3.20 *The NAND logic function*

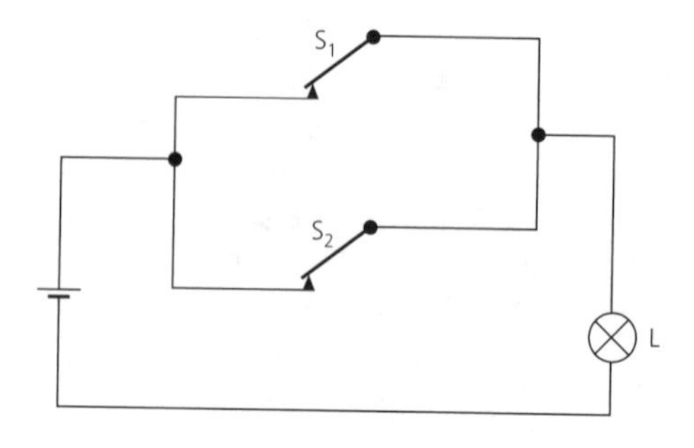

Table 3.15 *NAND function*

S_1	0	1	0	1
S_2	0	0	1	1
L	1	1	1	0

operated = 1 and lamp lit = 1, then Table 3.15 is the truth table for the circuit. Comparing Table 3.15 with Table 3.1 shows that the logical function of the circuit is NOT AND or NAND. The Boolean equation that describes the circuit is

$$L = \overline{S_1 \cdot S_2} \tag{3.8}$$

The NAND logic function performs the inverse of the AND logical function, so that its output is at logic 0 only when *all* of its inputs are at logic 1.

The NAND gate

The output F of a NAND gate is the same as that of an AND gate with its output inverted. If any one or more of the inputs is/are at logic 0 the output F will be at logic 1.

Two-input NAND gate

Table 3.16 gives the truth table of a 2-input NAND gate and its Boolean equation is given by

$$F = \overline{A \cdot B} \tag{3.9}$$

The output of the gate is LOW whenever both inputs are HIGH. The output is HIGH whenever either input or both is LOW.

Table 3.16 *Two-input NAND gate*

A	0	1	0	1
B	0	0	1	1
F	1	1	1	0

Three-input NAND gate

The truth table of a 3-input NAND gate is given in Table 3.17. Clearly the output of the NAND gate is only at logic 0 when all three of its inputs are at logic 1. This aetion can be described by the Boolean expression

$$F = \overline{ABC} \tag{3.10}$$

Table 3.17 *Three-input NAND gate*

A	0	1	0	1	0	1	0	1
B	0	0	1	1	0	0	1	1
C	0	0	0	0	1	1	1	1
F	1	1	1	1	1	1	1	0

Fig. 3.21 *The NAND function produced by (a) an AND gate followed by an inverter and (b) a NAND gate*

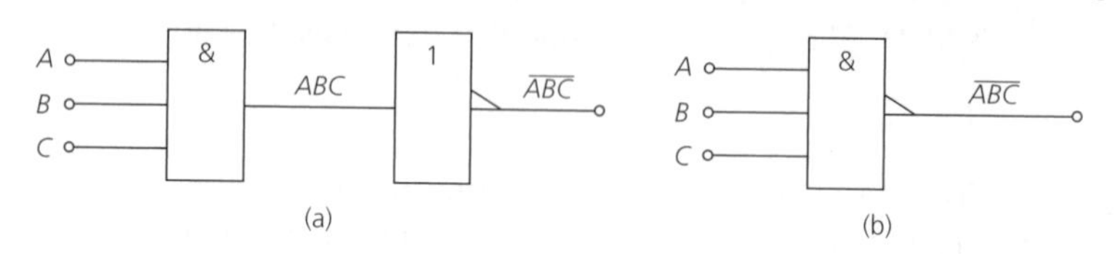

The NAND function can be produced by an AND gate followed by a NOT ga (Fig. 3.21(a)) but it is most often produced by a NAND gate (Fig. 3.21(b)). The NAN gate is readily available in an IC package.

EXAMPLE 3.10

Draw the output waveform of a 3-input NAND gate when the waveforms applied to its inputs are those shown in Fig. 3.22(a), (b) and (c).

Fig. 3.22

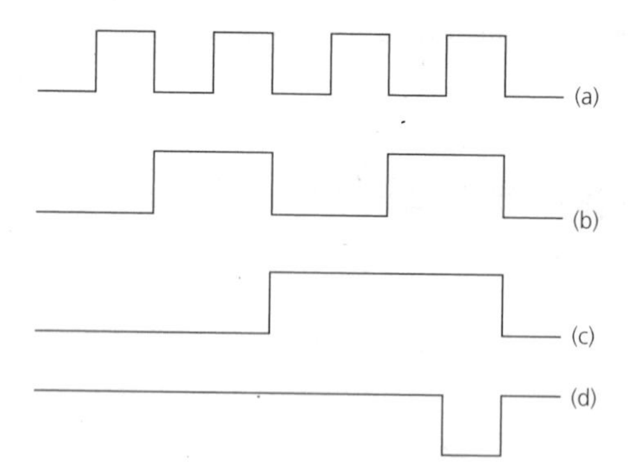

Solution

The output of the NAND gate goes LOW only when all its inputs are HIGH and so the output waveform is as shown in Fig. 3.22(d) (*Ans.*)

Fig. 3.23 *The NAND gate connected as an inverter*

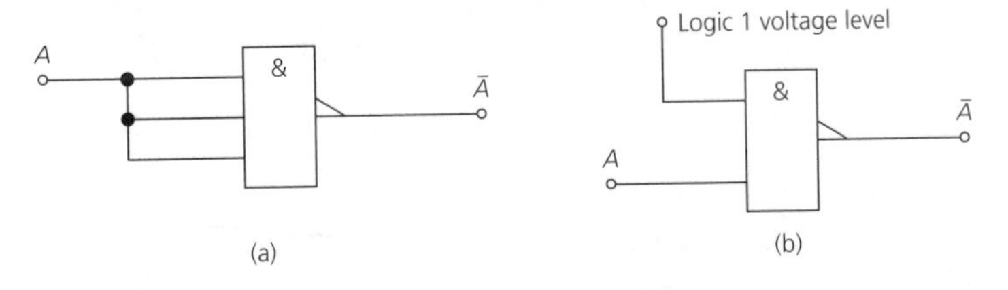

Fig. 3.24 *A 74LS00 quad 2-input NAND gate: (a) pinout and (b) logic symbol*

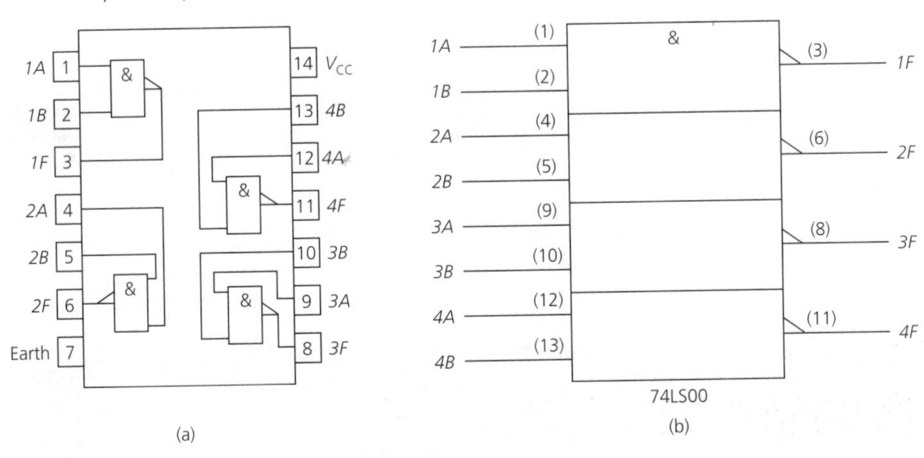

NOT function

The NOT function can be obtained using a NAND gate with its input terminals connected together as shown in Fig. 3.23(a). Alternatively, for a 2-input gate if one input is held at logical 1 (HIGH), any signal applied to the other input will be inverted (see Fig. 3.23(b)). This can readily be confirmed from the truth table of a 2-input NAND gate.

Figure 3.24(a) shows the pinout of the 74LS00 quad 2-input NAND gate and Fig. 3.24(b) gives the IEC symbol for the device. Since the 00 is a quad device the rectangular box has been divided into four parts and the input pin numbers are labelled on the left-hand side. The output pin numbers are labelled on the right-hand side. Other commonly employed NAND gates are the 74LS7410 triple 3-input and the 74LS20 dual 4-input. All these gates are available in the 74HC series.

PRACTICAL EXERCISE 3.2

To investigate the operation of the 74LS00 NAND gate (or the 74HC00 which has the same pinout).
Components and equipment: one 74LS00 quad 2-input NAND gate. One LED. One 270 Ω resistor. Power supply. Breadboard.

Fig. 3.25

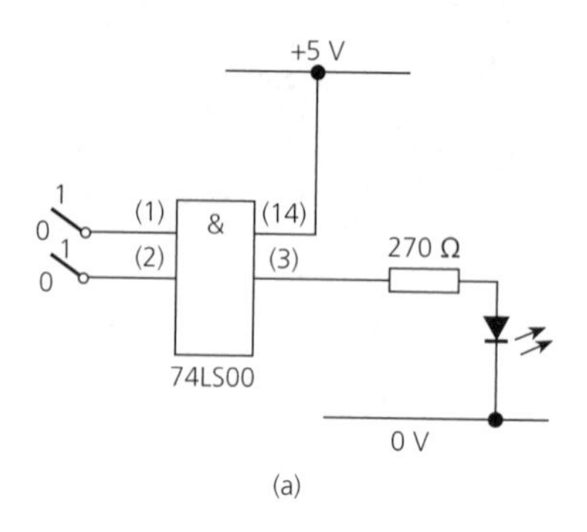

(a)

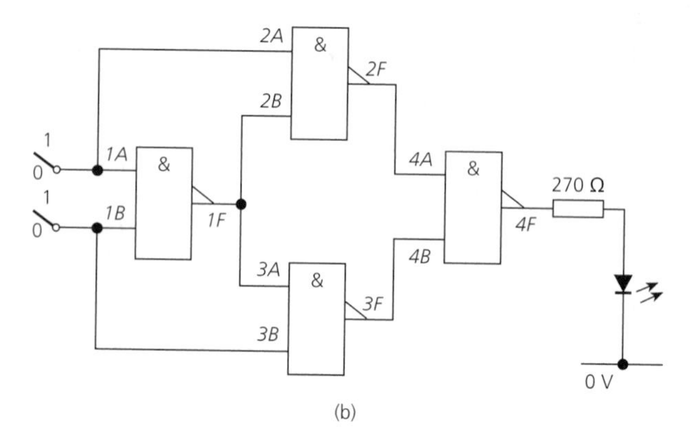

(b)

Procedure:

(a) Connect up the circuit shown in Fig. 3.25(a) with the 5 V power supply connected between pins 14 and 7.

(b) Connect both the inputs $1A$ and $1B$ to logic 0 and note the logical state of the LED. The LED should light to indicate that the output $1F$ is HIGH or at the logical 1 voltage level. Next, transfer $1A$ to logic 1 and note the logical state of the LED. Now reverse the connections to $1A$ and $1B$ and again note the logical state of the LED. Lastly, connect both $1A$ and $1B$ to logic 1 and observe the LED.

(c) Write down the truth table of the gate and compare it with Table 3.16.

(d) Build the circuit shown in Fig. 3.25(b) using all four gates in the IC.

(e) Apply, in turn, each of the four possible combinations of the input variables, i.e. (i) $A = B = 0$, (ii) $A = 1, B = 0$, (iii) $A = 0, B = 1$, and (iv) $A = B = 1$. Each time note the logical state of the LED.

(f) Write down the truth table for the circuit and use it to deduce the Boolean expression that describes the operation of the circuit. What logical function does the circuit perform?

Table 3.18 *NOR function*

S_1	0	1	0	1
S_2	0	0	1	1
L	1	0	0	0

Fig. 3.26 *The NOR logic function*

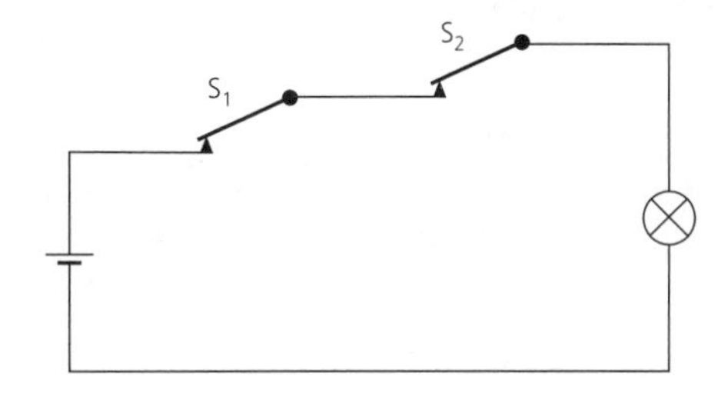

Fig. 3.27 *NOR function produced by OR gate with inverted output*

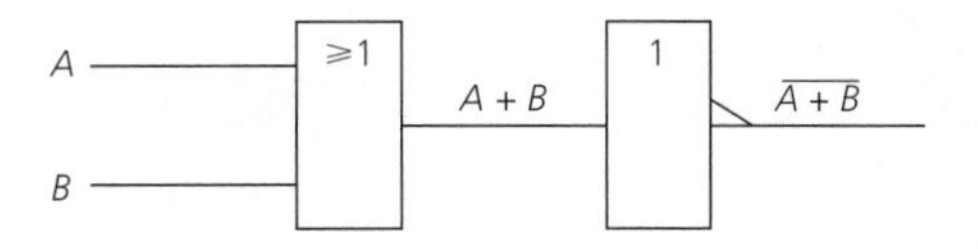

NOR logic function

The NOR (NOT OR) logical function is performed by the circuit given in Fig. 3.26. Clearly, only if both switches are non-operated will current flow in the circuit and the lamp light. Table 3.18 gives the truth table, and equation (3.11) the Boolean equation, describing the logical operation of the circuit

$$L = \overline{S_1 + S_2} \qquad (3.11)$$

The NOR gate

The output of a NOR gate is the same as that of an OR gate whose output has been inverted. The NOR logical function can be produced by an OR gate followed by a NOT gate as shown in Fig. 3.27, but more often a NOR gate IC is employed.

Two-input NOR gate

The truth table of a 2-input NOR gate is given in Table 3.19.

The output F of the 2-input NOR gate is at logic 1 only when both of its inputs are at logic 0. If either input, or both inputs, is/are at logic 1 the output will be at logic 0. This means that the output is only HIGH (1) when both inputs are LOW (0).

Table 3.19 *Two-input NOR gate*

A	0	1	0	1
B	0	0	1	1
F	1	0	0	0

Table 3.20 *Three-input NOR gate*

A	0	1	0	1	0	1	0	1
B	0	0	1	1	0	0	1	1
C	0	0	0	0	1	1	1	1
F	1	0	0	0	0	0	0	0

Three-input NOR gate

The truth table of a 3-input NOR gate is given in Table 3.20 and the Boolean expression that describes its action is given by

$$F = \overline{A + B + C} \quad (3.12)$$

NOT gate

The NOR gate can be used as an inverter or NOT gate by connecting its inputs together as shown in Fig. 3.28(a), or by connecting one input to logic 0 as in Fig. 3.28(b).

Fig. 3.28 *The NOR gate connected as an inverter*

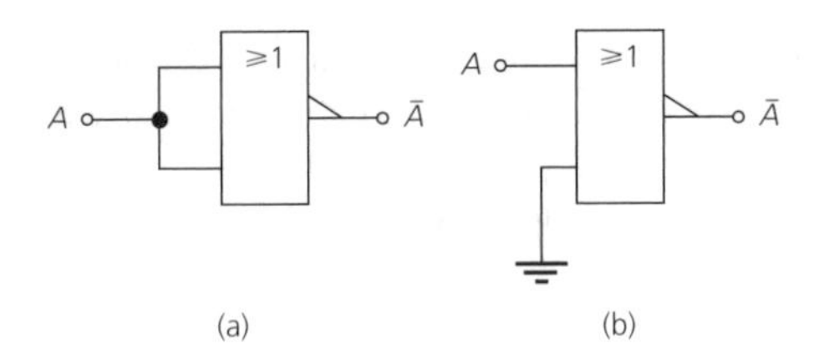

EXAMPLE 3.12

The waveforms shown in Fig. 3.29(a) and (b) are applied to (a) a 2-input AND gate, (b) a 2-input OR gate, (c) a 2-input NOR gate and (d) a 2-input NAND gate. For each case draw the output waveform of the gate.

Solution

The required waveforms are shown in Fig. 3.29(c) through to (f). (*Ans.*)

The pin connections of the 74LS02/74HC02 quad 2-input NOR gate are shown in Fig. 3.30(a) and its IEC symbol is shown in Fig. 3.30(b) (*Ans.*)

Fig. 3.29

Fig. 3.30 *A 74LS02 quad 2-input NOR gate: (a) pinout and (b) logic symbol*

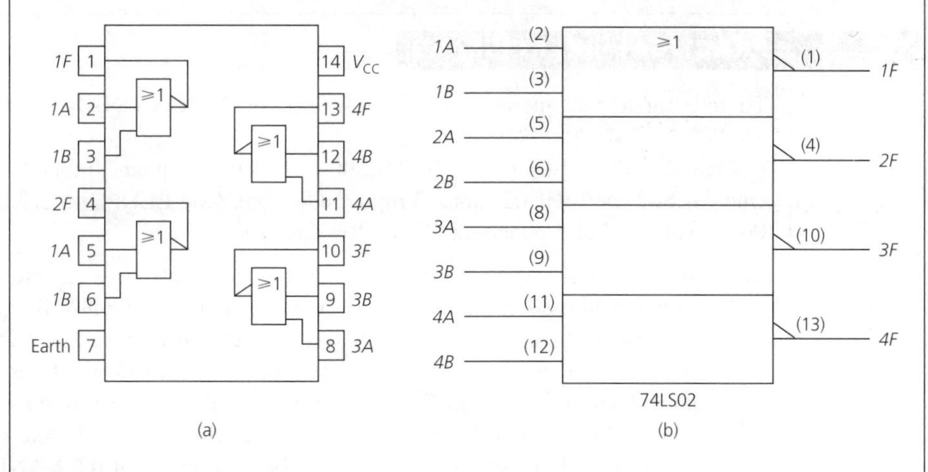

EXAMPLE 3.13

Draw the circuit that is described by the Boolean equation $F = (\overline{A + B}) \cdot (\overline{C + D})$. Use practical gates.

Solution

The two inverted OR functions can be obtained by using two 2-input NOR gates and their outputs can be applied to a 2-input AND gate. The logic diagram is shown in Fig. 3.31(a) and the practical diagram in Fig. 3.31(b) (*Ans.*)

Fig. 3.31

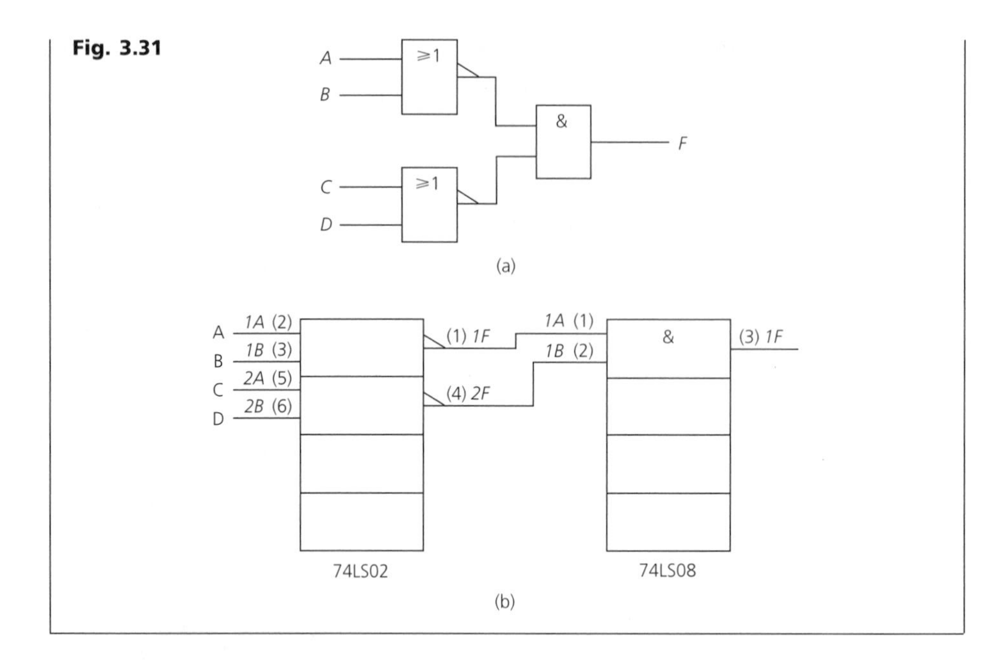

PRACTICAL EXERCISE 3.3

To investigate the operation of the 74LS00 (or 74HC00) NAND gate and the 74LS02 (or 74HC02) NOR gate.

Components and equipment: one 74LS00 (or 74HC00) quad 2-input NAND gate. One 74LS02 (or 74HC02) quad 2-input NOR gate. One LED. One 270 Ω resistor. Power supply. Pulse generator. CRO. Breadboard.

(a) Using the NAND gate connect the 5 V power supply between pins 7 and 14 with the positive terminal on pin 14. Connect pin 3 of the NAND gate to 0 V via a resistor R and a LED. The value of R should be about 270 Ω. Connect pin 1 to +5 V and pin 2 to 0 V and note if the LED glows. Reverse the connections to pins 1 and 2 and again observe the LED. Now try each of the other combinations of the input variables in turn and for each note whether or not the LED glows visibly. Write down the truth table for the NAND gate but note the pinout.

(b) Repeat procedure (a) using the NOR gate IC in place of the NAND gate.

(c) Connect one of the NOR gates to act as an inverter. Then apply first logic 1 and then logic 0 to its input terminals and each time note the logical state of the LED connected to its output. Write down the truth table for the inverter.

(d) Now connect the pulse generator to input *1A* and the CRO to output *1F*. The gate is to act to enable/disable the clock waveform produced by the pulse generator. Connect input *1B* to the appropriate logic level to disable the clock. Check that there is no output waveform displayed on the CRO. Now change the signal on *1B* and see if the CRO now displays the clock waveform.

Table 3.21 *Exclusive-OR gate*

A	0	1	0	1
B	0	0	1	1
F	0	1	1	0

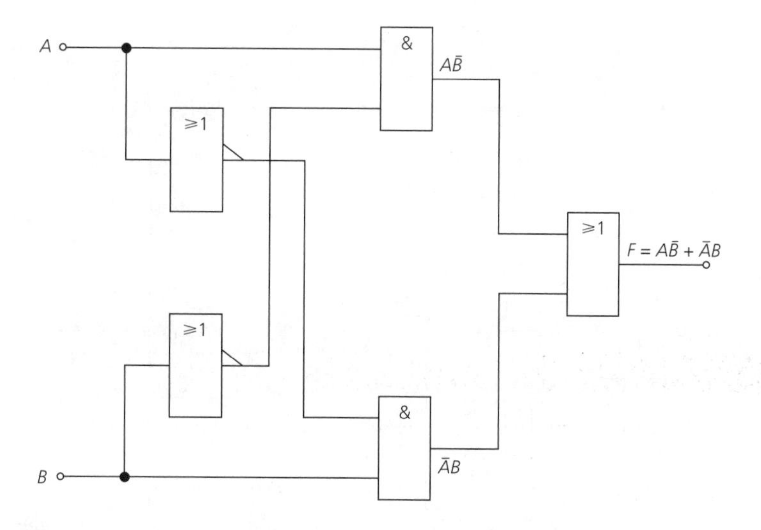

Fig. 3.32 *Exclusive-OR gate*

The exclusive-OR gate

The exclusive-OR gate has just two input terminals and one output terminal. The output F of the gate in HIGH only when either one, but not both, of its inputs is also HIGH. The truth table for the gate is given in Table 3.21. If both inputs are HIGH, or both inputs are LOW, the output F will also be LOW. The exclusive-OR gate performs the logical function

$$F = A\bar{B} + \bar{A}B = A \oplus B \tag{3.13}$$

The exclusive-OR gate can be made by suitably combining other types of gate and it can also be obtained in an IC package, e.g. the 74LS86 quad 2-input gate. Figure 3.32 shows how an exclusive-OR gate can be made using a mixture of AND, OR and NOT gates. Figure 3.33 shows another circuit which performs the exclusive-OR logic function.

Figure 3.34(a) shows the pinout of the 74LS86/74HC86 quad exclusive-OR gate IC and Fig. 3.34(b) shows its logic symbol.

Fig. 3.33 *Exclusive-OR function implemented using NAND gates*

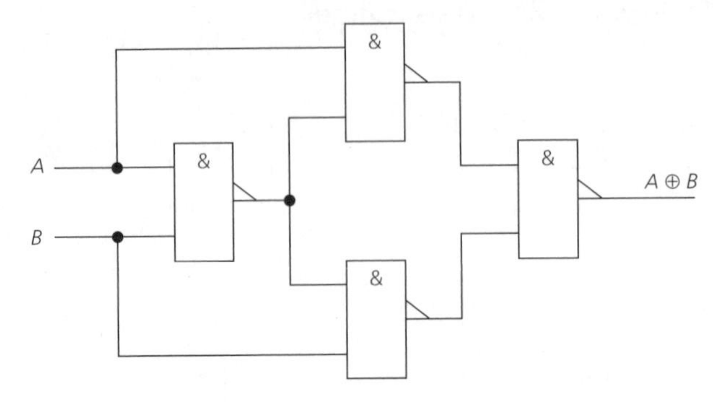

Fig. 3.34 *A 74LS86 quad exclusive-OR gate: (a) pinout and (b) logic symbol*

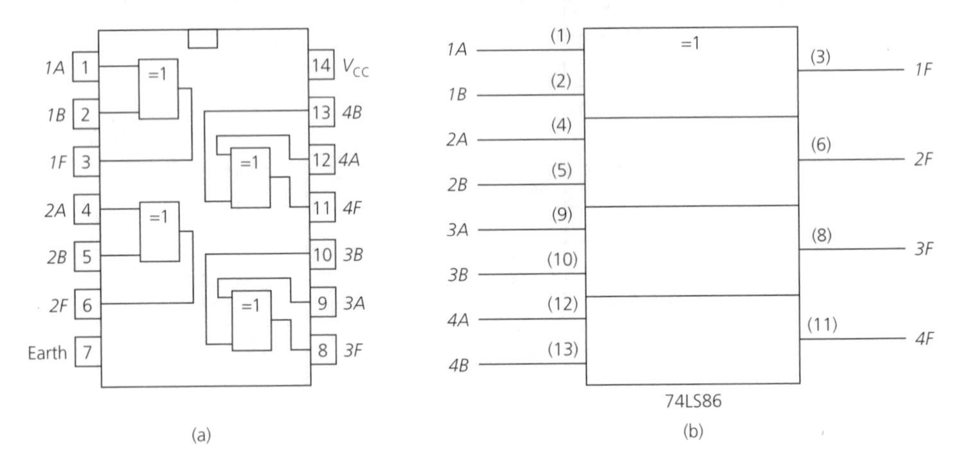

Table 3.22 *Exclusive-NOR gate*

A	0	1	0	1
B	0	0	1	1
F	1	0	0	1

The exclusive-NOR gate

The coincidence, or exclusive-NOR, gate is one which has two input terminals and one output terminal. It produces the logical 1 state at the output only when the two inputs are at the same logical state. The truth table of a coincidence gate is given in Table 3.22.

From the truth table the Boolean equation that describes an exclusive-NOR gate is

$$F = AB + \overline{AB} \quad (3.14)$$

It is shown on p. 90 that equation (3.14) is the complement of equation (3.13). This means that an exclusive-NOR gate performs the inverse function to an exclusive-OR gate.

Gates with inverted inputs

Each of the various logic functions can always be performed by two gates whose function is different from the function required. Two of these combinations have already been met; the AND gate followed by a NOT gate gives the NAND logical function (Fig. 3.21(a)), and an OR gate followed by a NOT gate gives the NOR function (Fig. 3.28). Various other gate combinations are also possible and any of them may sometimes be convenient to employ.

AND gate with inverted inputs

Table 3.23 shows that if the inputs A and B to a 2-input AND gate are inverted the output of the circuit is $\bar{A} \cdot \bar{B}$. This (see Table 3.9) is the same as the output of a NOR gate, i.e. the output is HIGH only when both the inputs are LOW (see Fig. 3.35(a)).

Table 3.23

A	B	$\bar{A}$	$\bar{B}$	$F = \bar{A} \cdot \bar{B}$
0	0	1	1	1
1	0	0	1	0
0	1	1	0	0
1	1	0	0	0

Fig. 3.35 *Gates with inverted inputs perform different logic functions*

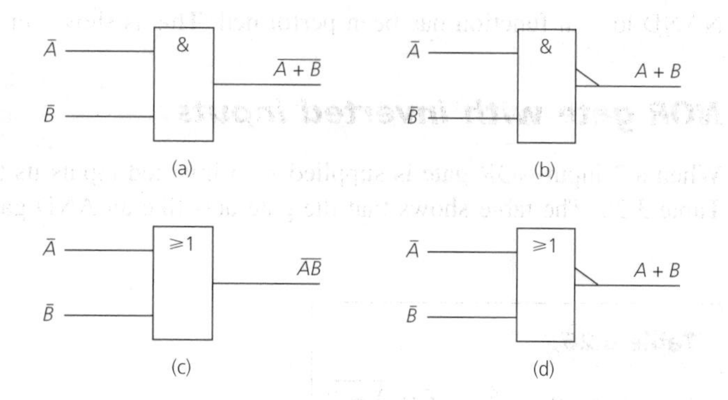

NAND gate with inverted inputs

The truth table of a NAND gate with inverted inputs is given in Table 3.24. The output of the gate is LOW only when both of its inputs are also LOW. If either, or both, input(s) is/are HIGH the output is also HIGH. Hence, a NAND gate with inverted inputs performs the OR logical function (see Fig. 3.35(b)).

Table 3.24

A	B	$\bar{A}$	$\bar{B}$	$F = \overline{\bar{A} \cdot \bar{B}}$
0	0	1	1	0
1	0	0	1	1
0	1	1	0	1
1	1	0	0	1

Table 3.25

A	B	$\bar{A}$	$\bar{B}$	$F = \bar{A} + \bar{B}$
0	0	1	1	1
1	0	0	1	1
0	1	1	0	1
1	1	0	0	0

OR gate with inverted inputs

Table 3.25 is the table of a 2-input OR gate both of whose inputs have been inverted. Comparing the last row of the table with the last row of Table 3.16, it is clear that the NAND logical function has been performed. This is shown in Fig. 3.35(c).

NOR gate with inverted inputs

When a 2-input NOR gate is supplied with inverted inputs its truth table is as given in Table 3.26. The table shows that the gate acts like an AND gate (see Fig. 3.35(d)).

Table 3.26

A	B	$\bar{A}$	$\bar{B}$	$F = \overline{\bar{A} + \bar{B}}$
0	0	1	1	0
1	0	0	1	0
0	1	1	0	0
1	1	0	0	1

PRACTICAL EXERCISE 3.4

To investigate various combinations of gates.
Components and equipment: one 74LS00 (or 74HC00) quad 2-input NAND gate IC. One 74LS02 (or 74HC02) quad 2-input NOR gate IC. One LED. One 270 Ω resistor. Power supply.
Procedure:

(a) Connect pins 4 and 5 of the 74LS00 NAND gate together to produce a NOT gate whose output is at pin 6. Do the same with pins 8, 9 and 10 to obtain another NOT gate.
(b) Connect pin 3 to pin 4 and connect pin 6 to 0 V via resistor R and the LED. Now connect each of the possible combinations of input voltages to pins 1 and 2 and for each one note whether the LED glowed or not. Write down the truth table of the circuit and state what logical function it represents.
(c) Connect the outputs (pins 6 and 8) of the two NAND gates that have been connected as NOT gates to pins 1 and 2, respectively. Connect pin 3 to 0 V via R and the LED. Apply each combination of input voltages to the NOT gate inputs (pins 4 and 9) and each time observe the LED. Write down the truth table of the circuit and state what logical function has been performed.
(d) Replace the NAND gate IC with the 74LS02 NOR gate IC and repeat procedures (a) through to (c).

Logic circuit from a Boolean expression

Combinational logic circuits are made by connecting the basic logic gates together to form a new, more complex, logic function. When a logic circuit is designed a logic expression is first obtained, simplified if possible and then a logic diagram that will implement the wanted logic operation is drawn. The basic process is simple; draw one gate for each logic operation in the Boolean equation and ensure that each gate is given the correct input(s).

EXAMPLE 3.14

Draw the logic diagram for the logic function $F = \overline{A + B} + A\overline{C}$.

Solution

The final stage of the required circuit must consist of an OR gate whose inputs are $\overline{A + B}$ and $A\overline{C}$. The first input can be obtained from a NOR gate with inputs A and B, and the second input can obtained from an AND gate with inputs A and $\overline{C}$. Lastly, an inverter is required to obtain the $\overline{C}$ term. The required circuit is shown in Fig. 3.36 (*Ans.*)

Fig. 3.36

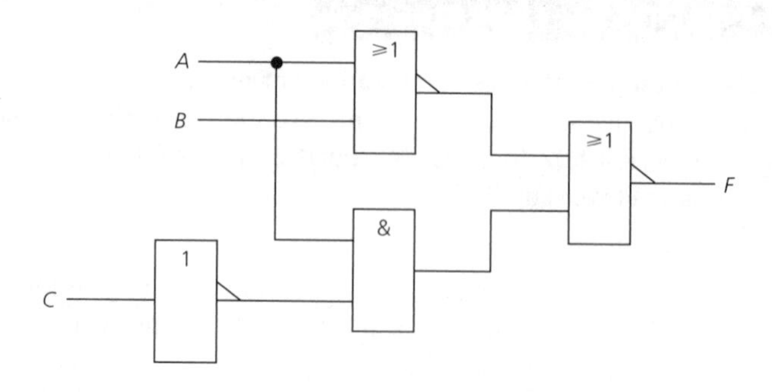

EXAMPLE 3.15

Draw the logic diagram of $F = (\overline{A+B})(\overline{C+D})(\overline{A+D})$. Show how the diagram can be re-drawn using IEC symbols.

Solution

Three 2-input NOR gates are required to give the three NOR functions, and one 3-input AND gate to produce the output. The logic diagram is shown in Fig. 3.37(a).

Fig. 3.37

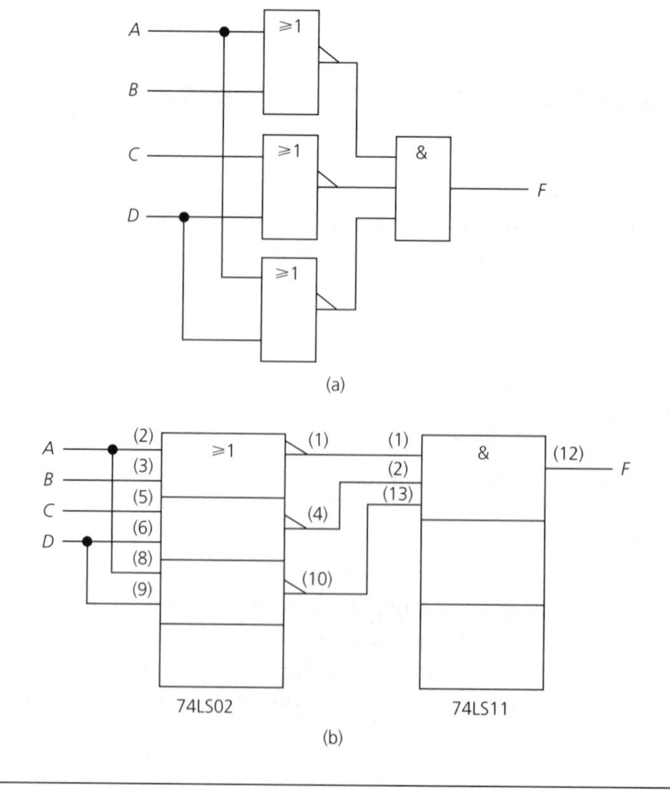

Using IEC symbols the diagram is redrawn as shown in Fig. 3.37(b). The 74LS11 is a triple 3-input AND gate but only one of its gates is needed, the gate employed has input pins 1, 2 and 3, and output pin 12 (*Ans.*)

EXAMPLE 3.16

Implement the logic function $F = (A + B)(A + \overline{B + C})(\overline{B + D})$.

Solution

$A + B$ requires one 2-input OR gate.
$A + \overline{B + C}$: $\overline{B + C}$ requires one 2-input NOR gate. The output of this is applied to the input of another 2-input OR gate along with input A.
$\overline{B + D}$: this function requires another 2-input NOR gate.
F: the outputs of the OR/NOR gates are connected to the inputs of a 3-input AND gate. The circuit is shown in Fig. 3.38. It requires the use of one quad 2-input OR gate, one quad 2-input NOR gate, and one triple 3-input AND gate. [The circuit can be reduced to a simpler form which uses fewer gates; see exercises (4.10) and (6.12).] (*Ans.*)

Fig. 3.38

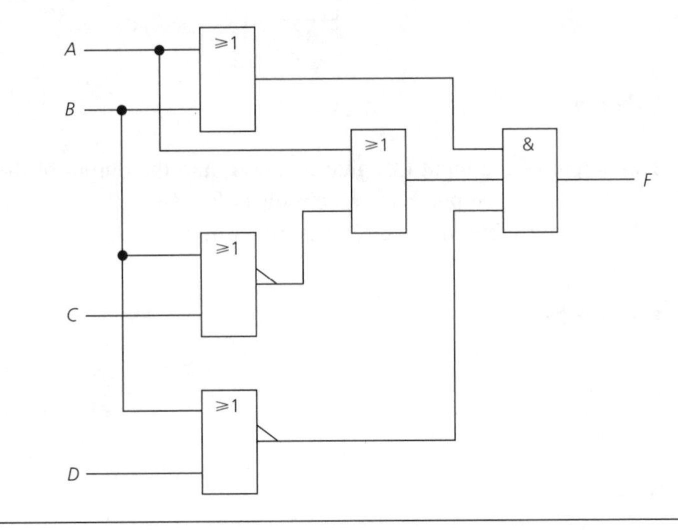

Analysis of circuits

A combinational logic circuit is analysed by proceeding from its input terminals through to its output terminal. The Boolean equation for the output of each gate is written down and combined to obtain an equation for the output F of the circuit. A truth table will then give the output of the circuit for each of the possible combinations of the input signals. The logic function at intermediate points in the circuit can be determined to simplify the determination of the output function.

EXAMPLE 3.17

Analyse the circuit given in Fig. 3.39. Write down the truth table for the circuit and use it to determine the output when (a) $A = B = C = 1$, and (b) $A = C = 0$ and $B = 1$. (c) What is the logic function of the circuit?

Fig. 3.39

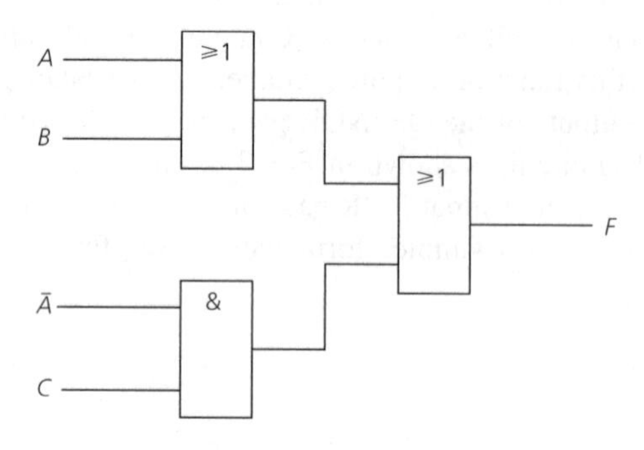

Solution

The output of the input OR gate is $A + B$, and the output of the input AND gate is $\bar{A}C$. Hence, the output F of the circuit is $F = (A + B) + \bar{A}C$ (*Ans.*)

The truth table of the circuit is given in Table 3.27.

Table 3.27

A	0	1	0	1	0	1	0	1
B	0	0	1	1	0	0	1	1
C	0	0	0	0	1	1	1	1
$A + B$	0	1	1	1	0	1	1	1
$\bar{A} \cdot C$	0	0	0	0	1	0	1	0
F	0	1	1	1	1	1	1	1

From Table 3.27:

(a) when $A = B = C = 1$, $F = 1$ (*Ans.*)
(b) when $A = C = 0$ and $B = 1$, $F = 1$ (*Ans.*)
(c) The output F of the circuit is at logic 1 whenever any one, or more, of the inputs is at 1, so that the circuit implements the OR logic function (*Ans.*)

Fig. 3.40

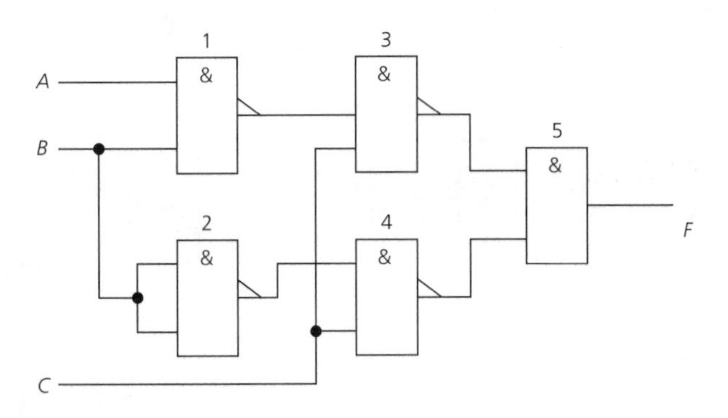

EXAMPLE 3.18

Determine the Boolean expression for the output of the circuit given in Fig. 3.40.

Solution

Gate 1: output = $\overline{A \cdot B}$. Gate 2: output = $\bar{B}$. Gate 3: output = $\overline{\overline{A \cdot B} \cdot C}$.
Gate 4: output = $\overline{\bar{B} \cdot C}$. Gate 5: output = $\overline{(\overline{\overline{A \cdot B} \cdot C})(\overline{\bar{B} \cdot C})}$ (*Ans.*)
[Note that this output expression can be reduced to a simpler form; see exercise (4.6b).]

Applications of logic gates

The number of possible applications of logic gates is extremely large and here some simple examples are given.

Door alarm

A system is required which will allow a door to be opened only when the correct combination of four push-buttons is pressed. Any incorrect combination is required to bring up an alarm.

Let the four buttons be labelled as A, B, C and D and suppose that the correct combination to open the door is $A = 1$, $B = 1$, $C = 0$ and $D = 0$. Then the output F of the system which opens the door is $F = AB\bar{C}\bar{D}$ and the system can be implemented using a 4-input AND gate and two NOT gates as shown in Fig. 3.41. The alarm requirement is easily satisfied by connecting the output of the 4-input AND gate via a NOT gate to the alarm circuit. Further circuitry would be needed to ensure that the alarm did not operate continually, and hence the NOT gate output is fed into an AND gate along with the output of an OR gate whose inputs are the four push-buttons. The alarm will now operate only when A or B or C or D and the output of the NAND gate are at 1.

Fig. 3.41

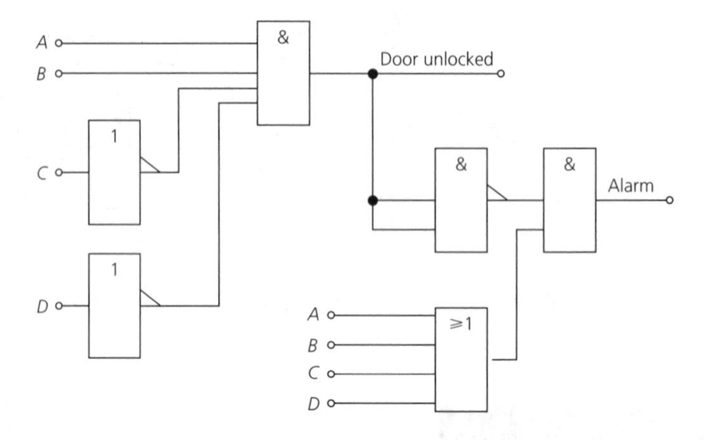

High-voltage power supply

Access (A) to the high-voltage section of a radio transmitter should be possible only if (i) the power (P) has been switched off, (ii) the door to the section has been unlocked with a special key (D) to ensure that the power cannot be turned on again by another person, (iii) the h.t. line (H) has been earthed.

The Boolean expression describing this action is $A = \bar{P}DH$, and this is easily implemented using one 3-input AND gate and one NOT gate.

Motor control

An electrical motor is to operate when (i) the power supply is connected (P), (ii) the current taken from the supply is less than some safety factor figure (I), (iii) the power supply should not be able to be switched on unless a safety guard (S) is in position, although this can be overridden by a maintenance technician by means of a special key (K).

The required logical function is $F = PIS\bar{K} + PI\bar{S}K$.

Implementation will require two 4-input AND gates, one 2-input OR gate and two NOT gates.

Self-service petrol pump

A self-service petrol pump is to provide the required grade of petrol (P) if the pump is switched on (S), and the grade selector is positioned to one of the 4-star, 5-star and green positions (4, 5, G) and a button (B) is pressed to alert the cashier that the pump is in operation.

The Boolean expression representing this action is $P = SB(4 + 5 + \text{G})$, and this can be implemented by one 3-input AND gate and one 3-input OR gate.

Water pump

A water pump is required to turn ON automatically whenever the water level in any two or more of three tanks *A*, *B*, and *C* falls below a pre-set level. Each water tank is provided with a level detector that generates a HIGH voltage whenever the water level in that tank is LOW. If a HIGH voltage turns ON the water pump, then $W = AB\overline{C} + A\overline{B}C + \overline{A}BC + ABC$. This expression could be implemented directly using AND, OR and NOT gates but it would, in practice, be first reduced to a simpler form to limit the number of gates required.

Deducing the truth table of a given circuit

A truth table can be obtained from a logic diagram by allocating a separated column in the table for the output of each gate in the circuit. Starting at the gate, or gates, to which the system inputs are applied write down one, or more, columns which show the first logic operations performed on the inputs. The output(s) obtained is then the input(s) to the next stage of gates, and new columns in the table are written down that show the results of the logic operations performed on the secondary inputs. The process is repeated until the output of the circuit is reached. At each stage the Boolean equation for that point can also be written down.

As an example, consider the circuit shown in Fig. 3.42. The inputs *A* and *B* are applied to an OR gate and the output of the OR gate is applied to one input of an AND gate along with the third input *C*. The truth table for the circuit is given in Table 3.28.

Since there are three inputs there should be $2^3 = 8$ columns in the table.

The Boolean expression for the circuit is $F = (A + B)\overline{C}$.

Fig. 3.42

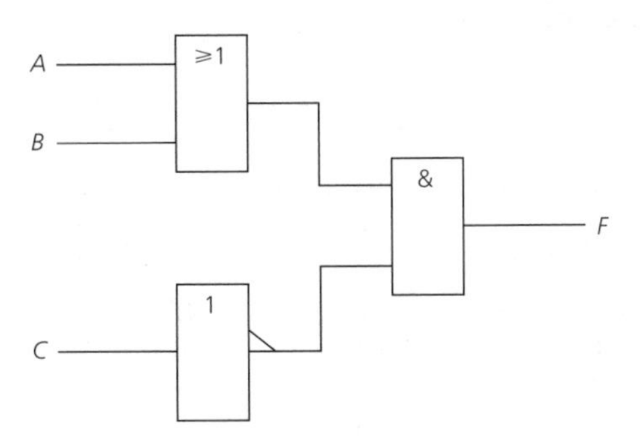

Table 3.28

A	0	1	0	1	0	1	0	1
B	0	0	1	1	0	0	1	1
C	0	0	0	0	1	1	1	1
$A + B$	0	1	1	1	0	1	1	1
$(A + B)\overline{C}$	0	1	1	1	0	0	0	0

EXAMPLE 3.19

Obtain the truth table for the circuit shown in Fig. 3.43.

Fig. 3.43

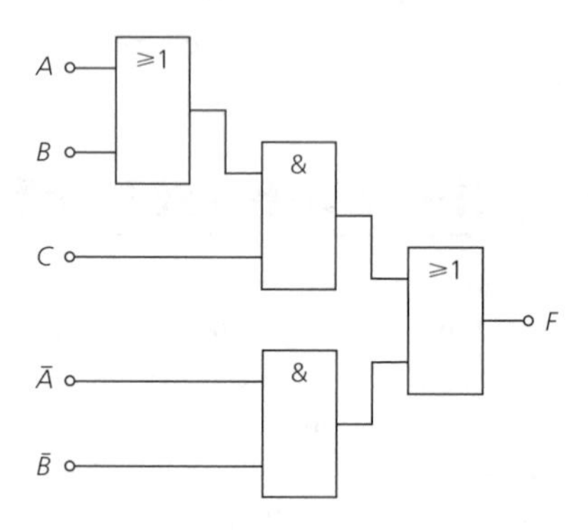

Solution

The output of the input OR gate is $A + B$ and the outputs of the two AND gates are (top) $(A + B)C$ and (bottom) $\bar{A}\bar{B}$. The output F of the circuit is $F = (A + B)C + \bar{A}\bar{B}$. The truth table of the circuit is shown in Table 3.29. (*Ans.*)

Table 3.29

A	B	C	$A + B$	$(A + AB)C$	$\bar{A}$	$\bar{B}$	$\bar{A}\bar{B}$	F
0	0	0	0	0	1	1	1	1
1	0	0	1	0	0	1	0	0
0	1	0	1	0	1	0	0	0
1	1	0	1	0	0	0	0	0
0	0	1	0	0	1	1	1	1
1	0	1	1	1	0	1	0	1
0	1	1	1	1	1	0	0	1
1	1	1	1	1	0	0	0	1

The truth table can be employed to determine the output of the circuit for any particular combination of the input variables. For example, if $A = B = C = 1$ then the output F is 1. If the requirement is merely to find the output of a circuit for a particular set of input variables it is not necessary first to obtain the truth table. The logic state of each input can be marked on the circuit and the state of each gate output found until the output F is determined.

EXAMPLE 3.20

The circuit in Fig. 3.44 has inputs $A = B = C = 1$. Find the output state of the circuit.

Solution

See Fig. 3.45.

Fig. 3.44

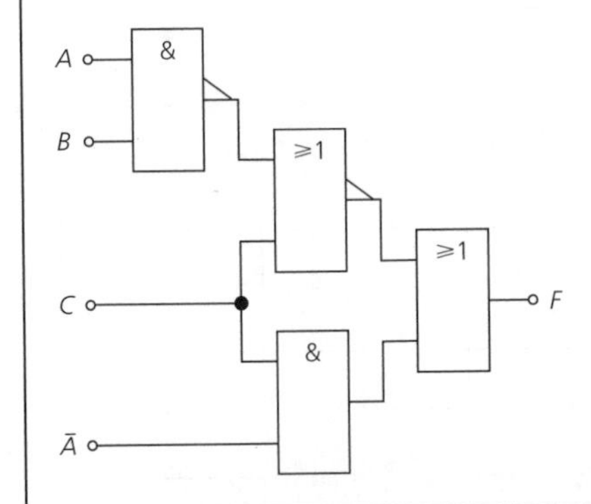

Fig. 3.45

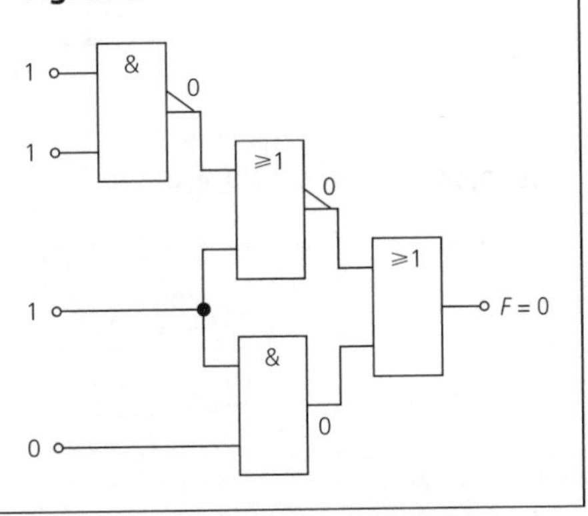

EXERCISES

3.1 Obtain the truth table for the circuit shown in Fig. 3.46.

3.2 Obtain the truth table and the Boolean equation for the circuit given in Fig. 3.47.

3.3 Obtain the Boolean equation and the truth table for the logic diagram of Fig. 3.48.

3.4 Write down the truth table of
(a) A 4-input AND gate.
(b) A 4-input OR gate.

Fig. 3.46

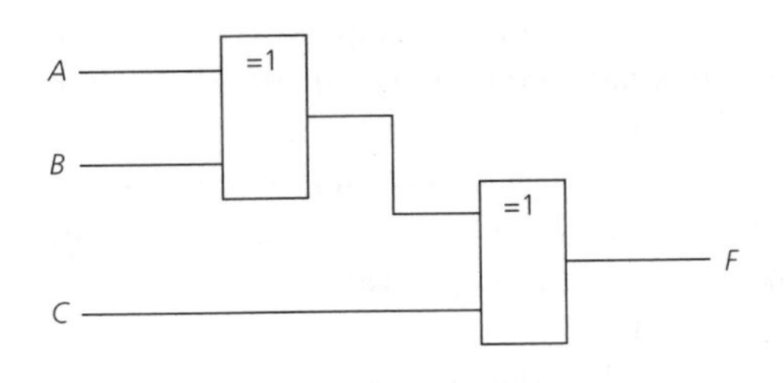

Fig. 3.47

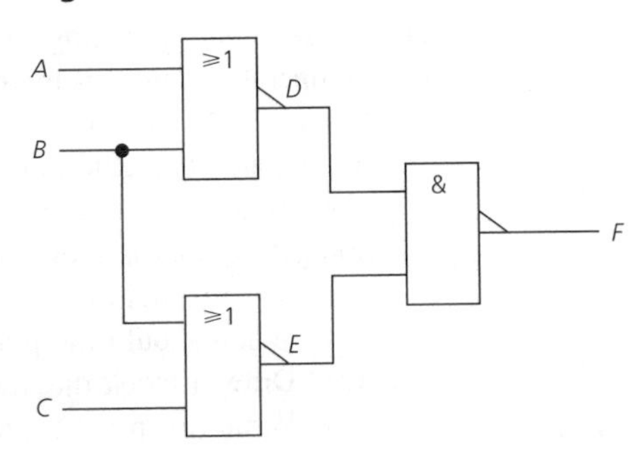

Fig. 3.48

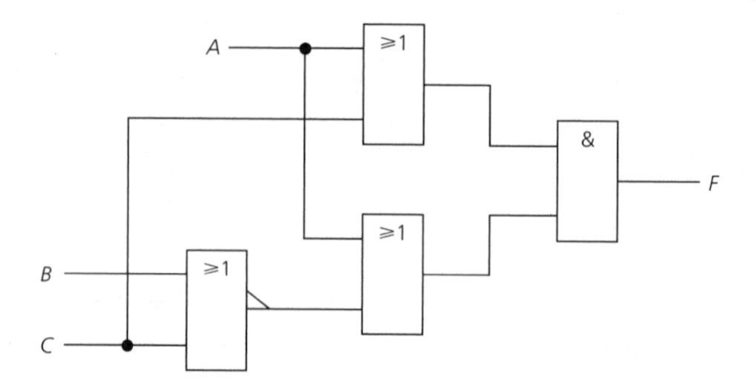

Fig. 3.49

3.5 Waveforms A, B and C of Fig. 3.49 are applied to logic circuits D, E and F. The output waveforms of D, E and F are as shown. Determine the Boolean expressions describing the operation of each circuit.

3.6 A gas central-heating system has a thermostat with two inputs, (i) the room temperature, and (ii) the manually set temperature control. The thermostat output signal, and a signal from a sensor mounted near the pilot flame, are applied to a logic gate. The gate output causes a valve to open and shut and thereby control the gas supplied to the boiler. When the two signals applied to the gate are HIGH the gate output is required to go HIGH also to open the gas valve.

(a) When should the thermostat output be HIGH?

(b) When should the pilot sensor provide a HIGH output?

(c) Draw a block diagram of the system and say what kind of gate is required.

(d) Write down the truth table for the system.

3.7 Design a heater system for a greenhouse. The system is to come into action whenever the temperature falls below a pre-set value, it is dark, and a master switch has been turned on.

(a) Draw a block diagram of a suitable arrangement bearing in mind that the output of a logic gate will not be able to directly turn an electrical heater on and off.

(b) Describe the operation of the system stating whether the inputs to the logic circuit are HIGH or LOW when active.

3.8 A 2-input NOR gate has inputs $\bar{A}$ and $\bar{B}$.

(a) Use a truth table to determine the logic function performed.

(b) Repeat for a 2-input NAND gate.

3.9 The circuit shown in Fig. 3.3 has all its AND gates replaced by NOR gates. Write down the truth table of the new circuit and determine the logic function now performed.

3.10 The circuit shown in Fig. 3.3 has all its gates replaced by NAND gates. Use a truth table to determine the new logic function performed.

3.11 (a) Write down the truth table of a logic circuit which will have a HIGH output whenever the 3-bit input number is even. Take 0 as an even number.

(b) Deduce the Boolean expression that describes the operation of the circuit.

3.12 A logic circuit is required in which the output is HIGH if the input A is HIGH together with either input B or C but not both, OR if all three inputs are LOW.

(a) Write down the truth table for the circuit.

(b) Write down the Boolean equation that describes the circuit.

(c) Draw the circuit.

3.13 Write down the truth table of a 2-input NAND gate and use it to show that a NAND gate with one input held HIGH will act as an inverter to signals applied to the other input.

3.14 Determine the logic function produced by $\bar{A}$ and $\bar{B}$ when applied to

(a) A NAND gate.

(b) A NOR gate.

(c) An exclusive-OR gate.

3.15 Inputs A and B are applied to an AND gate. The output of the gate is applied to one input of an OR gate. Input C is applied to the other input of the OR gate. What is the output of the OR gate?

3.16 A temperature sensor T_1 produces a logic 0 voltage when the temperature at its location is greater than 25 °C. Another temperature sensor T_2 produces a logic 1 voltage when the temperature at its location is above 10 °C. A heater H is to be turned ON when the temperature is between 10 and 25 °C. If the temperature falls below 10 °C an alarm is to be sounded. Obtain the Boolean equation for the required circuit.

3.17 Three pressure sensors each produce a logic 0 output when the pressure at their location falls below a pre-set pressure P_1, P_2, or P_3. A pump is to be turned on whenever the pressure at all three locations falls below the pre-set value. Obtain the Boolean equation describing the required circuit.

3.18 (a) Draw the logic diagram that implements the function $F = (\overline{A + B})(\overline{C + D})$.
(b) Re-draw using the IEC symbols for the 74LS02 and the 74LS11 ICs.

3.19 (a) Inputs A and B are applied to an AND gate. The output is applied to one input of an OR gate. Input C is applied to the other OR gate input. What logic function is performed?
(b) Draw logic diagrams to implement
(i) $F = A + BCD$.
(ii) $F = AB(C + D)$.
(iii) $F = AB(CD + EF)$.

3.20 Inputs A and B are applied to a NOR gate. Inputs B and C are applied to a second NOR gate. The outputs of the two NOR gates are then applied to a NAND gate. Determine the Boolean equation of the output F of the circuit.

3.21 Show how (a) an exclusive-OR gate and (b) an exclusive-NOR gate can be used as an inverter.

4 Simplification of Boolean equations

After reading this chapter you should be able to:

(a) Write the Boolean equation for a combinational logic circuit.
(b) Use Boolean algebra rules to simplify Boolean equations.
(c) Use De Morgan's rules to convert an equation from one form to another.
(d) Derive both SOP and POS expressions from a truth table.
(e) Obtain the truth table that corresponds to a given SOP or POS expression.
(f) Convert a Boolean equation from SOP to POS form, or from POS to SOP form.

When a combinational or random logic circuit is to be designed to solve a logic problem a Boolean equation and/or a truth table describing the required circuit operation may be produced. Very often the equation first obtained is not in its minimal form and it can then be simplified or *reduced*. If it is possible to reduce a Boolean equation to a simpler form fewer gates will be needed to implement the design. Usually this will mean that fewer integrated circuits (ICs) are employed with a consequent reduction in costs, weight, and physical dimensions. There are, however, some other factors that also have to be kept in mind; these include the question of IC availability, the hazard problem (p. 119), and the desirability of using only one kind of gate in a circuit (p. 126).

It is necessary to be able to express the output of a logical network in terms of Boolean algebra and also, given the Boolean expression for a required output, to be able to design the logic circuitry needed to produce this output. When designing a logic network, the required function is generally first simplified in an effort to minimize the number of gates required, although it must be noted that the simplest Boolean expression does not necessarily give the minimum number of ICs.

Simplification of a Boolean equation may be carried out using either the rules of Boolean algebra or a mapping method. The *Karnaugh map* is considered in chapter 5. Boolean equations often are, or can be rearranged to be, in either the *product-of-sums* (POS) form or the *sum-of-products* (SOP) form. A POS expression is of the form $F = (A + B)(C + D)$ and a SOP expression is in the form $F = AB + CD$. The SOP form is the more commonly employed because it is derived directly from a truth table and it is easier to enter on to a Karnaugh map. The output F of a SOP equation is 1 whenever any single product term is 1; conversely, the output of a POS equation is 0 whenever any one of its sum terms is 0. If an equation is in either the SOP or POS form it can then be implemented using only two levels of gates (not including inverters). Also, each expression requires only AND or OR gates plus inverters for its implementation. Equations can be rearranged so that a SOP expression can be implemented by two levels of NAND gates only, and a POS expression by two levels of NOR gates only. The SOP form of Boolean equations is used in *programmable logic devices* (p. 292).

EXAMPLE 4.1

(a) Write the equation $F = AB + C(D + B)$ in its SOP form.
(b) Write the equation $F = AB + AD + BC + CD$ in its POS form.

Solution

(a) $F = AB + BC + CD$ (*Ans.*)
(b) $F = (A + C)(B + D)$ (*Ans.*)

Logic rules

Boolean equations can be simplified by either algebraic or mapping methods. The variables A and B may be written in any order: thus, $A + B = B + A$ and $AB = BA$. Also, the grouping of three or more variables can be done in any convenient way: $(A + B) + C = A + (B + C) = (A + C) + B$, and $A \cdot (BC) = (AB) \cdot C = B(AC) = ABC$. Lastly, equations can be multiplied out: $A(B + C) = AB + AC$ and $(A + B)(C + D) = AC + AD + BC + BD$. Boolean algebra uses some of the same rules as ordinary algebra. In addition there are a number of *logic rules* which allow variables to be combined or eliminated to give a simpler version of an equation. These logic rules are:

1 $A + \bar{A} = 1$
2 $A + A = A$
3 $A + 0 = A$
4 $A + 1 = 1$
5 $A \cdot \bar{A} = 0$
6 $A \cdot A = A$
7 $A \cdot 0 = 0$
8 $A \cdot 1 = A$
9 $A \cdot (B + \bar{B}) = A$
10 $A + A \cdot B = A \cdot (1 + B) = A$
11 $\bar{A} + A \cdot B = \bar{A} + B$
12 $A \cdot (A + B) = A$
13 $A + \bar{A} \cdot B = A + B$
14 $A \cdot B + \bar{B} \cdot C + A \cdot C = A \cdot B + \bar{B} \cdot C$
15 $\overline{A + B} = \bar{A} \cdot \bar{B}$
16 $\overline{A \cdot B} = \bar{A} + \bar{B}$

Table 4.1 ***Rule 13***

B	A	$\bar{A}$	$\bar{A}B$	$A + \bar{A}B$	$A + B$
0	0	1	0	0	0
0	1	0	0	1	1
1	0	1	1	1	1
1	1	0	0	1	1

Table 4.2 ***Rule 14***

C	B	A	$\bar{B}$	AB	$\bar{B}C$	AC	$AB + \bar{B}C + AC$	$AB + \bar{B}C$
0	0	0	1	0	0	0	0	0
0	0	1	1	0	0	0	0	0
0	1	0	0	0	0	0	0	0
0	1	1	0	1	0	0	1	1
1	0	0	1	0	1	0	1	1
1	0	1	1	0	1	1	1	1
1	1	0	0	0	0	0	0	0
1	1	1	0	1	0	1	1	1

The logic rules can be employed to rearrange a Boolean equation into a simpler form and so allow a circuit to be implemented using fewer gates.

The accuracy of any of the rules can be confirmed. Thus:

Rule 1 If $A = 1$ then $\bar{A} = 0$: hence, $1 + 0 = 1$
If $A = 0$ then $A = 1$: hence, $0 + 1 = 1$
Rule 4 If $A = 1$ then $1 + 1 = 1$, if $A = 0$ then $1 + 0 = 1$
Rule 2 If $A = 1$ then $1(1 + B) = 1 \cdot 1 + 1 \cdot B = 1 + B = 1$
If $A = 0$ then $0(0 + B) = 0 + 0 \cdot B = 0$
Rule 13 The truth table is given in Table 4.1
Rule 14 Table 4.2 gives the truth table for this rule

EXAMPLE 4.2

Confirm logic rules 10 and 11.

Solution

(a) Rule 11; the truth table for the equation is given in Table 4.3 (*Ans.*)
(b) Rule 10; the truth table is given in Table 4.4 (*Ans.*)

Table 4.3

B	A	$\bar{A}$	$A \cdot B$	$\bar{A} + A \cdot B$	$\bar{A} + B$
0	0	1	0	1	1
0	1	0	0	0	0
1	0	1	0	1	1
1	1	0	1	1	1

Table 4.4

B	A	$A \cdot B$	$1 + B$	$A(1 + B)$
0	0	0	1	0
0	1	0	1	1
1	0	0	1	0
1	1	1	1	1

Table 4.5

C	B	A	$\bar{A}$	$\bar{B}$	$\bar{A} \cdot \bar{B}$	$B \cdot C$	$\bar{A} \cdot C$	$\bar{A} \cdot \bar{B} + B \cdot C + \bar{A} \cdot C$	$\bar{A} \cdot \bar{B} + B \cdot C$
0	0	0	1	1	1	0	0	1	1
0	0	1	0	1	0	0	0	0	0
0	1	0	1	0	0	0	0	0	0
0	1	1	0	0	0	0	0	0	0
1	0	0	1	1	1	0	1	1	1
1	0	1	0	1	0	0	0	0	0
1	1	0	1	0	0	1	1	1	1
1	1	1	0	0	1	1	0	1	1

Rule 14 allows a variable in the second term of an equation to be eliminated. Table 4.5 confirms $\bar{A}\bar{B} + BC + \bar{A}C = \bar{A}\bar{B} + BC$.

The symbol for the AND logical function, the dot, is often omitted and this will be done throughout the remainder of this book.

EXAMPLE 4.3

Confirm rule 14, i.e. $AB + \bar{B}C + AC = AB + \bar{B}C$.

Solution

$$
\begin{aligned}
AB + \bar{B}C &= AB(1 + C) + \bar{B}C(1 + A) && \text{(rule 7)} \\
&= AB + ABC + \bar{B}C + A\bar{B}C \\
&= AB + \bar{B}C + AC(B + \bar{B}) \\
&= AB + \bar{B}C + AC \; (Ans.) && \text{(rule 1)}
\end{aligned}
$$

EXAMPLE 4.4

Simplify $F = B(A + C) + C$.

Solution

$$F = AB + BC + C$$
$$= AB + C(1 + B) = AB + C \; (Ans.) \qquad \text{(rule 4)}$$

EXAMPLE 4.5

Simplify $F = (A + B)BC + A$

Solution

$$F = ABC + BC + A \qquad \text{(rule 6)}$$
$$= A(1 + BC) + BC$$
$$= A + BC \; (Ans.) \qquad \text{(rule 4)}$$

EXAMPLE 4.6

Simplify $F = B[(A + \bar{B})(B + C)]$.

Solution

$$F = B[AB + AC + B\bar{B} + \bar{B}C]$$
$$= AB + ABC + \bar{B}BC \qquad \text{(rule 6)}$$
$$= AB + ABC \qquad \text{(rule 5)}$$
$$= AB(1 + C)$$
$$= AB \; (Ans.) \qquad \text{(rule 4)}$$

EXAMPLE 4.7

Simplify $F = \overline{AB}(B + C)$.

Solution

$$F = (\bar{A} + \bar{B})(B + C) \qquad \text{(rule 16)}$$
$$= \bar{A}B + \bar{A}C + \bar{B}C \; (Ans.) \qquad \text{(rule 5)}$$

EXAMPLE 4.8

Solve the equation $F = \overline{A\bar{B}(A + C)} + \bar{A}B\overline{(A + \bar{B} + C)}$.

Solution

$$\begin{aligned}
F &= \overline{A\bar{B}} + \overline{A\bar{B}C} + \bar{A}B\overline{(A + \bar{B} + \bar{C})} && \text{(rule 6)}\\
&= (\overline{A\bar{B}})(\overline{A\bar{B}C}) + \bar{A}B(\bar{A}BC) && \text{(rule 15)}\\
&= (\bar{A} + B)(\bar{A} + B + \bar{C}) + \bar{A}BC && \text{(rule 16)}\\
&= \bar{A} + \bar{A}B + \bar{A}\bar{C} + \bar{A}B + B + B\bar{C} + \bar{A}BC && \text{(rule 6)}\\
&= \bar{A}(1 + B + \bar{C} + B + BC) + B(1 + \bar{C}) = \bar{A} + B \ (Ans.) && \text{(rule 4)}
\end{aligned}$$

EXAMPLE 4.9

Draw the logic circuit that will implement the Boolean expression $F = (A + B)(B + C)$. Also, simplify the expression and draw the circuit that will implement the simplified equation.

Solution

$$\begin{aligned}
F &= (A + B)(B + C)\\
&= AB + AC + BB + BC\\
&= AB + AC + B + BC && \text{(rule 6)}\\
&= AB + AC + B(1 + C)\\
&= A(B + C) + B && \text{(rule 4)}\\
&= AC + B(1 + A)\\
&= AC + B \ (Ans.) && \text{(rule 4)}
\end{aligned}$$

Figures 4.1(a) and (b) shows, respectively, the logical circuitry required to implement the equation $F = (A + B)(B + C)$ and its simplified version $F = AC + B$. Clearly, the saving in the number of gates needed is small, the number being reduced from three to two.

Fig. 4.1

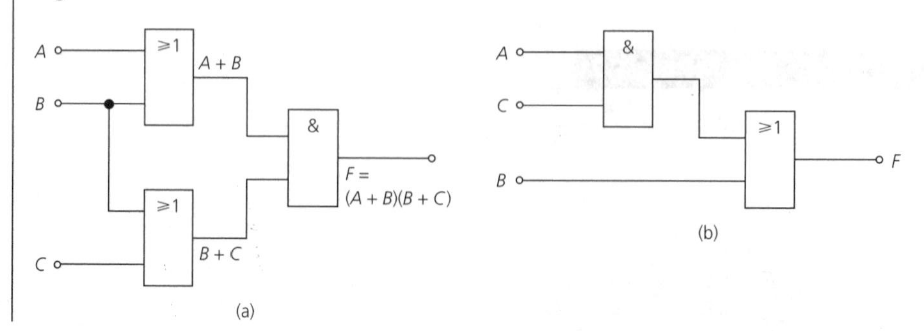

EXAMPLE 4.10

Simplify $F = \bar{A}(\bar{B} + \bar{C}) + BC + A\bar{C}$.

Solution

$$
\begin{aligned}
F &= \bar{A}\bar{B} + \bar{A}\bar{C} + BC + A\bar{C} \\
&= \bar{A}\bar{B} + \bar{C}(A + \bar{A}) + BC && \text{(rule 1)} \\
&= \bar{A}\bar{B} + B + \bar{C} && \text{(rule 13)} \\
&= \bar{A} + B + \bar{C} \ (Ans.) && \text{(rule 13)}
\end{aligned}
$$

De Morgan's rules

Rules 15 and 16 are known as *De Morgan's rules* and are particularly useful. The truth table for rule 15 is given in Table 4.6. Obviously, the fourth and the last columns are identical so that, for all values of A and B, $\overline{A + B} = \bar{A}\bar{B}$.

Table 4.7 gives the truth table for the De Morgan's second rule, i.e. rule 16. Again, the fourth and last columns are identical and the rule is confirmed.

De Morgan's rules can be stated in the following manner. The complement of a function F can be obtained by replacing each variable by its complement and then interchanging all AND (·) and OR (+) signs.

Both of De Morgan's rules can be extended to deal with three, or more, input variables. Thus

$$F = \overline{A + B + C} = \bar{A}\bar{B}\bar{C} \quad \text{(rule 15)}$$
$$F = \overline{ABC} = \bar{A} + \bar{B} + \bar{C} \quad \text{(rule 16)}$$

Table 4.6 ***De Morgan's first rule***

B	A	$A + B$	$\overline{A + B}$	$\bar{A}$	$\bar{B}$	$\bar{A}\bar{B}$
0	0	0	1	1	1	1
0	1	1	0	0	1	0
1	0	1	0	1	0	0
1	1	1	0	0	0	0

Table 4.7 ***De Morgan's second rule***

B	A	AB	$\overline{AB}$	$\bar{A}$	$\bar{B}$	$\bar{A} + \bar{B}$
0	0	0	1	1	1	1
0	1	0	1	0	1	1
1	0	0	1	1	0	1
1	1	1	0	0	0	0

EXAMPLE 4.11

If $F = A\bar{B} + C$, find $\bar{F}$.

Solution

Using rule 15, $\bar{F} = \overline{A\bar{B}}\bar{C}$
Using rule 16, $\bar{F} = (\bar{A} + B)\bar{C} = \bar{A}\bar{C} + B\bar{C}$ (*Ans.*)

EXAMPLE 4.12

Simplify $F = \overline{(A + B)(\bar{A} + C)}$.

Solution

$F = \overline{A + B} + \overline{\bar{A} + C}$ (rule 15)
$= \bar{A}\bar{B} + A\bar{C}$ (*Ans.*) (rule 16)

EXAMPLE 4.13

Use De Morgan's rules to prove that a coincidence gate provides the inverse function to an exclusive OR gate.

Solution

The output of an exclusive OR gate is $F = AB + AB$. Inverting,

$$\begin{aligned}\bar{F} &= \overline{\bar{A}\bar{B} + AB}\\ &= (\overline{\bar{A}\bar{B}})(\overline{AB})\\ &= (A + B)(\bar{A} + \bar{B})\\ &= A\bar{A} + A\bar{B} + B\bar{A} + B\bar{B}\\ &= A\bar{B} + \bar{A}B \qquad \text{(rule 5)}\end{aligned}$$

which is the equation for an exclusive-OR gate (*Ans.*)

EXAMPLE 4.14

Simplify the Boolean equation $F = \overline{\bar{A}(B + \bar{C}})(A + \bar{B} + C)(\bar{A}\bar{B}\bar{C})$ and draw the circuit which will implement the simplified equation.

Solution

$$
\begin{aligned}
F &= (A + \overline{B + \bar{C}})(A + \bar{B} + C)(A + B + C)\\
&= (A + \bar{B}C)(AA + AB + AC + A\bar{B} + \bar{B}B + \bar{B}C + AC + BC + CC)\\
&= (A + \bar{B}C)(A + C)\\
&= AA + AC + A\bar{B}C + \bar{B}C\\
&= A(1 + C) + \bar{B}C(1 + A)\\
&= A + \bar{B}C
\end{aligned}
$$

Figure 4.2 shows the required circuit (*Ans.*)

Fig. 4.2

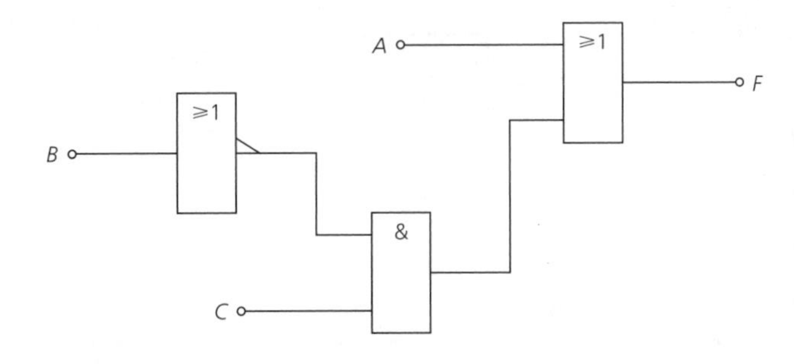

EXAMPLE 4.15

Simplify $F = \overline{A(B + C)} + \bar{A}B + \bar{C}(\overline{A + B})$.

Solution

$$
\begin{aligned}
F &= \bar{A} + \overline{B + C} + \bar{A}B + \bar{C}(\bar{A}\bar{B}) && \text{(rules 15 and 16)}\\
&= \bar{A} + \bar{B}\bar{C} + \bar{A}B + \bar{A}\bar{B}\bar{C} && \text{(rule 15)}\\
&= \bar{A}(1 + B) + \bar{B}\bar{C}(1 + \bar{A})\\
&= \bar{A} + \bar{B}\bar{C}\ (Ans.) && \text{(rule 4)}
\end{aligned}
$$

PRACTICAL EXERCISE 4.1

To confirm the accuracy of De Morgan's rules.

Components and equipment: one 74LS00 (or 74HC00) quad 2-input NAND gate IC, one 74LS08 (or 74HC08) quad 2-input AND gate IC, one 74LS32 (or 74HC32) quad 2-input OR gate IC. Two LEDs. Two 270 Ω resistors. Power supply. Breadboard.

Procedure:

(a) Connect up the circuit shown in Fig. 4.3. For each IC connect the 5 V power supply between pins 7 and 14.

Fig. 4.3

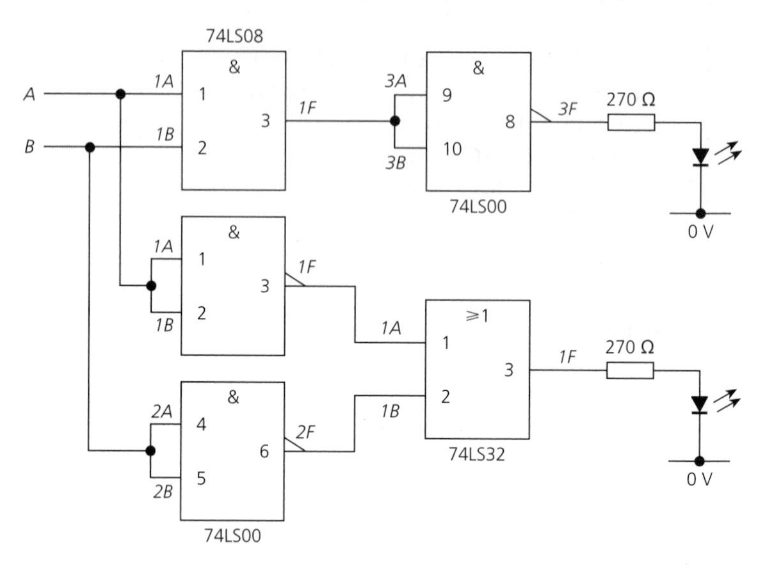

Fig. 4.4

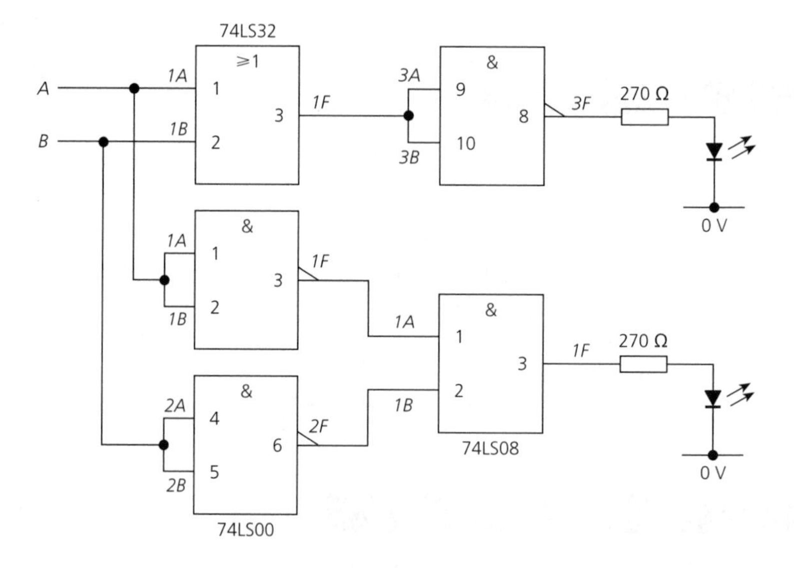

(b) Apply each of the four possible combinations of HIGH/LOW voltages to the input terminals of the circuit. Each time note whether or not the LED glows visibly. Write down the truth table for the circuit and then state the logic function performed. Which of De Morgan's rules has been confirmed?

(c) Build the circuit shown in Fig. 4.4.

(d) Repeat procedure (b). What logic function is now performed and which of De Morgan's rules has been followed?

PRACTICAL EXERCISE 4.2

To confirm that the OR logical function is performed when the inputs A and B to a 2-input NAND gate are both inverted.
Components and equipment: one 74LS00 (or 74HC00) quad 2-input NAND gate IC. One 74LS02 (or 74HC02) quad 2-input NOR gate IC. One LED. One 270 Ω resistor. Power supply. Breadboard.

Fig. 4.5

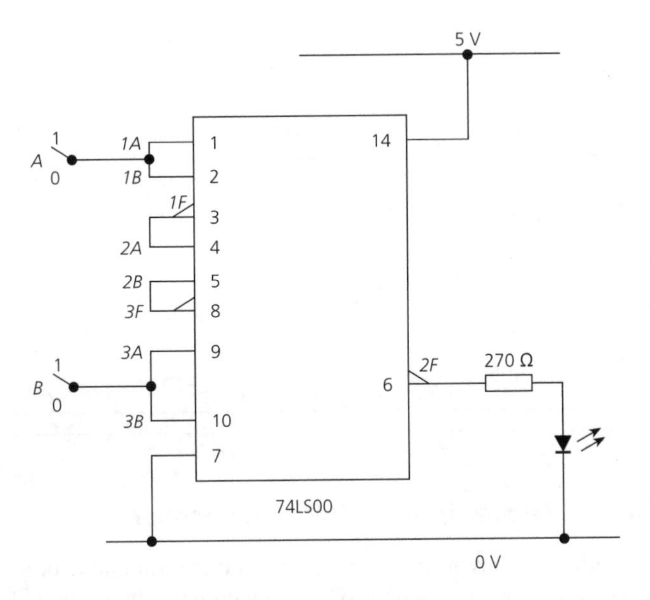

(a) Connect up the circuit shown in Fig. 4.5. Gate 2 acts as an AND gate and gates 1 and 3 invert the variables applied to this AND gate.
(b) Apply each of the four possible combinations of HIGH/LOW voltage to the inverter inputs and for each combination note whether or not the LED glows visibly.
(c) Write down the truth table for the circuit and from it deduce the logic function that has been performed.
(d) Replace the NAND gate IC with the NOR gate IC and repeat procedures (b) and (c). Check the pinout of the device first.

Use of truth tables

SOP expression from a truth table

Write each term of the input variables as a product and sum all the terms to specify the input combinations that exist. Consider the truth table given in Table 4.8.

The Boolean expression for this circuit is

$$F = \bar{A}\bar{B}\bar{C}D + \bar{A}\bar{B}\bar{C}\bar{D} + A\bar{B}\bar{C}D + A\bar{B}\bar{C}\bar{D} + A\bar{B}CD + A\bar{B}C\bar{D} + AB\bar{C}\bar{D}$$

Table 4.8

A	0	1	0	1	0	1	0	1	0	1	0	1	0	1	0	1
B	0	0	1	1	0	0	1	1	0	0	1	1	0	0	1	1
C	0	0	0	0	1	1	1	1	0	0	0	0	1	1	1	1
D	0	0	0	0	0	0	0	0	1	1	1	1	1	1	1	1
F	0	1	1	1	0	1	0	0	1	1	0	0	0	1	0	0

Table 4.9

A	B	C	ABC	$\bar{A}B\bar{C}$	$A\bar{B}\bar{C}$	$\bar{A}\bar{B}C$	F
0	0	0	0	0	0	0	0
1	0	0	0	0	1	0	1
0	1	0	0	1	0	0	1
1	1	0	0	0	0	0	0
0	0	1	0	0	0	1	1
1	0	1	0	0	0	0	0
0	1	1	0	0	0	0	0
1	1	1	1	0	0	0	1

Truth table from SOP expression

Sometimes it may be convenient to obtain the truth table of a SOP expression. The truth table should have a number of combinations of input variables equal to 2^n, where n is the number of variables. Consider the expression $F = ABC + \bar{A}B\bar{C} + A\bar{B}\bar{C} + \bar{A}\bar{B}C$; its truth table is given in Table 4.9.

The input combination for this expression is 111, 010, 100 and 001. The output for each combination is entered as 1 in the F column of the table, while all the remaining combinations are entered as 0 in the F column.

EXAMPLE 4.16

Obtain the truth table for the function $F = AB + \bar{B}\bar{C} + \bar{A}C$.

Solution

$$\begin{aligned} F &= AB(C + \bar{C}) + \bar{B}\bar{C}(A + \bar{A}) + \bar{A}C(B + \bar{B}) \\ &= ABC + AB\bar{C} + A\bar{B}\bar{C} + \bar{A}\bar{B}\bar{C} + \bar{A}BC + \bar{A}\bar{B}C \\ &= 111, 110, 100, 000, 011, 001. \end{aligned}$$

Table 4.10 gives the truth table (*Ans.*)

Table 4.10

A	0	1	0	1	0	1	0	1
B	0	0	1	1	0	0	1	1
C	0	0	0	0	1	1	1	1
F	1	1	0	1	1	0	1	1

Table 4.11

A	0	1	0	1
B	0	0	1	1
F	1	0	0	1

POS expression from a truth table

To obtain a Boolean expression in its product-of-sums form from a truth table the following steps must be followed:

- When a variable is 1 write it down in the equation in its complemented form.
- When a variable is 0 write it down in its uncomplemented form.
- Write each term of the individual input combinations that give $F = 0$ as a sum.
- AND all terms which specify an input combination to obtain the POS equation.

The POS expression gives the active-LOW equation of a logic function and it is the product of all input variable combinations that give an output F of 0. It *does not* give the complement of the equivalent SOP equation. Consider the POS equation $F = (A + \bar{B})(\bar{A} + B) = AB + \bar{A}\bar{B}$. The complement of this is $\bar{F} = \overline{AB + \bar{A}\bar{B}} = (\overline{AB}) \cdot (\overline{\bar{A}\bar{B}}) = (\bar{A} + \bar{B})(A + B) = \bar{A}B + A\bar{B}$.

The truth table for F is given in Table 4.11.

From this table the active-LOW POS equation is $F = (\bar{A} + B)(A + \bar{B})$ which is not the same as $\bar{F}$.

EXAMPLE 4.17

Obtain (a) the SOP and (b) the POS equation for the logic function whose truth table is given in Table 4.12.

Solution

(a) $F = \bar{A}\bar{B}\bar{C} + \bar{A}\bar{B}C + \bar{A}BC + ABC = \bar{A}\bar{B}(C + \bar{C}) + BC(A + \bar{A}) = \bar{A}\bar{B} + BC$ (*Ans.*)

(b) $F = (\bar{A} + B + C)(A + \bar{B} + C)(\bar{A} + B + \bar{C})(\bar{A} + \bar{B} + C)$ (*Ans.*)

Table 4.12

A	0	1	0	1	0	1	0	1
B	0	0	1	1	0	0	1	1
C	0	0	0	0	1	1	1	1
F	1	0	0	0	1	0	1	1

Table 4.13

A	0	1	0	1
B	0	0	1	0
F	1	0	0	1

Truth table from a POS equation

Consider the POS expression $F = (A + \bar{B})(\bar{A} + B)$. There are two variables, so the truth table must contain four combinations. Each ORed term in the equation gives a 0 output. The input combinations represented by the equation are: 01 and 10. These two output states are entered in the F row of the truth table as 0. All the other combinations are represented by 1. Thus, the truth table is given in Table 4.13.

EXAMPLE 4.18

Obtain the truth table for the POS expression $F = (A + B + \bar{C})(\bar{A} + B + \bar{C})(\bar{A} + \bar{B} + C)$.

Solution

The input combinations are: 001, 101 and 110 and these give a 0 output in the F row of the truth table. All other combinations give a 1. The truth table is shown in Table 4.14 (*Ans.*)

Table 4.14

A	0	1	0	1	0	1	0	1
B	0	0	1	1	0	0	1	1
C	0	0	0	0	1	1	1	1
F	1	1	1	0	0	0	1	1

Conversion from SOP to POS equations

The conversion of a SOP equation into the corresponding POS form is more difficult than the reverse process which merely requires simple multiplying out of brackets. The steps involved are as follows:

- Simplify the SOP equation if possible.
- Invert the SOP equation.
- Multiply out and simplify the inverse equation.
- Invert the result.

EXAMPLE 4.19

Convert the SOP equation $F = A\bar{B}C + \bar{A}B\bar{C}$ into its POS form.

Solution

(a) $\bar{F} = \overline{A\bar{B}C + \bar{A}B\bar{C}} = (\overline{A\bar{B}C})(\overline{\bar{A}B\bar{C}}) = (\bar{A} + B + \bar{C})(A + \bar{B} + C)$

(b) $\bar{F} = \bar{A}\bar{B} + \bar{A}C + AB + BC + A\bar{C} + \bar{B}\bar{C} = \bar{A}\bar{B} + \bar{A}C + AB + A\bar{C}$ (rule 14)

(c) $F = \overline{\bar{A}\bar{B} + \bar{A}C + AB + A\bar{C}} = (\overline{\bar{A}\bar{B}})(\overline{\bar{A}C})(\overline{AB})(\overline{A\bar{C}})$

$= (A + B)(A + \bar{C})(\bar{A} + \bar{B})(\bar{A} + C)$ (*Ans.*)

EXAMPLE 4.20

Convert $F = \bar{A}\bar{B}\bar{C} + \bar{A}\bar{B}C + \bar{A}B\bar{C} + A\bar{B}\bar{C} + AB\bar{C} + ABC$ into its POS form.

Solution

(a) $F = \bar{A}\bar{C}(B + \bar{B}) + A\bar{C}(B + \bar{B}) + \bar{A}\bar{B}C + ABC$

$= \bar{A}\bar{C} + A\bar{C} + \bar{A}\bar{B}C + ABC$ (rule 1)

$= \bar{C} + C(\bar{A}\bar{B} + AB)$ (rule 1)

$= \bar{C} + \bar{A}\bar{B} + AB$ (rule 11)

(b) $\bar{F} = \overline{\bar{C} + \bar{A}\bar{B} + AB} = C(\overline{\bar{A}\bar{B}})(\overline{AB})$

$= C(A + B)(\bar{A} + \bar{B}) = C(A\bar{B} + \bar{A}B)$

$= A\bar{B}C + \bar{A}BC$

(c) $F = \overline{A\bar{B}C + \bar{A}BC} = (\overline{A\bar{B}C})(\overline{\bar{A}BC})$

$= (\bar{A} + B + \bar{C})(A + \bar{B} + \bar{C})$ (*Ans.*)

EXERCISES

4.1 Use De Morgan's rules to show that
(a) A NOR gate with inverted inputs acts like an AND gate.
(b) A NAND gate with inverted inputs acts like an OR gate.
(c) An AND gate with inverted inputs acts like a NOR gate.

4.2 Simplify the Boolean equation $F = \overline{(\bar{A}\bar{B})C + D}$.

4.3 Inputs A and B are applied to the inputs of a 2-input NAND gate. Inputs C and B are applied to the inputs of another 2-input NAND gate. The outputs of the two NAND gates are applied to a NOR gate. Show that the circuit could be replaced by one 3-input AND gate.

4.4 (a) Show that the A term in the expression $F = (\overline{AB})(\overline{B + C})$ is redundant. How can the simplified circuit be implemented?
(b) Simplify $F = \overline{(\bar{A}C + B\bar{C})(A + \bar{B} + D)}$

4.5 (a) The function $F = AB + AC$ is to be implemented. How many gates are required? Show how the function could be implemented using only two gates.
(b) Simplify the equations (i) $F = ABC + \overline{ABC}$ and (ii) $F = AB + ABC$.
(c) Simplify $F = \overline{\overline{\overline{ABC}}} + A$.
(d) Simplify $F = BC + \bar{C}D + BD$.

4.6 Simplify
(a) $F = (\overline{AC} + BC)(\overline{A + B + D})$.
(b) $F = \overline{\overline{(\overline{\bar{A}\bar{B}C})(\overline{\bar{B}C})}}$.

4.7 Simplify
(a) $F = \overline{A + \bar{B} + \bar{C} + \bar{D}} + \bar{A}BC\bar{D}$.
(b) $F = (A + B)(\bar{A} + B)(A + \bar{B})$.
(c) $F = (A + \bar{B} + \bar{C})(\bar{A} + \bar{B} + \bar{C})$.
(d) $F = \overline{\bar{A}B(CD + \bar{A}\bar{C})}$.

4.8 Obtain the truth table for the POS expression $F = (A + \bar{B} + C)(\bar{A} + B + \bar{C})(\bar{A} + \bar{B} + C)$.

4.9 Show that the answer to example (4.17(b)) is *not* the complement of the answer to example (4.17(a)).

4.10 The output of the circuit given in Fig. 3.38 is $F = (A + B)(A + \overline{B + C})(\overline{B + D})$. Simplify the expression and implement the simplified circuit. How many ICs are needed?

4.11 Reduce to its simplest form

$$F = \overline{(\overline{A + \bar{B} + C}) + (\overline{\bar{A} + \bar{B} + C + D}) + (\overline{\bar{A} + B + \bar{C} + \bar{D}}) + (\overline{A + C})}.$$

4.12 Simplify $F = \bar{A}C + BC + \bar{A}\bar{B} + \bar{A}\bar{B}\bar{C}$.

4.13 Determine the expression for the output F of the circuit in Fig. 4.6.

4.14 Determine the Boolean equation representing each of the circuits shown in Fig. 4.7. Redraw each circuit using AND and/or OR gates.

4.15 In the circuit shown in Fig. 4.8 the output is 0. Determine the logical states of each of the inputs A through to G.

4.16 For the circuit shown in Fig. 4.8 obtain a Boolean expression for the output F of the circuit. Simplify the equation if possible and then draw the simplified circuit.

Fig. 4.6

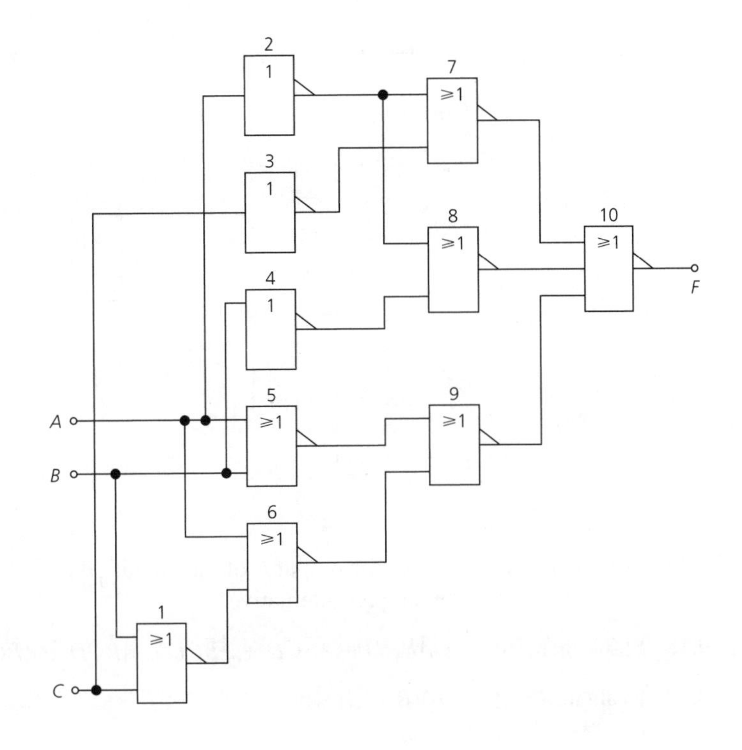

Fig. 4.7

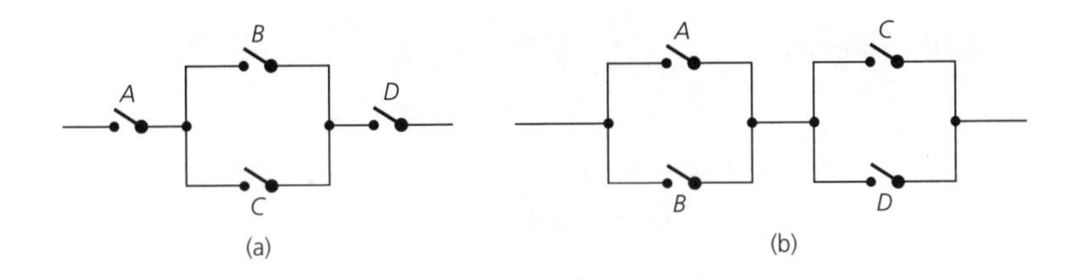

Fig. 4.8

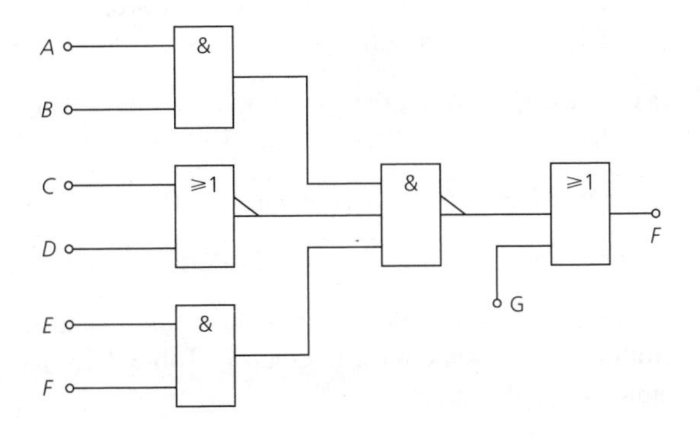

Fig. 4.9

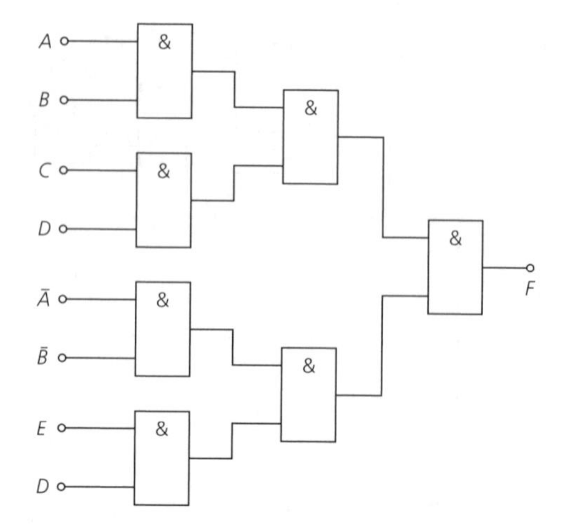

4.17 Obtain an expression for the output F of the circuit given in Fig. 4.9. Simplify the circuit and draw the simpler alternative.

4.18 Show that $\bar{A}BC\bar{D} + \bar{A}BCD + A\bar{B}\bar{C}D + A\bar{B}CD + AB\bar{C}D + ABCD = AD + \bar{A}BC$.

4.19 Evaluate $F = (A + B)(B + C)$ when
(a) $A = B = C = 0$.
(b) $A = B = C = 1$.
(c) $A = 1, B = C = 0$.

4.20 Simplify
(a) $F = \overline{A + B + \overline{A + B}}$.
(b) $F = \overline{\bar{A}\bar{B} + \overline{A + B}}$.
(c) $F = (AB + \bar{C})(A + BC)$.
(d) $F = (\overline{AC + D})(\overline{AC} + B)$.

4.21 (a) Show that $(AB + \bar{C}) + (\bar{A} + \bar{B})C = 1$.
(b) Show that $B(\overline{AD + \overline{BC}})(\bar{A} + \bar{D})\bar{C}(\bar{A} + \bar{D}) = 0$.

4.22 (a) Express $F = AB\bar{C} + A\bar{B}C + \bar{A}\bar{B}\bar{C}$ in product-of-sums form.
(b) Express $F = (A + B + \bar{C})(A + \bar{B} + C)(\bar{A} + \bar{B} + \bar{C})$ in sum-of-products form.

4.23 (a) Show that $(B + D)(D + C)(A + D) = D + ABC$.
(b) Show that $(B + D)(A + D)(B + C)(A + C) = AB + CD$.

4.24 Minimize $F = (\overline{\bar{A}\bar{B} + AB})\overline{(\bar{A} + \bar{B})(A + B)}$.

4.25 Minimize $F = \overline{\overline{AB(B + C)} + \overline{AB}(1 + C)}$.

4.26 A digital circuit has three inputs A, B, and C and three outputs X, Y, and Z. The truth table for this circuit is given in Table 4.15. Determine the Boolean expressions for X, Y, and Z.

Table 4.15

A	0	1	0	1	0	1	0	1
B	0	0	1	1	0	0	1	1
C	0	0	0	0	1	1	1	1
X	1	0	0	0	0	0	1	0
Y	0	1	1	0	1	1	0	0
Z	0	0	0	1	0	0	0	1

Table 4.16

A	0	0	0	0	1	1	1	1
B	0	0	1	1	0	0	1	1
C	0	1	0	1	0	1	0	1
F	1	1	1	0	1	0	1	0

Table 4.17

A	0	1	0	1	0	1	1	0
B	0	0	0	1	1	0	1	1
C	1	0	0	0	0	1	1	1
F	1	0	0	0	0	1	1	1

4.27 Obtain the Boolean equation for the output *F* of the circuit whose truth table is given in Table 4.16.

4.28 Table 4.17 gives the truth table of a logic circuit.

Obtain the Boolean equation describing the circuit and reduce it to its simplest form.

5 The Karnaugh map

After reading this chapter you should be able to:

(a) Use a Karnaugh map to reduce a 2-, 3-, or 4-variable SOP Boolean equation to its simplest form.
(b) Use a Karnaugh map to reduce a POS equation to its simplest form.
(c) Make use of don't care/can't happen states when simplifying a Boolean equation.
(d) Use a Karnaugh map to convert a SOP equation into its POS form.
(e) Use a Karnaugh map to convert a POS equation into its SOP form.
(f) Design a combinational logic circuit from a truth table.
(g) Understand static hazards.

When Boolean algebra is employed to simplify a logic expression it is not always clear if the result obtained is the simplest expression possible. The procedure may be lengthy and error-prone and it is not easy to see if a *hazard* (p. 119) exists. The *Karnaugh map* gives a method of obtaining the simplest possible version of a Boolean expression and shows up any possible static hazards that may exist. It provides a convenient method of simplifying Boolean equations in which the function to be simplified is displayed diagrammatically on a set of squares or *cells*. Each cell maps one term of the function. The number of cells in a map is equal to 2^n, where n is the number of variables in the equation to be simplified. A map must be arranged so that a single change occurs in a mapped term when a move is made from any one cell to an adjacent cell. Usually, the equation to be mapped is written in its sum-of-products (SOP) form, but a product-of-sums (POS) expression can also be mapped.

The use of the Karnaugh map is applicable only to simple Boolean expressions with up to five variables, more complex equations can be solved using a tabular method. Tabular methods are lengthy and rather error-prone when solved on paper but can be programmed for solution by computer. One popular software package for this purpose is known as *Expresso*.

Sum-of-products equations

Two variables

Consider the equation $F = A\bar{B} + \bar{A}B$. The number of variables n is 2 and hence four cells are needed. The rows and columns of the map are labelled as shown so that each cell represents a different combination of the two input variables. Thus the four cells represent, respectively, $\bar{A}\bar{B}$, $A\bar{B}$, $\bar{A}B$, and AB.

B \ A	0	1
0	$\bar{A}\bar{B}$	$A\bar{B}$
1	$\bar{A}B$	AB

A 1 written in a cell indicates the presence, in the SOP equation being mapped, of the term represented by that cell. A 0 in a cell means that that particular term is not present in the SOP equation being mapped. When the variable A only is to be mapped there is no B or $\bar{B}$ term and they must be eliminated by putting 1s in two cells as shown:

B \ A	0	1
0	0	1
1	0	1

Then $F = A\bar{B} + AB = A(B + \bar{B}) = A$.

Similarly, the mapping of $\bar{A}$ is

B \ A	0	1
0	1	0
1	1	0

When the variable B is to be mapped, A and $\bar{A}$ must be eliminated and then the mapping is

B \ A	0	1
0	0	0
1	1	1

This mapping gives $F = \bar{A}B + AB = B(A + \bar{A}) = \mathrm{B}$.

Similarly, the mapping of $\bar{B}$ is

$B \backslash A$	0	1
0	1	1
1	0	0

This means that the mapping for the equation $F = A\bar{B} + \bar{A}B$ is

$B \backslash A$	0	1
0	0	1
1	1	0

To simplify an equation using the Karnaugh map, adjacent squares containing a 1 are looped together. When any two squares have been looped together, it means that the corresponding terms in the equation being mapped have been combined; and any terms of the form $A\bar{A}$ have been eliminated.

Consider the equation $F = AB + \bar{A}B + \bar{A}\bar{B}$. The mapping is

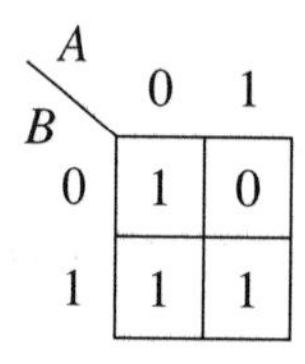

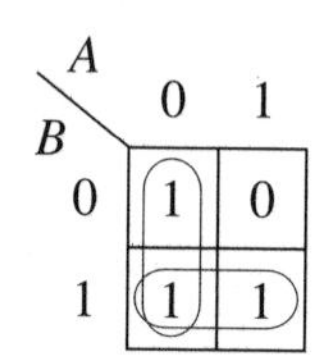

Looping adjacent squares in the map as shown simplifies the equation to $F = \bar{A}(B + \bar{B}) + B(A + \bar{A}) = \bar{A} + B$.

EXAMPLE 5.1

Use a Karnaugh map to confirm the Boolean rule numbers 11, and 13, i.e. $\bar{A} + AB = \bar{A} + B$, and $A + \bar{A}B = A + B$.

Solution

The mapping for $\bar{A} + AB$ is

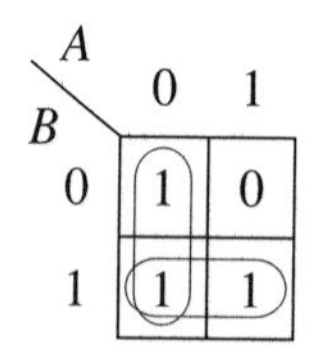

Looping the adjacent cells that contain 1s gives $\bar{A} + B$ (*Ans.*)

The mapping for $A + \bar{A}B$ is

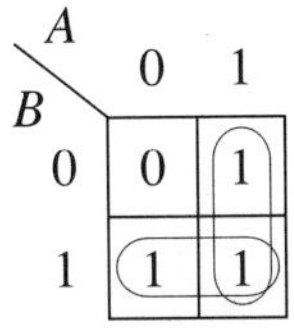

B \ A	0	1
0	0	1
1	1	1

From the looped cells, $A + AB = \bar{A} + B$ (*Ans.*)

Three variables

The Karnaugh map is easily extended for use with three variables A, B, and C. Now $n = 3$ and the number of cells required in the Karnaugh map is $2^3 = 8$. The labelling is as shown:

C \ AB	00	01	11	10
0	$\bar{A}\bar{B}\bar{C}$	$\bar{A}B\bar{C}$	$AB\bar{C}$	$A\bar{B}\bar{C}$
1	$\bar{A}\bar{B}C$	$\bar{A}BC$	ABC	$A\bar{B}C$

The steps to be taken in the Karnaugh map reduction of a Boolean equation are:

- Transform the equation into its SOP form.
- Map the terms of the SOP equation.
- Loop adjacent cells in groups of one, two, four or eight. The two outer (extreme left and extreme right) cells in each row are taken as being adjacent to one another, and so are the top and bottom cells in a column. The looping should be carried out to form as few groups of cells as possible. Every 1 entered on the map must be included in at least one group, although in some cases a group may contain only a single cell. Groups may overlap. The larger a group of 1s, the simpler will be the product term it represents.
- Write down the terms of the simplified SOP expression by determining which variables are included within each loop. Terms of the type $A\bar{A}$ cancel out.

EXAMPLE 5.2

Use a Karnaugh map to simplify the logic expression $F = A(B\bar{C} + \bar{B}\bar{C}) + A\bar{B}C$.

Solution

(a) Transform the expression into its SOP form. Thus, $F = AB\bar{C} + A\bar{B}\bar{C} + A\bar{B}C$.

(b) Mapping gives

C \ AB	00	01	11	10
0	0	0	1	1
1	0	0	0	1

(c) Looping adjacent cells

C \ AB	00	01	11	10
0	0	0	(1	1)
1	0	0	0	1

(d) The simplified expression is $F = A\bar{B} + A\bar{C}$ (*Ans.*)

The answer can be checked using the logic rules of Chapter 4.

$$F = AB\bar{C} + A\bar{B}\bar{C} + A\bar{B}C = A\bar{B}(C + \bar{C}) + AB\bar{C} = A\bar{B} + AB\bar{C}$$
$$= A(\bar{B} + B\bar{C}) = A(\bar{B} + \bar{C}) = A\bar{B} + A\bar{C}.$$

EXAMPLE 5.3

(a) Use a Karnaugh map to simplify the Boolean equation $F = AC + \bar{A}BC + \bar{B}C$.
(b) Check the answer using Boolean algebra.

Solution

The Karnaugh mapping of the equation is

AC:

C \ AB	00	01	11	10
0	0	0	0	0
1	0	0	1	1

$\bar{A}BC$:

C \ AB	00	01	11	10
0	0	0	0	0
1	0	1	0	0

$\bar{B}C$:

C \ AB	00	01	11	10
0	0	0	0	0
1	1	0	0	1

F:

C \ AB	00	01	11	10
0	0	0	0	0
1	1	1	1	1

The term AC has no B variable in it and so it fills two cells ABC and $A\bar{B}C$ to give $AC(B + \bar{B}) = AC$. Looping the 1 cells in the lower row gives $F = C$ (*Ans.*)

Normally, the complete mapping would be written down directly to obtain the fourth map.

(b) $F = AC + \bar{A}BC + \bar{B}C = C(A + \bar{A}B) + \bar{B}C = C(A + B) + \bar{B}C$ (rule 13)
$= C(A + B + \bar{B}) = C(1 + A)$ (rule 1)
$= C$ (*Ans.*) (rule 4)

EXAMPLE 5.4

Simplify the Boolean equation $F = ABC + \bar{A}\bar{B}\bar{C} + AB\bar{C} + \bar{A}\bar{C}$ using (a) a Karnaugh map and (b) Boolean algebra.

Solution

The mapping of the expression is

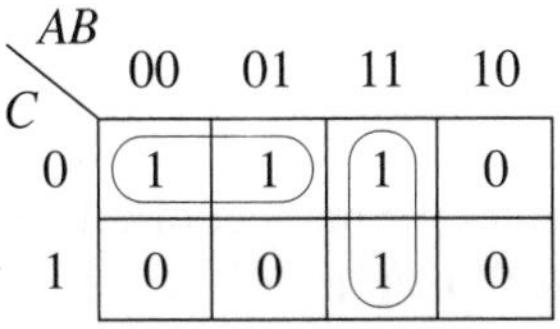

C \ AB	00	01	11	10
0	1	1	1	0
1	0	0	1	0

The cells can be looped together in two groups of two, as shown. From the mapping, $F = AB + \bar{A}\bar{C}$ (*Ans.*)

(b) $F = ABC + \bar{A}\bar{B}\bar{C} + AB\bar{C} + \bar{A}\bar{C}$
$= AB(C + \bar{C}) + \bar{A}\bar{C}(1 + B)$
$= AB + \bar{A}\bar{C}$ (*Ans.*) (rule 1 and rule 4)

EXAMPLE 5.5

Use a Karnaugh map to simplify (a) $F = ABC + A\bar{B}\bar{C} + \bar{A}\bar{B}\bar{C}$, (b) $F = ABC + \bar{A}\bar{B}\bar{C} + A\bar{B}\bar{C} + A\bar{B}C$, and (c) $F = \bar{A}B\bar{C} + AB\bar{C} + \bar{A}\bar{B}C + \bar{A}BC$.

Solution

(a) The mapping is

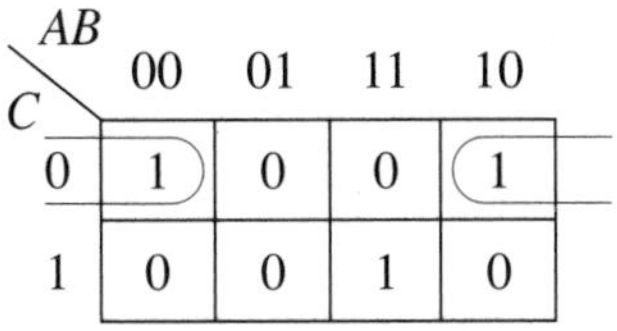

C \ AB	00	01	11	10
0	1	0	0	1
1	0	0	1	0

From the looped cells, $F = \bar{B}\bar{C} + ABC$ (*Ans.*)

(b) The mapping is

C \ AB	00	01	11	10
0	1	0	0	1
1	0	0	1	1

From the looped cells, $F = AC + \bar{B}\bar{C}$ (*Ans.*)

(c) The mapping is

C \ AB	00	01	11	10
0	0	1	1	0
1	1	1	0	0

From the looped cells, $F = B\bar{C} + \bar{A}C$ (*Ans.*)

Four variables

When a Boolean expression contains four variables the number of cells in the Karnaugh map must be $2^4 = 16$. The labelling of the 16-cell map is

CD \ AB	00	01	11	10
00	$\bar{A}\bar{B}\bar{C}\bar{D}$			$A\bar{B}\bar{C}\bar{D}$
01				
11			$ABCD$	
10	$\bar{A}\bar{B}C\bar{D}$			

EXAMPLE 5.6

Simplify the Boolean equation $F = ACD + AB\bar{C}D + \bar{A}BD + A\bar{B}\bar{C}D$.

Solution

The mapping of the expression is

CD \ AB	00	01	11	10
00	0	0	0	0
01	0	1	1	1
11	0	1	1	1
10	0	0	0	0

The cells can be looped in two groups each containing four 1s. Hence, the simplified equation is $F = AD + BD$ (*Ans.*)

EXAMPLE 5.7

Use a Karnaugh map to simplify $F = \bar{A}\bar{B}\bar{C}\bar{D} + \bar{A}\bar{B}C\bar{D} + \bar{A}B\bar{C}D + \bar{A}BCD + \bar{A}B\bar{C}\bar{D} + \bar{A}BC\bar{D} + A\bar{B}\bar{C}\bar{D} + A\bar{B}C\bar{D}$.

Solution

The mapping of the expression is

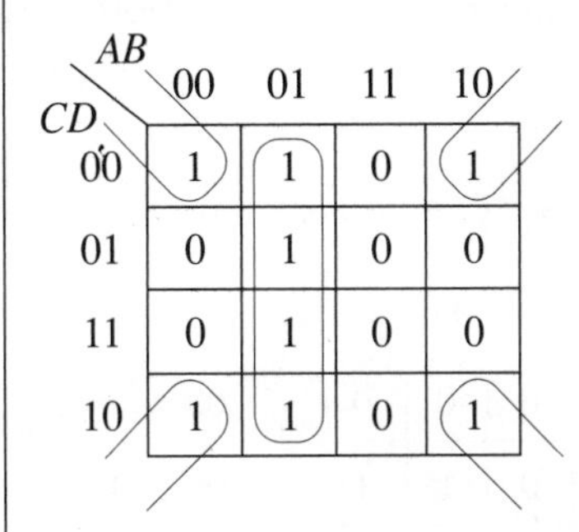

The four 1s in the second column can be looped together to give the term $\bar{A}B\,(C + \bar{C} + D + \bar{D}) = \bar{A}B$. The four corner cells can also be looped together to give $\bar{B}\bar{D}$. Hence, $F = \bar{A}B + \bar{B}\bar{D}$ (*Ans.*)

EXAMPLE 5.8

Map and then simplify $F = \bar{A}\bar{B}\bar{C}D + AB\bar{C}D + A\bar{B}\bar{C}D + \bar{A}\bar{B}CD + A\bar{B}CD$.

Solution

CD \ AB	00	01	11	10
00	0	0	0	0
01	1	0	1	1
11	1	0	0	1
10	0	0	0	0

From the map, $F = \bar{B}D + A\bar{C}D$ (*Ans.*)

Product-of-sums equations

Combining individual maps

The Karnaugh map can also be used to simplify an equation in its POS form. Each of the sum terms can be mapped individually in the same way as for a SOP equation. The individual maps can then be combined to obtain the mapping of F. Note that $1.1 = 1$ and $1.0 = 0$.

EXAMPLE 5.9

Use a Karnaugh map to simplify the equation $F = (AC + A\bar{C}D)(AD + AC + BC)$.

Solution

The mappings of $(AC + A\bar{C}D)$ and $(AD + AC + BC)$ are

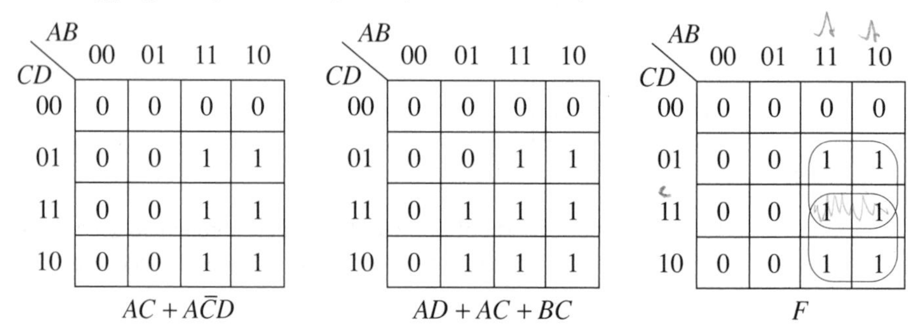

ANDing the two maps gives the mapping for F. Looping the 1 cells gives $F = AC + AD$ (*Ans.*)

EXAMPLE 5.10

Minimize $F = (A + B)(B + C)$ using a Karnaugh map.

Solution

$F = (A + B)(B + C)$ is mapped as

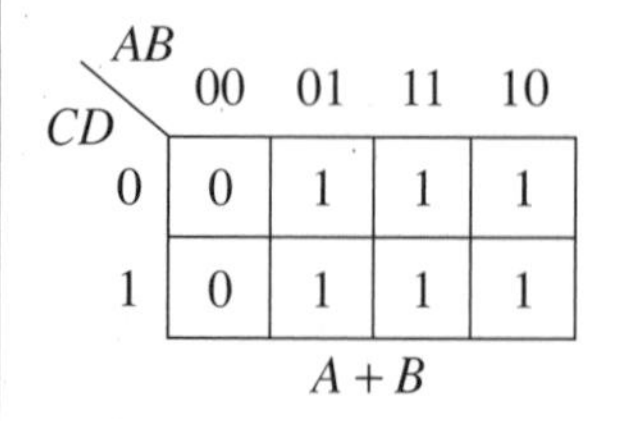

C \ AB	00	01	11	10
0	0	1	1	0
1	1	1	1	1

$B + C$

C \ AB	00	01	11	10
0	0	1	1	0
1	0	1	1	1

$(A + B)(B + C)$

From the map, $F = B + AC$ (*Ans.*)

EXAMPLE 5.11

Use a Karnaugh map to simplify $F = (\bar{A} + \bar{B} + C + D)(\bar{A} + \bar{B} + C + \bar{D})(\bar{A} + \bar{B} + \bar{C} + \bar{D})$.

Solution

The mapping is

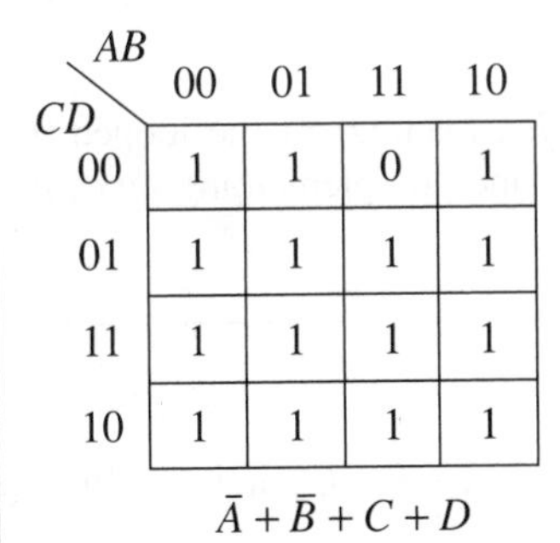

CD \ AB	00	01	11	10
00	1	1	0	1
01	1	1	1	1
11	1	1	1	1
10	1	1	1	1

$\bar{A} + \bar{B} + C + D$

CD \ AB	00	01	11	10
00	1	1	1	1
01	1	1	0	1
11	1	1	1	1
10	1	1	1	1

$\bar{A} + \bar{B} + C + \bar{D}$

CD \ AB	00	01	11	10
00	1	1	1	1
01	1	1	1	1
11	1	1	0	1
10	1	1	1	1

$\bar{A} + \bar{B} + \bar{C} + \bar{D}$

CD \ AB	00	01	11	10
00	1	1	0	1
01	1	1	0	1
11	1	1	0	1
10	1	1	1	1

F

From the final map, $F = \bar{A} + \bar{B} + C\bar{D}$ (*Ans.*)

Direct mapping

The need to draw a separate map for each term in a POS Boolean expression before deducing the composite map can be avoided and the mapping written down directly. Since the POS expression is the active LOW representation of a logic function, the following method can be employed.

- Convert each term in a product by changing all AND signs to OR and all OR signs to AND, and also complementing each variable, and enter a 0 into the corresponding cells in the map. This means that (i) a term such as $A + B$ must be read as $\bar{A}\bar{B}$ and (ii) $A + \bar{B}$ read as $\bar{A}B$. This is shown in the map.

$C \backslash AB$	00	01	11	10
0	0	0	1	1
1	0	0	1	1

$(A+B)(A+\bar{B})$

- Enter 1s into all the remaining cells.
- Loop the 0 cells.
- Read the groups of 0 cells with the complements of each term and with + and • signs interchanged.

Thus, for the mapping of $(A + B)(A + \bar{B})$, reading off the looped 0 cells in the usual way gives $F' = \bar{A}$, and changing signs and complementing gives $F = A$.

EXAMPLE 5.12

Map $F = (A + B + C)(A + \bar{B} + C)(A + \bar{B} + \bar{C})(\bar{A} + B + C)$ and simplify the equation.

Solution

(a) Enter 0s into the cells $\bar{A}\bar{B}\bar{C}$, $\bar{A}B\bar{C}$, $\bar{A}BC$, and $A\bar{B}\bar{C}$.
(b) Loop the 0 cells and read off as a SOP equation to give $F' = \bar{A}B + \bar{B}\bar{C} + \bar{A}\bar{C}$.
(c) Changing signs and complementing gives $F = (A + \bar{B})(B + C)(A + C)$ (*Ans.*)

[To check: multiplying out the original expression gives

$$A + A\bar{B} + AC + AB + BC + AC + \bar{B}C + C)(A + \bar{B} + \bar{C})(\bar{A} + B + C)$$
$$= (A + C)(A + \bar{B} + \bar{C})(\bar{A} + B + C) = (A + A\bar{B} + A\bar{C} + AC + \bar{B}C)(\bar{A} + B + C)$$
$$= (A + \bar{B}C)(\bar{A} + B + C) = AB + AC + \bar{B}C$$

Multiplying out the reduced expression gives: $(AB + AC + \bar{B}C)(A + C) = AB + AC + \bar{B}C$.

Inverse of a function

When a Boolean equation has been mapped on a Karnaugh map the inverse of the function can easily be obtained by looping the cells that contained a 0. Thus, referring to

(a) Example (5.4), $\bar{F} = \bar{A}C + A\bar{B}$.
(b) Example (5.6), $\bar{F} = \bar{D} + \bar{A}\bar{B}$.
(c) Example (5.8), $\bar{F} = \bar{D} + \bar{A}B + BC$.

The minimized form of a Boolean equation can be obtained by looping the 0 cells to obtain $\bar{F}$ and then inverting to obtain F.

EXAMPLE 5.13

Use a Karnaugh map to simplify the Boolean equation $F = \bar{A}B\bar{C} + A\bar{C}D + BC + \bar{A}\bar{B}D + A\bar{B}CD$.

Solution

The mapping is

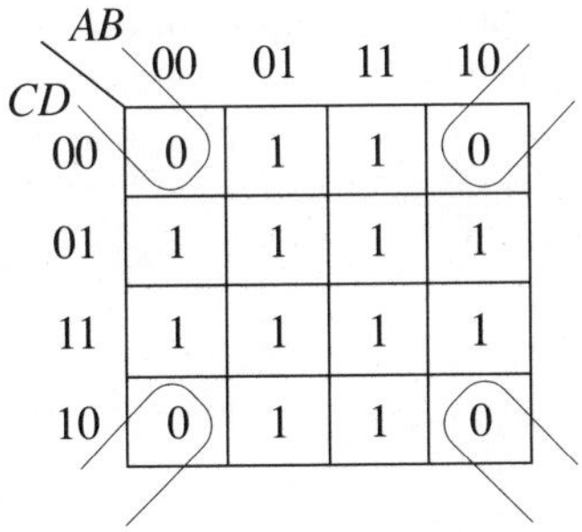

CD \ AB	00	01	11	10
00	0	1	1	0
01	1	1	1	1
11	1	1	1	1
10	0	1	1	0

Looping the 0 cells gives $\bar{F} = \bar{B}\bar{D}$. Therefore, $F = \overline{\bar{B}\bar{D}} = B + D$ (*Ans.*)

EXAMPLE 5.14

The mapping of a Boolean equation is

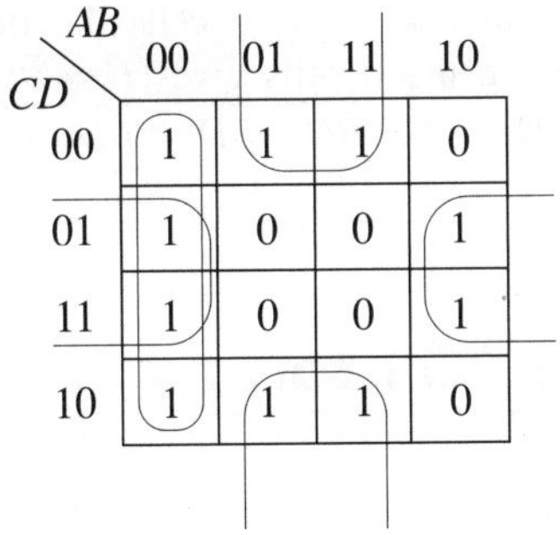

CD \ AB	00	01	11	10
00	1	1	1	0
01	1	0	0	1
11	1	0	0	1
10	1	1	1	0

Obtain the simplified expression for F (a) directly and (b) by finding $\bar{F}$ and then inverting.

Solution

(a) Looping the 1 cells gives $F = \bar{A}\bar{B} + B\bar{D} + \bar{B}D$ (*Ans.*)

(b) Looping the 0 cells gives $\bar{F} = BD + AB\bar{D}$.

Therefore, $F = \overline{BD + A\bar{B}\bar{D}} = (\overline{BD})(\overline{A\bar{B}\bar{D}}) = (\bar{B} + \bar{D})(\bar{A} + B + D)$

$= \bar{A}\bar{B} + \bar{A}\bar{D} + B\bar{D} + \bar{B}D = \bar{A}\bar{B} + B\bar{D} + \bar{B}D$ (*Ans.*)

Don't care/can't happen states

In some logic circuits, particularly coding circuits, certain combinations of the input variables will never occur. One example of this which has already been met is the BCD code in which only the input variables representing decimal numbers 0 through to 9 can occur. The combinations representing decimal 10 through to 15 *can't happen.* Since it does not matter whether such combinations are treated as a 1 or as a 0, these combinations are known as *don't care states.* When a Boolean equation is mapped, any don't care terms are indicated by a ×. When the equation is simplified, each cell containing a × may be looped with *either* 1 cells *or* with 0 cells. It is not necessary for all the × cells to be included in a loop. Only those × cells whose inclusion assists in the simplification of the equation should be used.

Suppose the mapping of a function is

CD \ AB	00	01	11	10
00	×	×	×	×
01	0	1	1	0
11	0	1	1	0
10	1	0	0	1

Looping the 1 cells gives $F = BD + \bar{B}C\bar{D}$.

If the don't care cells are looped together with the 1 cells the function further reduces to $F = BD + \bar{B}\bar{D}$. Similarly, looping the 0 cells gives $\bar{F} = \bar{B}D + BC\bar{D}$, and including the × cells, gives $\bar{F} = \bar{B}D + B\bar{D}$. Then, $F = \overline{\bar{B}D + B\bar{D}} = (\overline{\bar{B}D})(\overline{B\bar{D}}) = (B + \bar{D})(\bar{B} + D) = BD + \bar{B}\bar{D}$, as before.

Converting from SOP to POS form

A Karnaugh map provides a convenient method of converting a SOP equation into its corresponding POS form. The procedure to be followed is:

- Map the SOP equation.
- Loop the 0 cells to obtain the simplified equation.
- For each group of 0 cells read off the POS term, interchanging • and + signs and complementing all variables.

EXAMPLE 5.15

Use a Karnaugh map to obtain the POS form of $F = \bar{B}\bar{C} + \bar{A}B + ABC$.

Solution

The mapping is

C \ AB	00	01	11	10
0	1	1	0	1
1	0	1	1	0

The looped 0 cells give $F' = \bar{B}C + AB\bar{C}$. Complementing and changing signs, $F = (B + \bar{C})(\bar{A} + \bar{B} + C)$ (*Ans.*)

EXAMPLE 5.16

Use a Karnaugh map to solve example (4.20), i.e. convert $F = \bar{A}\bar{B}\bar{C} + \bar{A}\bar{B}C + \bar{A}B\bar{C} + A\bar{B}\bar{C} + AB\bar{C} + ABC$ into its POS form.

Solution

The mapping is

C \ AB	00	01	11	10
0	1	1	1	1
1	1	0	1	0

From the 0 cells, $F' = \bar{A}BC + A\bar{B}C$. Interchanging • and + signs and complementing each variable gives $F = (A + \bar{B} + \bar{C})(\bar{A} + B + \bar{C})$ (*Ans.*)

Converting from POS to SOP form

It is also possible to employ a Karnaugh map to convert an equation from its POS form to its SOP form. The procedure is as follows:

- Map the POS equation inserting 0s into the relevant cells.
- Loop the cells containing a 1.
- Read off the simplified SOP equation.

EXAMPLE 5.17

Use a Karnaugh map to convert $F = (A + B)(\bar{B} + C)$ into its SOP form.

Solution

C \ AB	00	01	11	10
0	0	0	0	1
1	0	1	1	1

(a) Map $\bar{A}\bar{B}$ and $B\bar{C}$ inserting 0s into the relevant cells.
(b) Loop the 1 cells and read off to get $F = A\bar{B} + BC$ (*Ans.*)

EXAMPLE 5.18

Convert the answer to example (5.16) into its SOP form.

Solution

(a) Map $A\bar{B}C$ and $\bar{A}BC$ using 0s. This is shown in the map.

C \ AB	00	01	11	10
0	1	1	1	1
1	1	0	1	0

(b) Looping the 1 cells gives $F = \bar{C} + \bar{A}\bar{B} + AB$ (*Ans.*)

EXAMPLE 5.19

Convert the answer to example (4.19), i.e. $F = (A + B)(A + \bar{C})(\bar{A} + \bar{B})(\bar{A} + C)$ into the SOP form.

Solution

(a) Mapping, with 0s, $\bar{A}\bar{B}$, $\bar{A}C$, AB, and $A\bar{C}$ gives

C \ AB	00	01	11	10
0	0	1	0	0
1	0	0	0	1

(b) From the 1 cells, $F = \bar{A}B\bar{C} + A\bar{B}C$ (*Ans.*)

Table 5.1

A	0	0	0	0	0	0	0	1	1	1	1	1	1	1	1
B	0	0	0	1	1	1	1	0	0	0	0	1	1	1	1
C	0	0	1	0	0	1	1	0	0	1	1	0	0	1	1
D	0	1	0	1	0	1	0	1	0	1	0	1	0	1	0
F	0	1	0	0	1	0	0	1	1	1	1	0	1	0	0

Designing a circuit from a truth table

The Boolean equation that describes the operation of a combinational logic circuit can be derived from the truth table of the circuit. If a SOP equation is wanted each 1 that appears in the output column of the table must be represented by a term in the equation. This term must include each variable that is at 1, and the complement of each variable that is at 0.

As an example of the technique consider the truth table given in Table 5.1.

The Boolean expression for this circuit is

$$F = \bar{A}\bar{B}\bar{C}D + \bar{A}B\bar{C}\bar{D} + A\bar{B}\bar{C}D + A\bar{B}\bar{C}\bar{D} + A\bar{B}CD + A\bar{B}C\bar{D} + AB\bar{C}\bar{D}$$

The Karnaugh mapping of the function F is shown with adjacent squares looped together. From the map the simplified equation representing the logical operation given by the truth table is

$$F = A\bar{B} + B\bar{C}\bar{D} + \bar{B}\bar{C}D$$

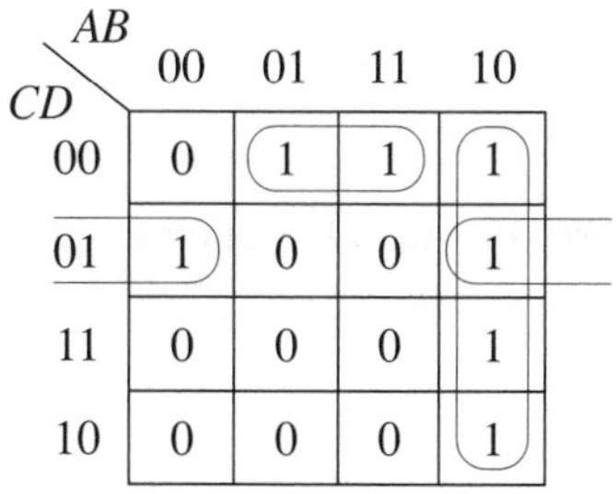

CD \ *AB*	00	01	11	10
00	0	1	1	1
01	1	0	0	1
11	0	0	0	1
10	0	0	0	1

If a POS equation for the output signal is required each 0 appearing in the output column of the truth table must be represented by a term in the Boolean equation describing the circuit. This term must contain each input variable that is at 0 and the complement of each input variable that is at 1.

Consider the truth table of a circuit that is given in Table 5.2.

The SOP expression for the circuit is

$$F = \bar{A}\bar{B}\bar{C} + \bar{A}\bar{B}C + \bar{A}BC + ABC$$

Table 5.2

A	0	1	0	1	0	1	0	1
B	0	0	1	1	0	0	1	1
C	0	0	0	0	1	1	1	1
F	1	0	0	0	1	0	1	1

The POS expression is

$$F = (\bar{A} + B + C)(A + \bar{B} + C)(\bar{A} + \bar{B} + C)(\bar{A} + B + \bar{C})$$

[Both equations reduce to $F = \bar{A}\bar{B} + BC$.]

EXAMPLE 5.20

The truth table of a logic circuit is given in Table 5.3.

All other variables are don't cares. Obtain an expression for the output F of the circuit if the output is (a) active HIGH, and (b) active LOW.

Table 5.3

A	0	1	0	1	0	1	0	1	0	1
B	0	0	1	1	0	0	1	1	0	0
C	0	0	0	0	1	1	1	1	0	0
D	0	0	0	0	0	0	0	0	1	1
F	0	1	0	0	0	1	0	1	0	0

Solution

(a) From the truth table, $F = A\bar{B}\bar{C}\bar{D} + A\bar{B}C\bar{D} + ABC\bar{D}$. Mapping gives

CD \ AB	00	01	11	10
00	0	0	0	1
01	0	×	×	0
11	×	×	×	×
10	0	0	1	1

Looping the cells that contain either a 1 or a × gives $F = AC + A\bar{B}\bar{D}$ (*Ans.*)

(b) Looping the 0s and the ×s gives $F' = \bar{A} + B\bar{C} + D$, $F = A\bar{D}(\bar{B} + C)$ (*Ans.*)

Table 5.4

A	0	1	0	1
B	0	0	1	1
Sum	0	1	1	0
Carry	0	0	0	1

Fig. 5.1 *Half-adder circuit*

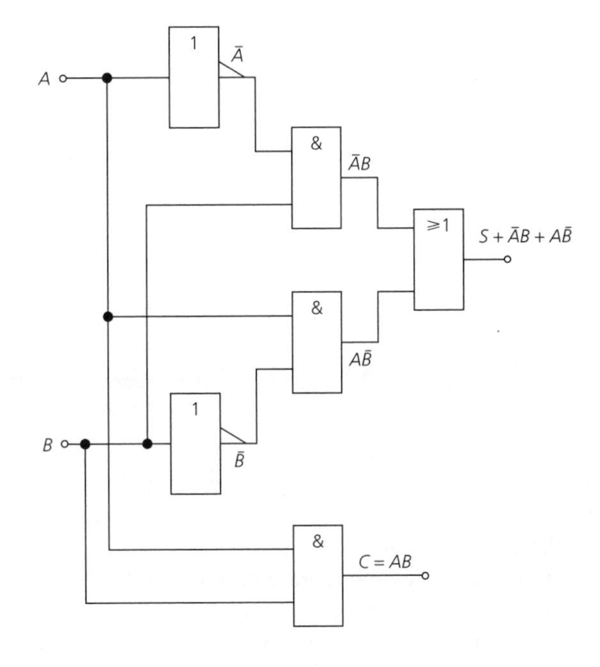

Half-adder

The half-adder is a circuit that adds two inputs A and B to produce a sum and a carry but cannot take account of any carry originating from a previous stage. The truth table of a half-adder is given in Table 5.4.

From the truth table the sum and the carry of A and B are $S = A\bar{B} + \bar{A}B$, $C = AB$.

Implementing these equations using AND, NOT, and OR gates gives the circuit shown in Fig. 5.1.

Static hazards

So far it has been assumed that all input variables change state instantaneously. However, a delay always occurs whenever a signal is passed through a gate. Different gates will have different delay times of a few milliseconds but this may lead to errors, known as *hazards*, in the output signal of a combinational logic circuit.

Fig. 5.2 *Circuit with a static hazard*

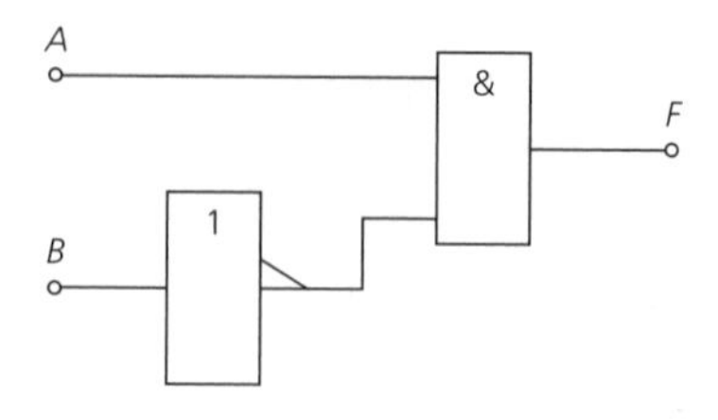

Fig. 5.3 *Waveforms for Fig. 5.2*

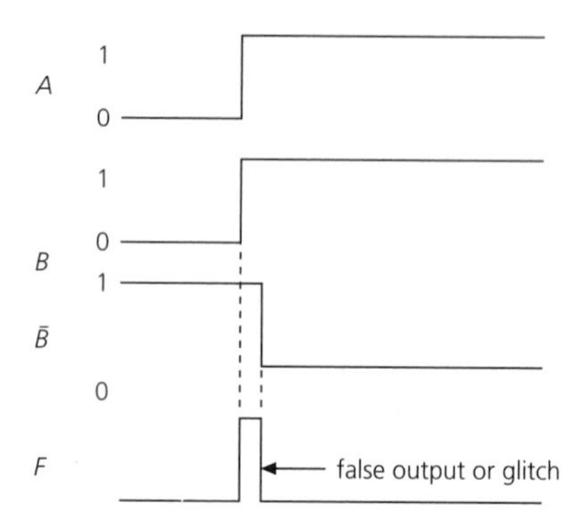

In many cases a hazard is generated when an input variable is complemented. Terms such as $A + \bar{A} = 1$ or $A\bar{A} = 0$ cannot occur instantaneously. Once the delay time has passed the correct output is obtained.

The Boolean expression representing Fig. 5.2 is $F = A\bar{B}$. When $A = B = 0$ the output of the circuit will be 0. If, now, A and B change simultaneously to 1 the output of the circuit should remain at 0. However, because of the time delay introduced by the inverting stage, the output will momentarily go to 1. The waveforms for the circuit are shown in Fig. 5.3. A hazard is always likely if adjacent 1 cells in a Karnaugh mapping are not looped together. The hazard can be removed by ensuring that all adjacent 1 cells *are* looped together. This will generally mean that the minimal solution to a mapped equation is not obtained.

Consider the equation $F = AC + B\bar{C}$. When $A = B = 1$, F should be of the form $C + \bar{C} = 1$. If C changes from 1 to 0, C will – after a short time delay – change from 0 to 1. For a short time, therefore, both C and $\bar{C}$ will be at the logical 0 state and the output of the circuit will be incorrect.

The mapping of $F = AC + B\bar{C}$ is

C \ AB	00	01	11	10
0	0	1	1	0
1	0	0	1	1

The static hazard can be eliminated by looping the ABC and $AB\bar{C}$ cells as well to give the term AB. Now $F = AB + B\bar{C} + AC$ and the output will be at logic 1 when $A = B = 1$ regardless of the state of C (and $\bar{C}$).

EXAMPLE 5.21

(a) Show that the expression $F = AB + C(\bar{B} + AB)$ contains a possible static hazard. (b) Explain how it arises. (c) Eliminate the hazard.

Solution

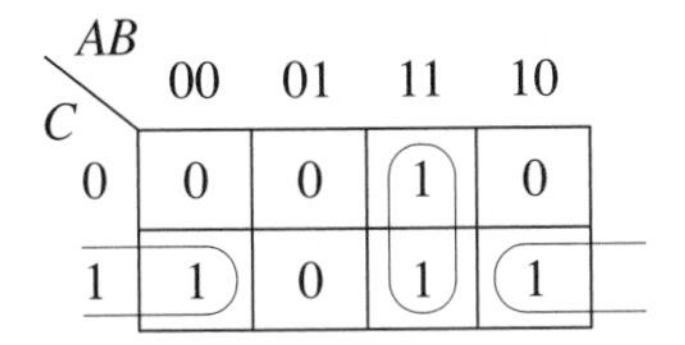

(a) Multiplying out, $F = AB + \bar{B}C + ABC$ and this function is mapped above. Looping the 1 cells gives the minimal expression as $F = \bar{B}C + AB$. This equation has a static hazard because the two loops are not linked together.
(b) When $B = 1, \bar{B} = 0$. If B changes from 1 to 0 there will be a small time interval during which $B = 0$ and $\bar{B} = 0$. During this time $F = A0 + C0 = 0$ instead of $F = 0 + C(1 + 0) = C$.
(c) To eliminate the hazard, the cells ABC and $A\bar{B}C$ must be linked together. This introduces a third term, AC, into the equation. Hence, the hazard-free equation is $F = AB + \bar{B}C + AC$ (*Ans.*)

EXAMPLE 5.22

Design a combinational logic circuit that has 10 inputs, numbered 0 through to 9, and one output. The output is required to go HIGH whenever any one, or more, of the inputs numbered 2, 5, 6 or 7 go HIGH. The circuit should be free of static hazards.

Solution

The truth table for the required circuit is given in Table 5.5.

From the truth table the Boolean equation describing the circuit is $F = \bar{A}BC\bar{D} + A\bar{B}C\bar{D} + \bar{A}BC\bar{D} + ABC\bar{D}$. This equation is mapped as shown. If the 1 and × cells are looped to minimize the equation, the result is $F = \bar{A}B + AC$, but a static hazard will exist. To eliminate the hazard, a third loop must be introduced to produce the redundant term BC. Hence, $F = \bar{A}B + AC + BC$ (*Ans.*)

Table 5.5

Denary	Input Binary D	C	B	A	Output
0	0	0	0	0	0
1	0	0	0	1	0
2	0	0	1	0	1
3	0	0	1	1	0
4	0	1	0	0	0
5	0	1	0	1	1
6	0	1	1	0	1
7	0	1	1	1	1
8	1	0	0	0	0
9	1	0	0	1	0

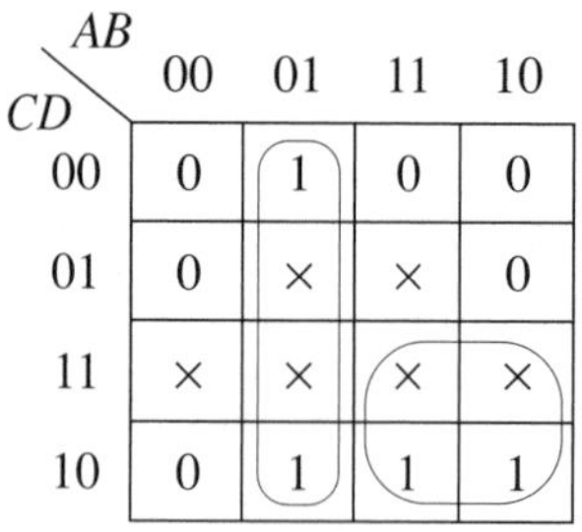

CD \ AB	00	01	11	10
00	0	1	0	0
01	0	×	×	0
11	×	×	×	×
10	0	1	1	1

EXAMPLE 5.23

The output of a BCD parity-checking circuit is to be at logic 1 whenever there is an even number of 1s at its input terminal. Write down the truth table for the circuit. Map the Boolean equation that describes the operation of the circuit and determine if a static hazard exists.

Solution

The truth table for the circuit is given in Table 5.6.

From the table, $F = AB\bar{C}\bar{D} + A\bar{B}C\bar{D} + \bar{A}BC\bar{D} + A\bar{B}\bar{C}D$. From the map, $F = AB\bar{C} + A\bar{B}C + AD + \bar{A}BC$. There are two adjacent 1 cells that are not looped, so a static hazard exists (*Ans.*)

Table 5.6

Inputs				Outputs
D	C	B	A	F
0	0	0	0	0
0	0	0	1	0
0	0	1	0	0
0	0	1	1	1
0	1	0	0	0
0	1	0	1	1
0	1	1	0	1
0	1	1	1	0
1	0	0	0	0
1	0	0	1	1

CD \ AB	00	01	11	10
00	0	0	1	0
01	0	×	×	1
11	×	×	×	×
10	0	1	0	1

PRACTICAL EXERCISE 5.1

To design a circuit whose output goes HIGH when its inputs A and B are both HIGH AND either input C OR input D is also HIGH.

Components and equipment: one 74LS10 (or 74HC10) triple 3-input NAND gate IC. One 74LS27 (or 74HC27) triple 3-input NOR gate IC. One LED. One 270 Ω resistor. Power supply. Breadboard.

Procedure:

(a) Write down the Boolean expression that describes the performance of the required circuit. Rearrange the expression into its SOP form.

(b) Construct the circuit using the 74LS10 IC with the output terminal connected to 0 V via the resistor and the LED.

(c) Apply the following combinations of input variables to the circuit and each time note the logical state of the LED. (i) $A = B = 0$, $C = D = 1$, (ii) $A = B = 1$, $C = D = 0$, (iii) $A = B = C = 1$, $D = 0$, (iv) $A = B = D = 1$, $C = 0$, (v) $A = B = C = D = 1$, and (vi) $A = 1$, $B = C = D = 0$.

(d) Put the equation into its POS form and construct an active-LOW circuit using the 74LS27 NOR gate IC. Now connect the output terminal via the resistor and the LED to +5 V.

(e) Repeat procedure (c).

EXERCISES

5.1 Design a logic circuit that inputs a BCD number and gives a HIGH output whenever the input number is odd and less than 8.

5.2 Design a traffic light system to go through the normal red, red and yellow, green, yellow, red, etc. sequence for each of the following conditions.

(a) All four-colour periods are of the same duration.
(b) The red and green periods are of equal length and three times as long as the red and the yellow periods.
(c) The green period is five times as long as the yellow period, and the red period is three times the yellow period.

5.3 A voting system has four inputs A, B, C, and D. If any three, or all four inputs are HIGH the output must go HIGH also. If the voting is 2/2 then the output must take up the state of input D. If only one, or none, of the inputs is HIGH the output must be LOW. Design the system.

5.4 Design a decoder circuit which inputs a 4-bit digital number and lights the appropriate LED to indicate the decimal equivalent of the input number. Assume the highest input number is decimal 9.

5.5 Design an encoder circuit to perform the inverse function to the circuit of 5.4.

5.6 Design a logic circuit whose output will go HIGH whenever the 4-bit input digital number is odd.

5.7 Re-design the circuit in 5.6 so that the output goes HIGH when the input number is odd and in the decimal range 5–13.

5.8 The equipment of a car includes a buzzer (B) that sounds audibly if either the headlights (H) OR the sidelights (S) are ON AND the driver's door (D) is open, OR if the ignition is turned on (I) AND the door (D) is open. Determine the Boolean equation for the operation of the buzzer.

5.9 Convert $F = ABC + A\bar{B}CD + \bar{A}B\bar{C}D + ACD$ into its POS form.

5.10 Obtain the minimal solution for the complement of the SOP equations
(a) $F = ABC + A\bar{B}CD + AB\bar{C}D + \bar{A}BCD + A\bar{B}D + \bar{B}C\bar{D} + \bar{A}B\bar{C}D$.
(b) $F = ABC + A\bar{B}CD + \bar{A}B\bar{C}D + ACD$.

5.11 Determine the complement of $F = (A + \bar{B})(\bar{A} + C)(\bar{A} + B)$
(a) Using De Morgan's rules.
(b) Using a Karnaugh map.

5.12 Use a Karnaugh map to convert $F = C\bar{D} + AC + \bar{A}\bar{C}\bar{D}$ into its POS form.

5.13 Use a Karnaugh map to convert $F = (A + B + D)(\bar{A} + \bar{C} + \bar{D})(A + \bar{B} + C)$ into its SOP form.

5.14 Determine the complement of $F = A\bar{B}C + \bar{A}BD$
(a) Using De Morgan's rules.
(b) Using a Karnaugh map.

5.15 Use a Karnaugh map to simplify $F = A\bar{B}C + \bar{A}\bar{B}\bar{C} + \bar{A}\bar{B}C + A\bar{B}\bar{C} + ABC$.

5.16 Use a Karnaugh map to simplify $F = AB\bar{C} + ABC + \bar{A}B\bar{C} + \bar{A}\bar{B}\bar{C}$.
For both 5.15 and 5.16, loop the squares marked with 0 to obtain the minimal solution for $\bar{F}$. Invert $\bar{F}$ and check with the answer obtained in the first part of the question.

5.17 Use a Karnaugh map to minimize the equation $F = \bar{A}B\bar{C}\bar{D} + \bar{A}B\bar{C}D + A\bar{B}CD + A\bar{B}C\bar{D} + ABCD + \bar{A}\bar{B}\bar{C}\bar{D}$.

5.18 Use a Karnaugh map to minimize $F = (AB + \bar{B}\bar{C})(\bar{A}\bar{C} + \bar{A}\bar{B}\bar{C} + BC)$.

5.19 Use a Karnaugh map to minimize $F = A\bar{B} + \bar{C}D + C + \bar{B}\bar{C}D$.

5.20 Use a Karnaugh map to minimize $F = A\bar{B}\bar{C}\bar{D} + \bar{A}\bar{B}\bar{C}\bar{D} + AB\bar{C}D + \bar{A}BCD + ABD + \bar{B}C\bar{D} + \bar{A}B\bar{C}D$.

6 NAND and NOR logic

After reading this chapter you should be able to:

(a) Use De Morgan's rule to implement other gates using either NAND gates only, or NOR gates only.
(b) Understand the universal capability of NAND and NOR gates.
(c) Obtain the NAND gate only version of a combinational logic circuit using (i) double inversion and (ii) OR/AND level techniques.
(d) Obtain the NOR gate only version of a combinational logic circuit using (i) double inversion and (ii) AND/OR level techniques.

Most of the ICs employed in modern digital equipment belong to one or other of two logic families, namely the TTL and CMOS families. NAND and NOR gates are the most commonly employed since the use of one type of gate in a circuit often results in fewer ICs being needed. Also, the NAND and NOR gates generally have a faster operating speed and a lower power dissipation than AND and OR gates. Very often, a random logic circuit is made up using *only* NAND or NOR gates. The AND function can be obtained by connecting two NAND gates as shown in Fig. 6.1(a), the second gate being employed as an inverter. Similarly, the OR function is easily obtained by connecting two NOR gates as shown in Fig. 6.1(b). The use of one type of gate in a circuit often results in fewer ICs being required even if there is a larger number of gates.

Fig. 6.1 *Implementation of (a) the AND function using NAND gates and (b) the OR function using NOR gates*

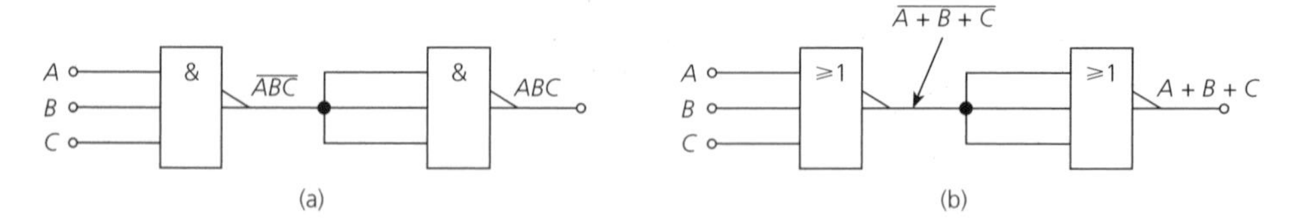

AND and OR functions

Implementation of the AND function using NOR gates, and of the OR function using NAND gates, is not quite as easy but the necessary connections can be deduced readily by the use of De Morgan's rules.

Fig. 6.2 *Implementation of (a) the AND function using NOR gates and (b) the OR function using NAND gates*

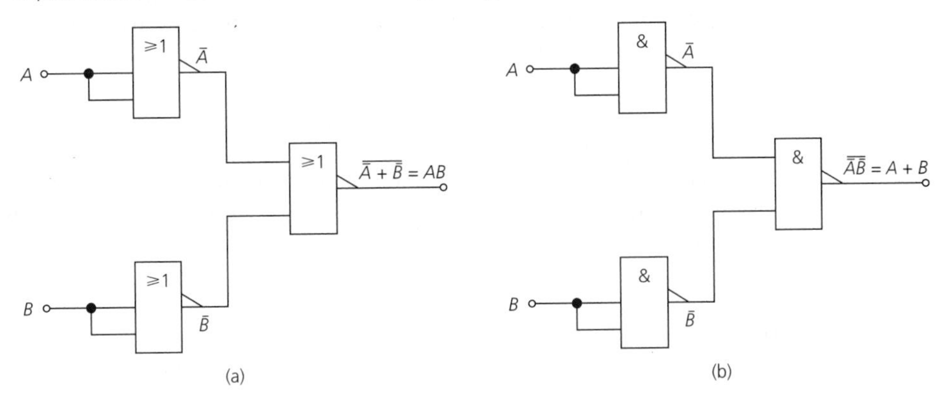

Rule (16) is

$$\overline{AB} = \bar{A} + \bar{B}$$

hence,

$$AB = \overline{\bar{A} + \bar{B}}$$

The right-hand expression is implemented easily using NOR gates as shown in Fig. 6.2(a). The other De Morgan rule is

$$\overline{A + B} = \bar{A}\bar{B}$$

hence,

$$A + B = \overline{\bar{A}\bar{B}}$$

and this expression can be implemented using NAND gates as shown in Fig. 6.2(b).

Clearly, more gates are needed to implement the AND/OR functions with NAND/NOR gates, but very often the apparent increase in the number of gates required is not as great as at first anticipated since consecutive stages of inversion need not be provided. This point is illustrated by the following example.

EXAMPLE 6.1

Implement the exclusive-OR function $F = A\bar{B} + \bar{A}B$ using (a) NAND gates only, and (b) NOR gates only.

Solution

The first step is to draw the logic diagram using AND, OR and NOT gates. When this has been done, replace each gate with its (a) NAND, (b) NOR equivalent circuit. Finally, if possible, simplify the resulting network by eliminating any redundant gates.

(a) Figure 3.32 shows the exclusive-OR gate built with AND and OR gates. Replacing each gate with the equivalent NAND gates gives the circuit shown in Fig. 6.3(a). It can be seen that this circuit includes two sets of two NAND gates in cascade. These are redundant since $\bar{\bar{A}} = A$. Figure 6.3(b) shows the simplified network which does not include the four redundant gates. It will be noticed that the number of NAND gates required (i.e. 5) to implement the exclusive-OR gate is the same as the number of AND and OR gates which are necessary. In this case there is no reduction in the number of ICs required.

Fig. 6.3 *Exclusive-OR function implemented using NAND gates*

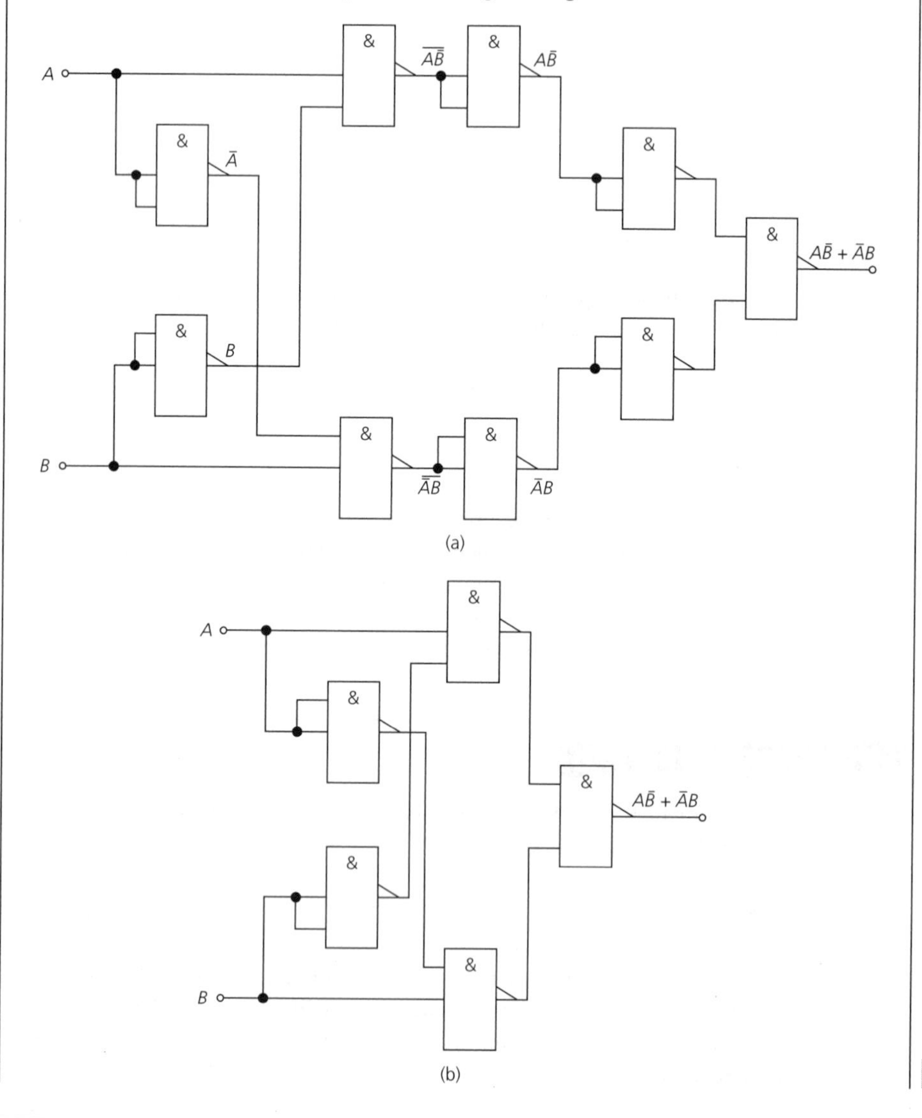

(b) Replacing each AND gate and each OR gate with the corresponding NOR gate version gives the circuit shown in Fig. 6.4(a). This circuit can be simplified by eliminating redundant gates to give the circuit of Fig. 6.4(b). Now five gates are needed requiring two ICs (*Ans.*)

Fig. 6.4 *Exclusive-OR function implemented using NOR gates*

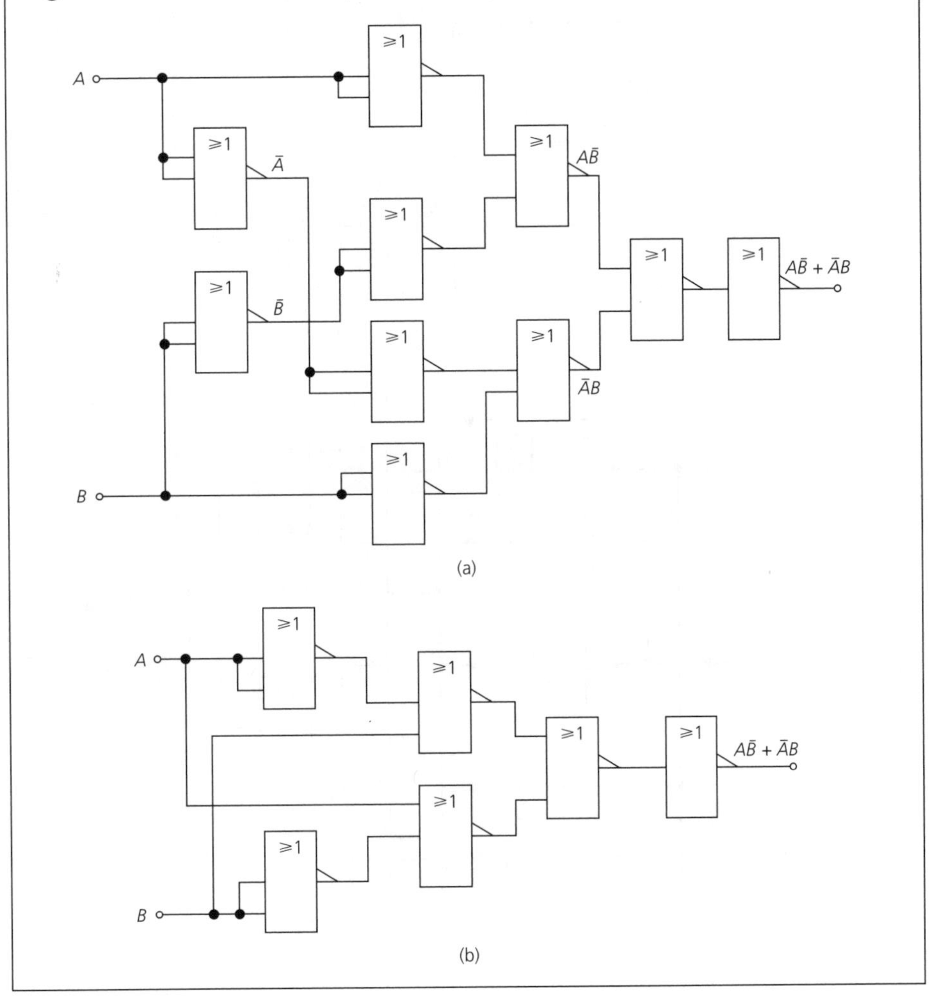

In general, Boolean equations in the POS form, e.g. $(A + B)(C + D)$, are best implemented using NOR gates, and SOP equations, e.g. $AB + CD$, are more easily implemented using NAND gates.

Figure 5.1 (p. 119) shows the circuit of a half-adder implemented using AND, OR and NOT gates.

The first step in obtaining the NAND version of this circuit is to replace each AND and each OR gate by its NAND equivalent. This has been done for the Fig. 5.1 circuit

Fig. 6.5 *Half-adder circuits implemented with (a) and (b) NAND gates only, and (c) and (d) NOR gates only*

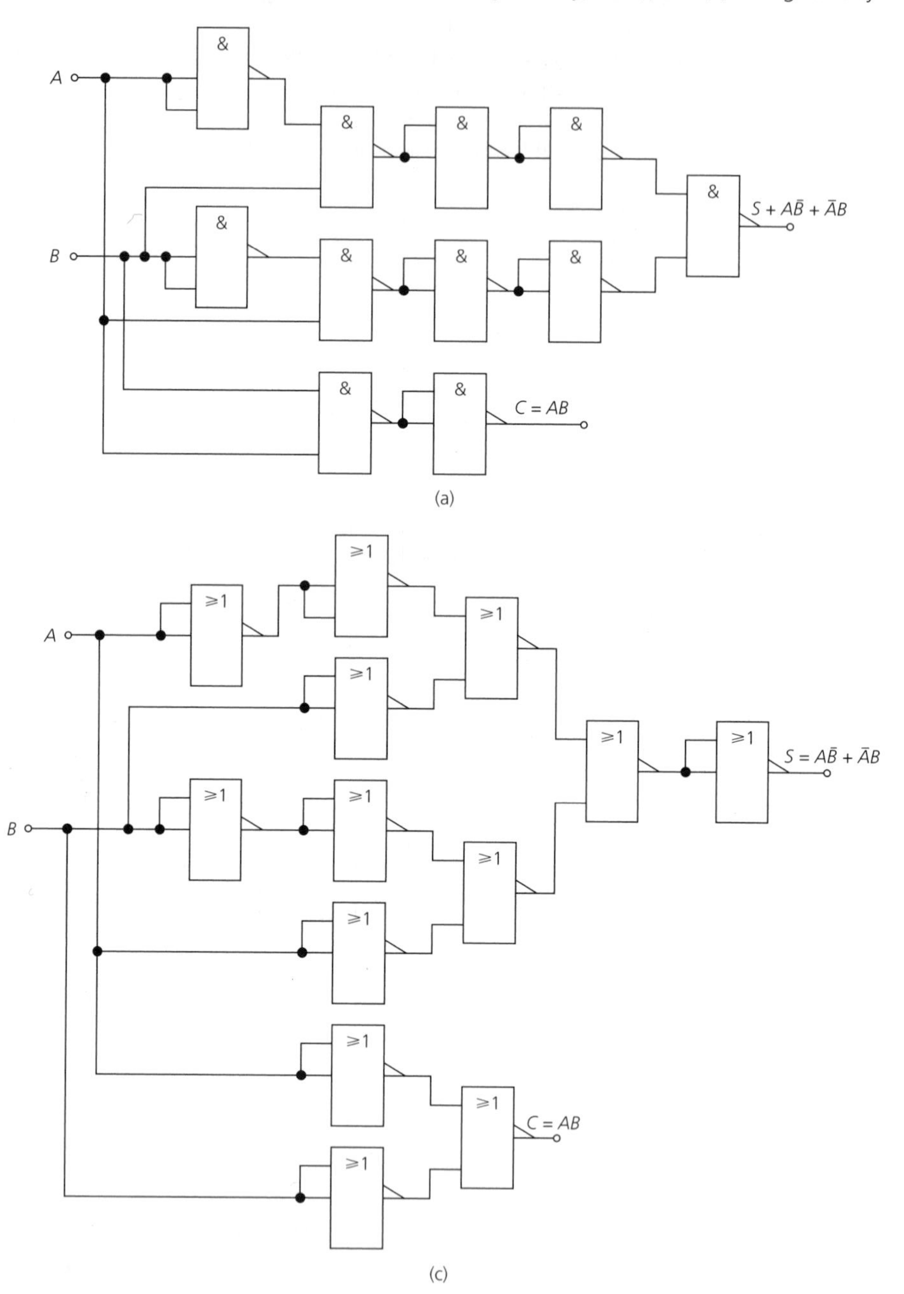

and the result is given in Fig. 6.5(a). The circuit can then be simplified by the removal of redundant gates to give the circuit of Fig. 6.5(b). Similarly, the NOR gate version of the 5.1 circuit is shown in Fig. 6.5(c). It can be seen that in this case there are four redundant gates. Removing these gives Fig. 6.5(d).

Fig. 6.5 *(cont.).*

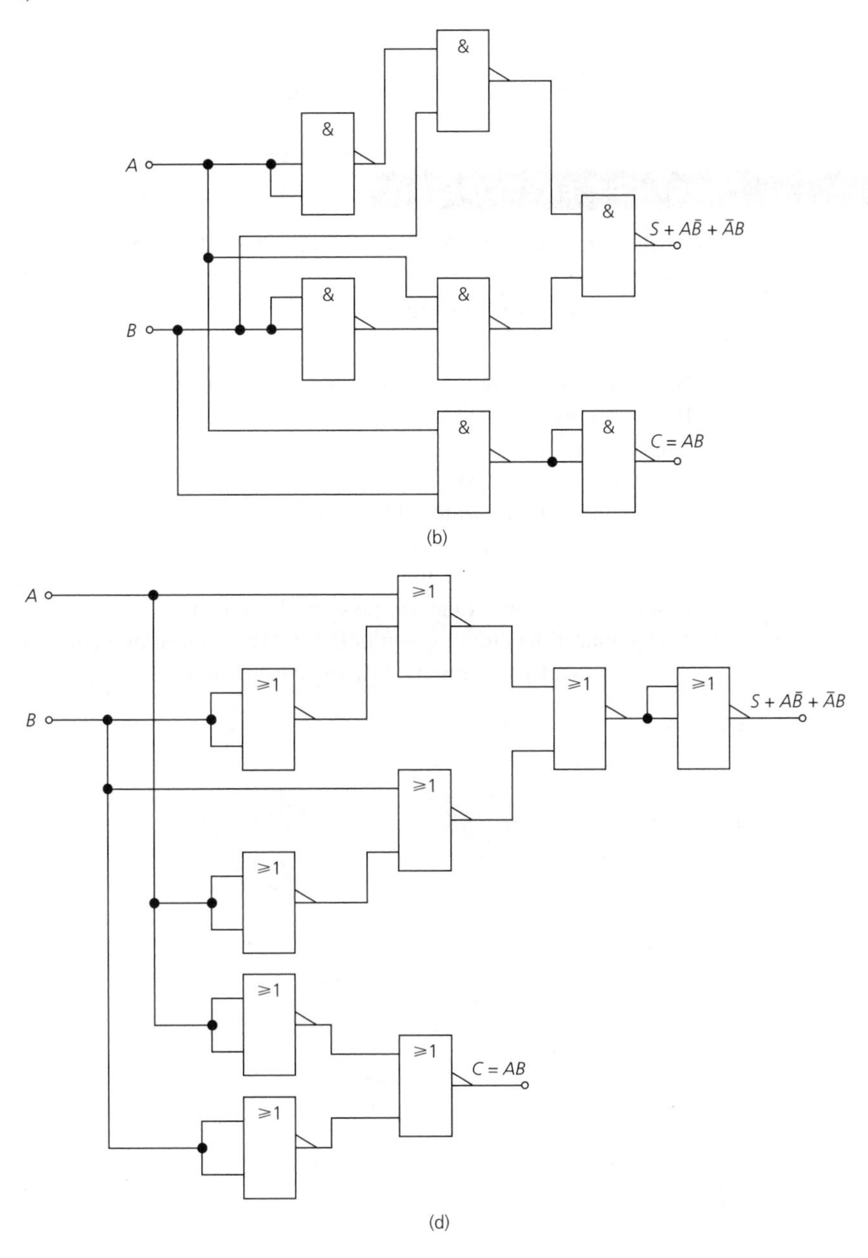

Comparing the half-adder circuits shown in Fig. 6.5(b) and (d), it is clear that the NOR version requires two more gates than the NAND equivalent. This is an indication that the NAND gate implementation of logical functions is more suited to those functions in SOP form. Conversely, NOR gate implementation is the most appropriate for a function in POS form.

It might appear that the NAND/NOR circuits will require more ICs than the circuit in Fig. 5.1. This, however, is not the case. Figure 5.1 uses two NOT, three AND and one OR gate which would require *three* ICs. Figure 6.5(b) uses seven NAND gates and Fig. 6.5(d) uses nine NOR gates; these are provided by just *two* or *three* ICs.

PRACTICAL EXERCISE 6.1

To investigate the use of NAND gates to implement the AND and OR logic functions.
Components and equipment: one 74LS00 (or 74HC00) NAND gate IC. One LED. One 270 Ω resistor. Breadboard. Power supply.
Procedure:

(a) Build the circuit shown in Fig. 6.6(a) and connect the 5 V power supply between pins 7 and 14.
(b) Apply each of the four possible combinations of 1 and 0 to the three input terminals of the circuit. For each combination note the logical state of the LED. Complete the truth table for the circuit started in Table 6.1 and confirm that the circuit has performed the AND logic function.
(c) Move the connection of the LED from 0 V to +5 V and repeat procedure (b). Determine the logic function now implemented.
(d) Now connect the circuit given in Fig. 6.6(b). Repeat procedure (b) and confirm that the circuit performs the OR logical function.

Fig. 6.6

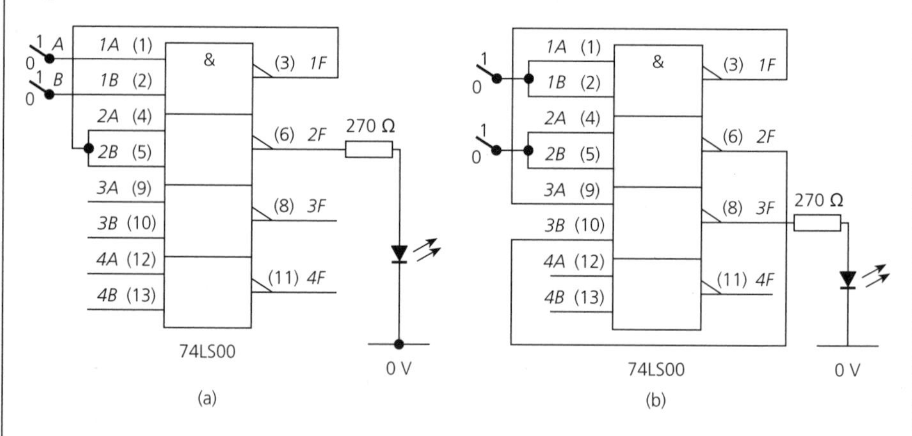

Table 6.1

B	A	F
0	0	
0	1	
1	0	
1	1	

NAND function implemented by NOR gates

The AND function can be produced by three NOR gates connected as shown in Fig. 6.2(a). If the output of this circuit is inverted the NAND logic function is obtained. One quad 2-input NOR gate is required.

NOR function implemented by NAND gates

To produce the NOR function using NAND gates it is necessary to invert the output of the NAND gate implementation of the OR function (see Fig. 6.2(b)). One quad 2-input NAND gate IC is required.

Function implementation using NAND gates

Any SOP equation can be realized easily using two levels of NAND gates. Each AND or OR gate in a logic circuit can be replaced by its NAND gate equivalent as shown in Figs 6.2 and 6.4. Two quicker methods of realizing a NAND gate only circuit are available: namely (a) double inversion and (b) OR/AND levels.

Double inversion

The NAND circuit can be obtained directly from the Boolean equation describing the circuit by employing the technique of *double inversion*. The procedure to be followed is:

- Invert twice the SOP expression to be implemented.
- Retain the top inversion bar and apply De Morgan's rule to the bottom bar to eliminate the OR operation(s).
- Implement the resulting expression using NAND gates only.

EXAMPLE 6.2

Implement the expression $F = ABC + \bar{A}BC + \bar{A}\bar{B}C$ using NAND gates only.

Solution

Inverting F twice gives

$$F = \overline{\overline{ABC + \bar{A}BC + \bar{A}\bar{B}C}}$$
$$= \overline{(\overline{ABC})(\overline{\bar{A}BC})(\overline{\bar{A}\bar{B}C})}$$

This result can be implemented directly using four 3-input NAND gates as shown in Fig. 6.7. This circuit requires the use of two triple 3-input NAND gate ICs as opposed to one triple 3-input AND gate IC and one triple 3-input OR gate IC, assuming that the complements of A and B are already available. Hence, there is no saving in the number of ICs needed. If, however, the complements of A and B are not available, then the unused gates in the second NAND gate IC can be connected as inverters so there is no increase in the number of ICs required. The AND/OR

Fig. 6.7

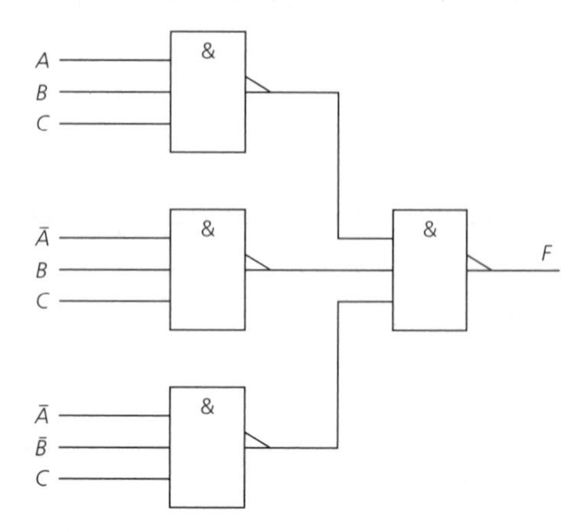

gate implementation, however, would need either a hex inverter IC or a NAND, or NOR, gate IC with two gates connected as inverters. In this case there is a saving of one IC (*Ans.*)

EXAMPLE 6.3

Implement the logic function $F = A + BD + C\bar{D}$ using NAND gates only.

Solution

$$F = \overline{\overline{A + BD + C\bar{D}}}$$
$$= \overline{\bar{A}\ (\overline{BD})\ (\overline{C\bar{D}})}$$

This equation can be implemented using four 2-input NAND gates and one 3-input NAND gate as shown in Fig. 6.8. Two ICs are needed (*Ans.*)

Fig. 6.8

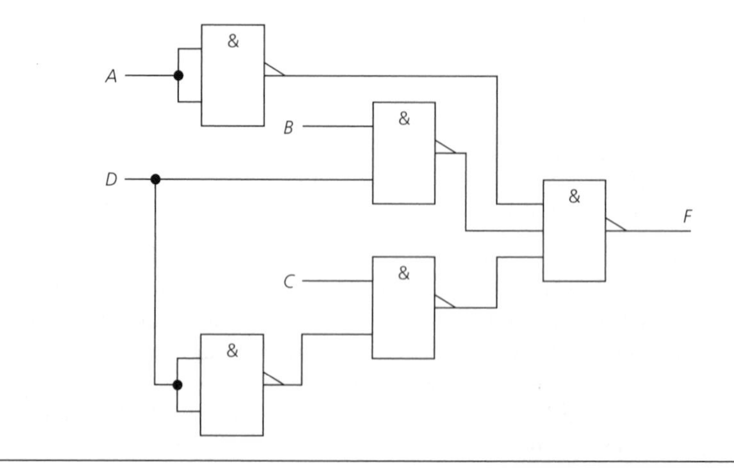

OR/AND levels

Considering the gate from which the output F of the circuit is taken as being level 1, the preceding gates are taken as being in levels 2, 3 and so on, not including any inverters.

- In odd numbered levels the NAND gate(s) perform the OR logical function. Any input variable entering at an odd level appears in complemented form at the output.
- In even numbered levels the NAND gate(s) perform the AND logical function. Any variables entering at an even level appear in true form (uncomplemented) at the output.

Consider example (6.3) again. The output NAND gate acts like an OR gate, so a 3-input NAND gate is required. The second level NAND gates acts like AND gates, so two 2-input NAND gates are needed. Input A enters at level 1 (odd) so it will appear complemented at the output; since A is wanted in its true form, the input A must be inverted. Inputs B, C and D all enter at an even level (2) and so must enter in the same form as they are to appear at the output. This means that only input D must be inverted. Hence, the required circuit is that shown in Fig. 6.8.

EXAMPLE 6.4

Implement the function $F = ABC + A\bar{B}\bar{C} + D$.

Solution

Using the OR/AND rule the circuit can be written down directly (see Fig. 6.9). The variables A, B and C enter the circuit at the second level and so enter uncomplemented. Variable D enters the circuit at the first level and so it must be complemented or inverted (*Ans.*)

Fig. 6.9

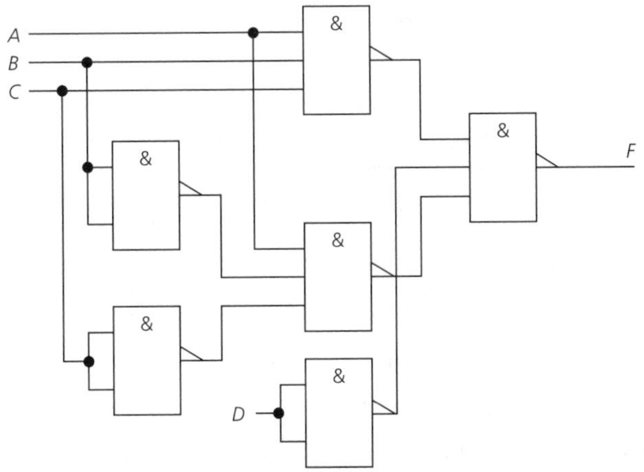

PRACTICAL EXAMPLE 6.2

To build an exclusive-OR circuit using (a) NAND gates only and (b) NOR gates only. Components and equipment: two 74LS00 (or 74HC00) quad 2-input NAND gate ICs. Two 74LS02 (or 74HC02) quad 2-input NOR gate ICs. One LED. One 270 Ω resistor. Power supply. Breadboard.
Procedure:

(a) Write down the Boolean expression for the exclusive-OR function and then draw the NAND gate only implementation. Build the circuit using the 74LS00 IC and connect its output terminal via the 270 Ω resistor and the LED to 0 V.
(b) Apply each of the four possible combinations of input levels to the inputs A and B and each time note the logical state of the LED. Confirm that the circuit implements the exclusive-OR function.
(c) Put the equation for the exclusive-OR function into suitable form for NOR gate only implementation and then draw the circuit. Build the circuit using the 74LS02 IC and then repeat procedure (b).
(d) Compare the two circuits. Is there any way the exclusive-OR circuit can be made using just one IC (other than an exclusive-OR IC!)?

Logic functions implemented by NOR gates only

The implementation of a SOP equation using NOR gates requires that the function is first converted into its POS form otherwise the resulting equation may be more complex than necessary. When in POS form the same methods used to determine the NAND gate implementation of a circuit can be employed, i.e. double inversion and AND/OR levels.

Double inversion

- Invert twice the POS expression that is to be implemented.
- Keep the upper inversion bar and apply De Morgan's rules to the lower bar to remove the AND operations.
- Implement the result directly using NOR gates only.

Converting the equation for the exclusive-OR function into its POS form

$$F = A\bar{B} + \bar{A}B = (A + B)(\bar{A} + \bar{B}) = \overline{\overline{(A + B)(\bar{A} + \bar{B})}}$$
$$= \overline{\overline{(A + B)} + \overline{(\bar{A} + \bar{B})}}$$

This equation for F is shown implemented, using three NOR gates, in Fig. 6.10(b). This circuit should be compared with Fig. 6.4(b). The two exclusive-OR implementations in Fig. 6.10(a) and (b) both require the use of one IC. Figures 6.11(a) and (b) show

Fig. 6.10 *Implementing the exclusive-OR function using (a) NAND and (b) NOR gates only*

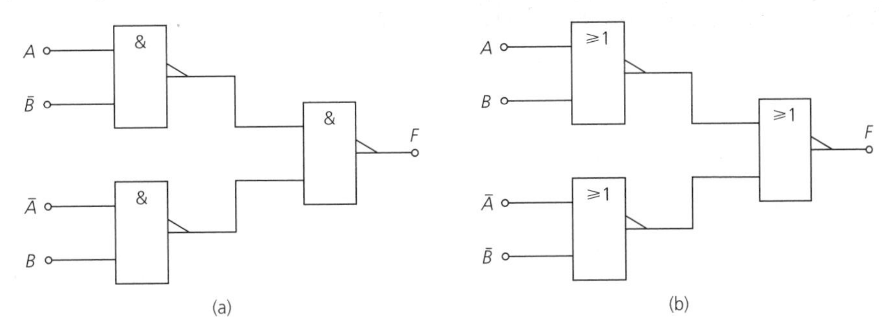

Fig. 6.11 *Exclusive-OR function implemented by (a) 74LS00 and (b) 74LS02 ICs*

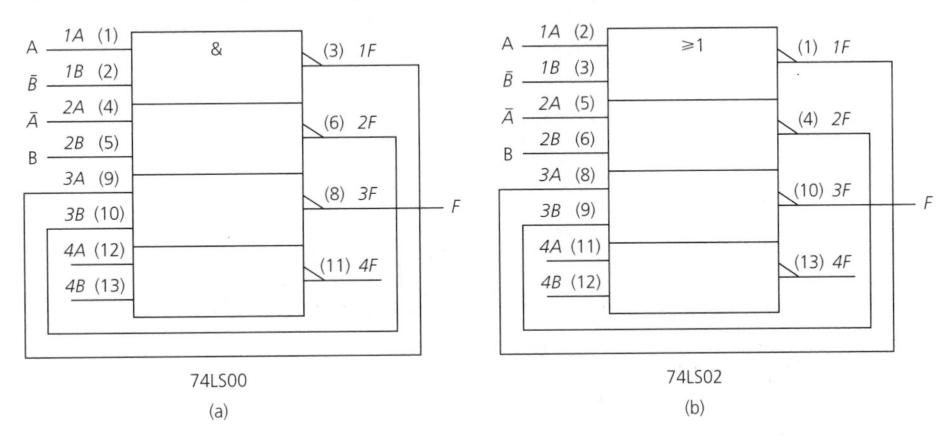

the required connections for the 74LS00 and the 74LS02 ICs, respectively, assuming $\bar{A}$ and $\bar{B}$ are available.

AND/OR levels

- At odd levels a NOR gate acts like an AND gate. Any variable entering at this level will appear at the output in its complemented form.
- At even levels each NOR gate acts like an OR gate. Any input variable will appear at the output uncomplemented.

EXAMPLE 6.5

Implement, using NOR gates only, the expression $F = AB + BD + BC + CD$, (a) using the double inversion method and (b) using the AND/OR level method.

Solution

The mapping of the equation is

CD \ AB	00	01	11	10
00	1	0	0	1
01	1	1	1	1
11	0	1	1	1
10	1	1	1	1

Looping the 0 cells gives $F' = \bar{A}\bar{B}CD + B\bar{C}\bar{D}$. Changing signs and complementing, $F = (A + B + \bar{C} + \bar{D})(\bar{B} + C + D)$.

(a) Double inverting, $\bar{\bar{F}} = \overline{\overline{(A + B + \bar{C} + \bar{D})(\bar{B} + C + D)}} = \overline{\overline{(A + B + \bar{C} + \bar{D})} + \overline{(\bar{B} + C + D)}}$. Implement this expression to get the circuit shown in Fig. 6.12 (*Ans.*)

(b) For a NOR implementation the output level 1 acts like an AND gate. The second level NOR gates act like OR gates. A variable entering at the second level must be complemented. Hence, the circuit is shown in Fig. 6.12 (*Ans.*)

Fig. 6.12

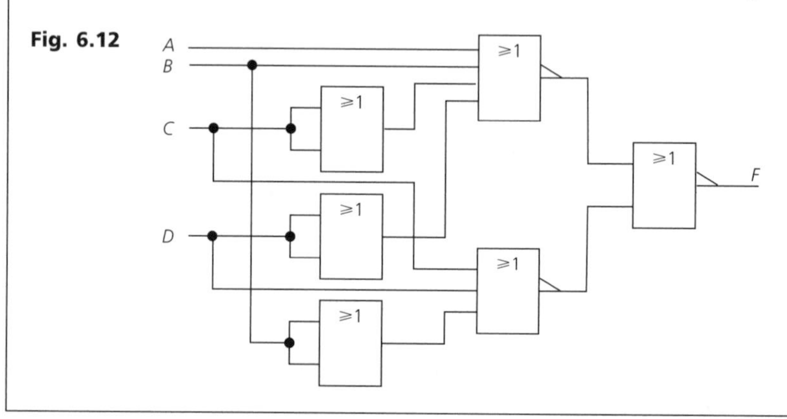

PRACTICAL EXERCISE 6.3

To design and implement, using NOR gate ICs only, a 3-bit combinational logic circuit that turns ON a LED whenever the input to the circuit is an even number. Take 0 as an odd number.

Components and equipment: one 74LS02 (or 74HC02) quad 2-input NOR gate IC. One 74LS04 (or 74HC04) hex inverter. One LED. One 270 Ω resistor. Breadboard. Power supply.

Procedure:

(a) Write down the truth table for the wanted circuit and from it obtain the Boolean equation that describes the operation of the circuit. Map the equation and simplify it including the don't care cells.

(b) Convert the simplified equation into its POS form (this could be read off the map), and construct the circuit.

(c) The equation to be implemented is $F = \bar{A}\ (B + C)$. The required circuit is shown in Fig. 6.13. Connect the output of the circuit to 0 V via the 270 Ω resistor and the LED.

(d) Apply each of the BCD numbers in turn to the circuit and each time note whether or not the LED glows. Write down the truth table of the circuit and compare it with the table used for the design. State whether the circuit performed its required function.

Fig. 6.13

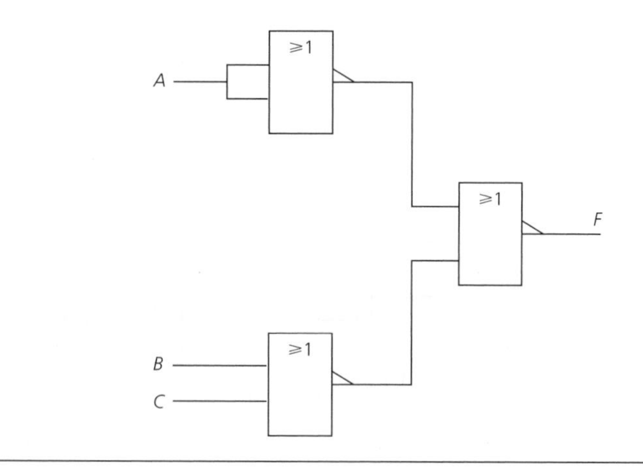

EXERCISES

6.1 (a) A 2-input NAND gate is operated with both of its inputs inverted. What logic function is performed?

(b) A 2-input NOR gate is operated with both of its inputs inverted. What logic function is performed?

(c) The inputs to a 3-input NAND gate are A, B and C. Determine the output of the gate.

6.2 The equation $F = A + \bar{B}\bar{C}$ is to be implemented using one NOR gate IC. Rearrange the equation and then draw the circuit.

6.3 Determine the logic function performed by the circuit shown in Fig. 6.14.

6.4 Implement, using (a) NAND gates only, and (b) NOR gates only, the function $F = A + BD + CD$.

6.5 Implement, using (a) NAND gates only, and (b) NOR gates only, the function $F = A + BC$.

Fig. 6.14

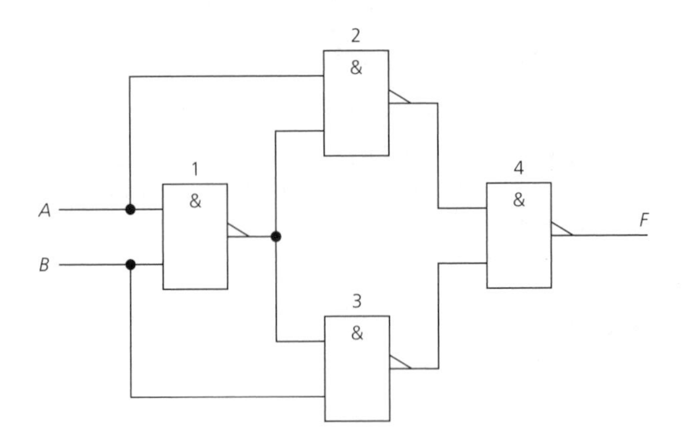

Fig. 6.15

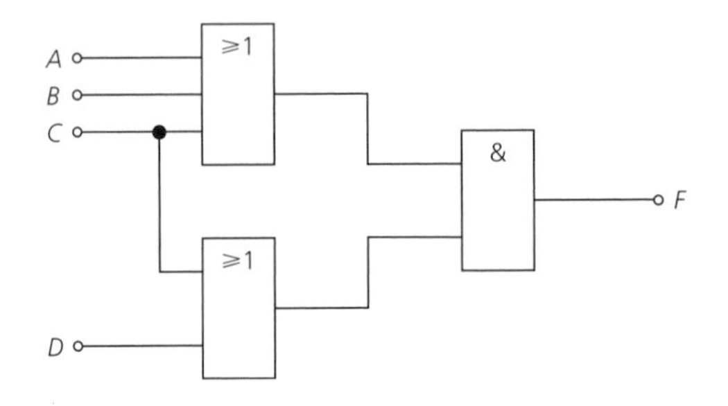

6.6 (a) Implement the circuit shown in Fig. 6.15 using NOR gates only.
(b) In the circuit shown in Fig. 6.16 replace each NAND gate by a NOR gate and then determine the logical function then performed by the new circuit.

6.7 (a) Implement the function $F = A + BC(A + C)$ using NAND gates only.
(b) Implement the function $F = \bar{A}\bar{B} + BC + \bar{D}$ using NOR gates only.

6.8 Implement, using NAND gates only, the function $F = (A + B + C)(\bar{A}\bar{C} + A\bar{B})$.

6.9 Implement, using NOR gates only, the function $F = ABC + \bar{A}\bar{B}\bar{C}$.

6.10 Implement the function $F = \bar{A}\bar{B}\bar{C}\bar{D} + \bar{A}B\bar{C}\bar{D} + \bar{A}\bar{B}C\bar{D} + A\bar{B}\bar{C}\bar{D} + \bar{A}\bar{B}\bar{C}D + \bar{A}B\bar{C}D + A\bar{B}\bar{C}D + A\bar{B}C\bar{D}$ using NOR gates only.

6.11 Re-write $F = ABC\bar{D} + B\bar{C}D + \bar{A}\bar{B}\bar{D}$ into suitable form for its implementation
(a) Using NAND gates only.
(b) Using NOR gates only.

6.12 Simplify the equation $F = (A + B)(A + \overline{B + C})(\overline{B + D})$ and implement the result using NAND gates only.

Fig. 6.16

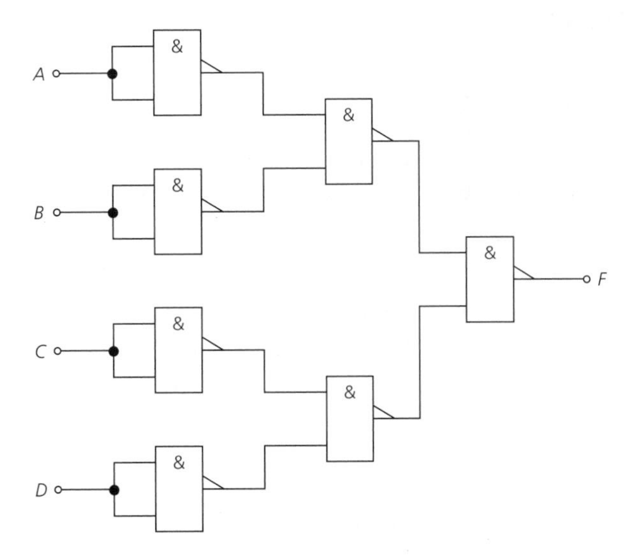

Fig. 6.17

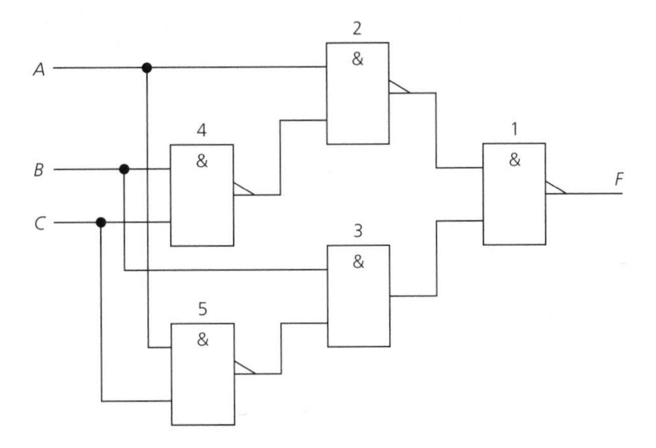

6.13 Implement, using NAND gates only, $F = ABC + \bar{A}\bar{B}\bar{C} + \bar{C}D$. How many ICs are needed?

6.14 Determine the output F of the circuit shown in Fig. 6.17

(a) By using the NAND gate rules.
(b) By writing down the output of each gate in turn.

6.15 The truth table of a logic circuit is given in Table 6.2.

(a) Obtain the minimal expression for F taking the don't cares into consideration.
(b) Put the expression into suitable form for implementing using
 (i) NAND gates only.
 (ii) NOR gates only.

Table 6.2

A	0	1	0	1	0	1	0	1	0	1	0	1	0	1	0	1
B	0	0	1	1	0	0	1	1	0	0	1	1	0	0	1	1
C	0	0	0	0	1	1	1	1	0	0	0	0	1	1	1	1
D	0	0	0	0	0	0	0	0	1	1	1	1	1	1	1	1
F	1	1	0	0	×	×	×	×	×	×	1	1	0	0	1	1

6.16 (a) Simplify $F = (AC + AB)(A + C)$.
(b) Implement the result using 2-input NAND gates only.

7 Logic technologies

After reading this chapter you should be able to:

(a) Understand the use of transistors and diodes as electronic switches.
(b) Explain the terms fan-in, fan-out, noise margin, propagation delay and power dissipation with reference to a gate.
(c) Understand the differences between the various sub-families in both the TTL and CMOS technologies.
(d) Know how to interface TTL and CMOS devices.
(e) Understand the difference between current sinking and current sourcing.
(f) Explain why open-collector (drain) circuits are sometimes employed.
(g) Use wired-AND logic.

The various kinds of gates described in chapter 3 can be obtained in a number of different technologies. The most popular logic technologies are *transistor–transistor logic* (TTL), of which several versions exist, and *complementary metal oxide semiconductor* (CMOS), although *emitter-coupled logic* (ECL) is available for very fast applications. These technologies provide both *small-scale integrated* (SSI) devices, *medium-scale integrated* (MSI) devices, *large-scale integrated* (LSI) devices, and *very-large-scale integrated* (VLSI) devices. The terms SSI and MSI refer to devices having fewer than 10 gates, or equivalent circuits, and fewer than 100 gates (or equivalent circuits), respectively. Similarly, LSI devices contain between 100 and 5000 gates and VLSI devices may contain several thousand gates.

Integrated circuits containing 1 to 4 gates are often referred to as *random logic* and have been the mainstay of most logic design in the past. In modern circuitry the use of MSI and LSI devices such as multiplexers and microprocessors have led to fewer applications for random logic but it is expected to continue to be used for such purposes as interfacing to, from, and between the LSI devices. In addition, standard SSI/MSI devices can provide simple solutions to many digital requirements.

The logic technology TTL incorporates several families: these are (a) standard TTL (74 series), low-power Schottky TTL (74LS series), advanced low-power Schottky TTL (ALS), and advanced Schottky (AS). The 74 series is rarely, if ever, used today and the 74LS series is not often used in new designs. The CMOS technology includes the standard 4000 series, also rarely used for new designs, high-speed CMOS (HC series), and advanced CMOS (AC series). Within both the HC and AC families there are two sub-families that are either with, or without, TTL compatible inputs. HC devices are compatible with CMOS and HCT devices are compatible with TTL circuitry; the AC family has CMOS compatible inputs and outputs while ACT devices are compatible with TTL. The AC and ACT devices combine the high speed of the AS and ALS TTL families with the low static power dissipation of 4000 CMOS. AHC/AHCT devices offer even greater speed with low noise and low power.

The power supply voltage for TTL devices was originally standardized as +5 V because it was just below the breakdown voltage of the multiple-emitter transistors used in standard TTL (p. 161). A need for lower voltages has since arisen because the desirability of minimizing power dissipation. The power dissipated within a digital integrated circuit (IC) increases linearly with the square of the power supply voltage and, hence, the best way of reducing power dissipation is to reduce the voltage as much as possible. A reduction in the power dissipated is obviously desirable for battery-operated equipment since it prolongs battery life, but it also leads to a reduction in the generation of unwanted heat. This, in turn, reduces the problem of heat removal, perhaps obviating the need for a cooling fan, and increases the reliability of components, both active and passive. Lower temperatures also allow components to be packed together more densely and so help to minimize the physical dimensions of the equipment. On the negative side, reducing the power supply voltage increases the propagation time (p. 159) and reduces the capability of one IC to drive another. Many LSI devices operate with a 3.3 V internal power supply voltage and have their input and output voltage levels made compatible with their external 5 V environment.

Logic families HC and AC can be operated with power supply voltages as small as 2 V for HC and 3 V for AC but, to overcome the speed and drive disadvantages, some new logic families have been developed. There are four low-voltage logic families: LV, LVC, ALVC, and LVT. The first three are CMOS devices and they have similar characteristics to CMOS; the fourth family, LVT, uses CMOS circuitry except for a TTL output stage. Devices in the LV family can operate with a power supply voltage as low as 2 V, and the other three families have a minimum power supply voltage of 2.7 V. All four logic families have inputs and outputs that are compatible with TTL circuitry. Both 3.3 and 5 V devices can be used in the same system using a voltage translator IC. Such a device offers bi-directional voltage translation between 5 V TTL and 3.3 V LVTTL devices, or uni-directional voltage shifting from 5 V CMOS ICs to 3.3 V LVTTL ICs.

The choice of logic family for a particular design is made after considering the relative importance of factors such as: (a) switching speed, (b) power dissipation, (c) cost, (d) availability, and (e) noise immunity, and (for modern equipment) power supply voltage.

The advantages to be gained by the use of integrated circuits instead of discrete circuitry are (a) a reduction in costs, (b) circuits are physically smaller and lighter in weight, (c) it is easier to replace a part of a circuit, particularly if IC holders are employed, (d) standardization is easier to achieve, and (e) power dissipation is much reduced. The main disadvantage is that it may be more difficult to locate faults.

With a large variety of MSI and VLSI devices readily available SSI devices are rarely used in modern circuitry other than for inter-connection purposes (*glue* logic).

Fig. 7.1 *(a) Generalized diode current/voltage characteristic, (b) equivalent circuit of a diode when conducting and (c) equivalent circuit of a diode when non-conducting*

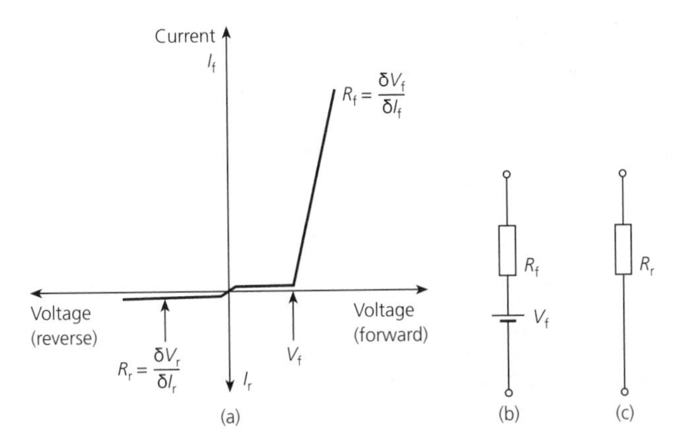

Semiconductor diodes and transistors as switches

The requirements for an electronic switch are (a) that it can be turned ON and OFF very rapidly, (b) that it has a very high resistance when OFF and a very low resistance when ON, and (c) that it dissipates very little power when either ON or OFF.

The Semiconductor Diode

A semiconductor diode is able to act as a switch because it can be turned ON and OFF by an applied voltage.

When the diode is OFF its resistance is high, usually several thousand ohms, and when it is ON the voltage across it is not zero but approximately 0.7 V. The time taken by a diode to turn ON is very small but it takes longer to turn OFF and the turn-OFF time limits the maximum frequency at which a diode can be switched.

A generalized diode characteristic is given in Fig. 7.1(a). The diode conducts only when the forward bias voltage is greater than a threshold value labelled as V_f.

When a diode is ON it can be replaced, on paper, by an equivalent circuit consisting of a battery of e.m.f. V_f connected in series with a resistance R_f (see Fig. 7.1(b)).

V_f is the voltage dropped across the diode when it is conducting and it has a value that varies between about 0.4 and 0.75 V depending upon the type of diode, the diode current, and the junction temperature. (A typical temperature coefficient is –2 mV/°C.) For silicon diodes V_f is very often assumed to be 0.6 V.

R_f is the a.c. resistance of the diode when it is forward biased. When a diode is OFF it can be represented by a resistance R_r (Fig. 7.1(c)), which is the reverse resistance of the diode.

The equivalent circuits of a diode are linear and they may assist in the determination of the currents and voltages in a diode circuit.

EXAMPLE 7.1

Calculate the output voltage of the circuit given in Fig. 7.2 when (a) $V_1 = +5$ V, $V_2 = 0$ V, (b) $V_1 = +5$ V, $V_2 = +5$ V, (c) $V_1 = 0$ V, $V_2 = +5$ V, and (d) $V_1 = 0$ V, $V_2 = 0$ V, if $V_f = 0.6$ V and $R_f = 10\ \Omega$.

Fig. 7.2

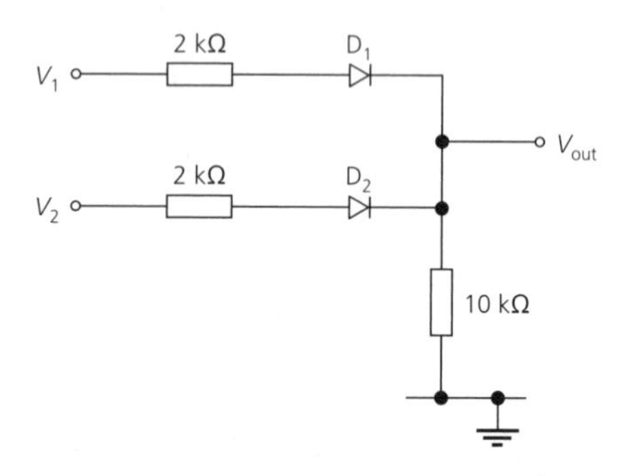

Solution

(a) Diode D_1 will be forward biased and ON while diode D_2 is OFF. Therefore, $V_{OUT} = [(5 - 0.6) \times 10\,000]/12\,010 = 3.66$ V (*Ans.*)
(b) Both diodes are now ON and, therefore, $V_{OUT} = [(5 - 0.6) \times 10\,000]/[(0.5 \times 2010) + 10\,000] = 4.0$ V (*Ans.*)
(c) This is really the same situation as in (a). Hence, $V_{OUT} = 3.66$ V (*Ans.*)
(d) Neither diode conducts and so $V_{OUT} = 0$ V (*Ans.*)

When the polarity of the voltage applied across a diode is reversed, the diode will not instantaneously change its state from ON to OFF or vice versa. However, the time taken for a diode to switch on is generally very small and can be neglected. Suppose a diode has a forward voltage V_F applied to it and is passing a current

$$I_F = (V_F - V_f)/(R_f + R_L) \simeq V_F/R_L$$

If the voltage is suddenly reversed to a new value, V_R, the current flowing in the circuit will also reverse its direction and have a magnitude of V_R/R_L for a time, shown in Fig. 7.3 as t_s, during which excess stored minority charge carriers are removed. The time period labelled as t_s is known as the *storage time*. The voltage across the diode does *not* reverse its direction for this period of time. When the excess charges have been removed, the diode voltage reverses its polarity and the diode current starts to fall taking a time t_f to reach zero (Fig. 7.3).

Manufacturers' data generally quote the *reverse recovery time* of a diode. This is the time that elapses from the moment the diode current first reverses its direction of flow to the time the diode current reaches a defined value. The reverse recovery time may vary from a few nanoseconds to a few microseconds.

Fig. 7.3 *Switching a diode: (a) applied voltage; (b) diode voltage and (c) diode current*

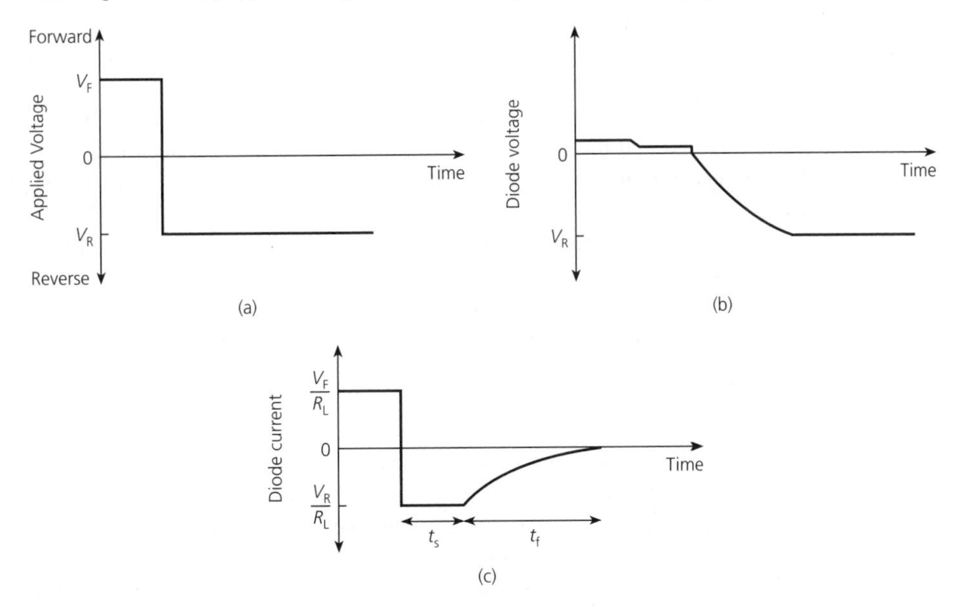

Fig. 7.4 *The transistor as a switch*

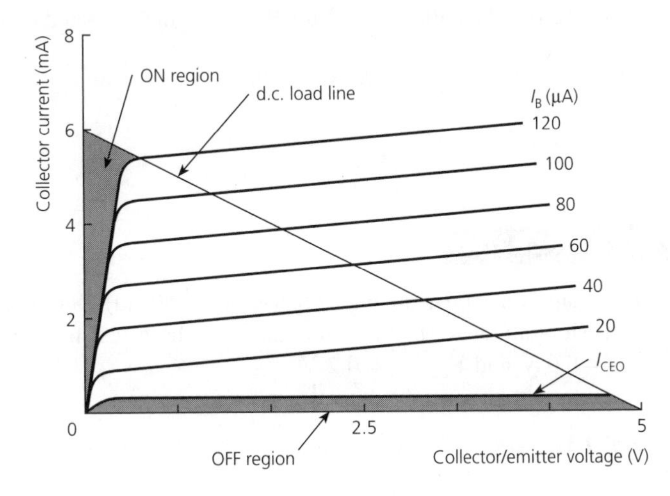

Schottky diodes are made in such a way that their forward voltage drop is reduced to about 0.3 V and their switching time is very small. Schottky diodes are used in 74LS family devices to increase their switching speed.

The bipolar transistor

Figure 7.4 shows a typical set of output characteristics for a bipolar transistor with a d.c. load line drawn between the points

$$V_{CE} = V_{CC} = 5 \text{ V}, I_C = 0 \text{ and } V_{CE} = 0 \text{ V}, I_C = V_{CC}/R_L = 6 \text{ mA}$$

where R_L is the total load on the transistor. When a transistor is used as a switch it is rapidly switched between two stable states, OFF and ON.

When the transistor is OFF, both its collector/base and emitter/base junctions are reverse biased and the collector current is only the small collector leakage current I_{CEO}. The collector/emitter voltage V_{CE} of the transistor is then equal to the supply voltage V_{CC}.

When the transistor is conducting current in its active region, the collector/base junction is reverse biased but the emitter/base junction is forward biased. As the base/emitter voltage V_{BE} is increased, the base current increases also and this produces an increase in the collector current since

$$I_C = h_{FE}I_B + I_{CEO} \simeq h_{FE}I_B$$

Eventually the point is reached at which the collector/base junction becomes forward biased and the transistor is said to be *saturated* or *bottomed*. The collector current now has its maximum, or ON, value $I_{C(SAT)}$ and the base current is $I_B = I_{C(SAT)}/h_{FE}$. The base/emitter voltage $V_{BE(SAT)}$ producing this base current usually has a value of about 0.75 V. Any further increase in the base current will *not* produce a corresponding increase in the collector current. The collector/emitter voltage $V_{CE(SAT)}$ of the transistor is then very low, being typically in the region of 0.1–0.2 V, because most of the supply voltage is dropped across the collector resistor R_L. Thus, the saturated collector current is equal to

$$I_{C(SAT)} = (V_{CC} - V_{CE(SAT)})/R_L \approx V_{CC}/R_L \quad (7.1)$$

Note that $V_{CC}/R_L > h_{FE}I_B$.

EXAMPLE 7.2

The transistor shown in Fig. 7.5 has $h_{FE} = 50$. (a) Determine whether or not the transistor saturates. (b) Find the maximum base resistance for saturation to occur. $V_{BE} = 0.6$ V and $V_{CE(SAT)} = 0.2$ V.

Fig. 7.5

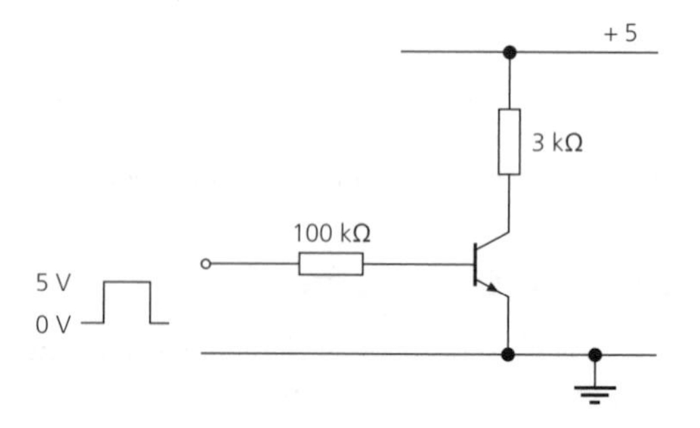

Solution

(a) $I_B = (5 - 0.6)/(100 \times 10^3) = 44\ \mu A$

and so $I_C = 50 \times 44 \times 10^{-6} = 22$ mA

$I_{C(SAT)} = (5 - 0.2)/(3 \times 10^3) = 1.6$ mA

So the transistor is saturated (*Ans.*)
(b) For saturation to occur

$I_B \geq (1.6 \times 10^{-3})/50 \geq 32\ \mu A$

The maximum base resistance is equal to

$(5 - 0.6)/(32 \times 10^{-6}) \approx 137.5\ k\Omega$ (*Ans.*)

A transistor can be rapidly switched ON and OFF by the application of a rectangular waveform of sufficient amplitude to its base. In either of the two stable states the power dissipated within the transistor is small because *either* V_{CE} *or* I_C is approximately equal to zero. The active (or amplifying) region of the transistor is rapidly passed through as the transistor switches from one state to the other, and so little power is dissipated.

A transistor is unable to change state instantaneously when the voltage applied to its base is changed, because of charges stored in (a) the base region; (b) the collector/base depletion layer, and (c) the base/emitter depletion layer. When the transistor is OFF both its base/emitter and its collector/base junctions are reverse biased and the two depletion layers are wide. When a voltage is applied to the base to turn the transistor ON, the base current supplies charge to both the p–n junctions and this reduces the widths of the two depletion layers. At some point the base/emitter depletion layer will be sufficiently narrow to allow charge carriers to move from the emitter into the base, and when these reach the collector a collector current starts to flow. As the collector current increases, the collector/emitter voltage falls and the width of the collector/base depletion layer decreases. If the base current is sufficiently large the collector current continues to increase until the transistor is saturated; the collector/base junction is then forward biased and so charge carriers are no longer swept into the collector region. This means that an *excess charge* is stored in the base region.

When a voltage is applied to the base to turn the transistor OFF, the collector current will not start to fall until all the excess base charge has been removed. This time delay is known as the *storage delay*.

Figure 7.6 shows how the collector/emitter voltage of an initially OFF bipolar transistor varies when a voltage pulse is applied to its base. The terms shown in the figure are defined thus:

t_d is the time that elapses between the application of the base signal and the collector/emitter voltage falling to 90% of its original value of V_{CC} volts.

t_f is the time taken for the collector/emitter voltage to fall from 90 to 10% of V_{CC} volts.

Fig. 7.6 *Switching a bipolar transistor circuit: (a) input voltage and (b) output voltage*

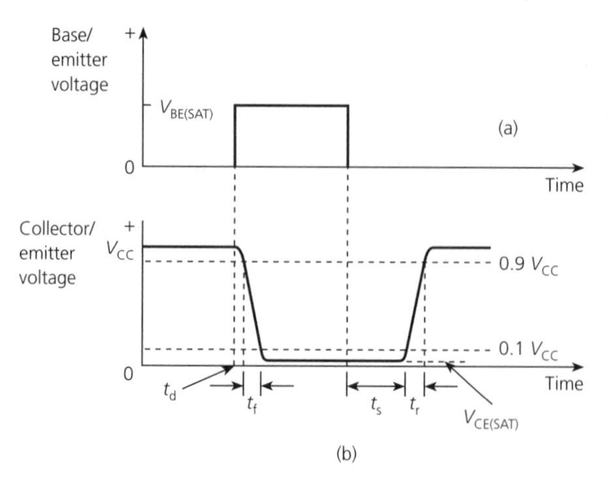

Fig. 7.7 *Use of a diode to increase switching speed*

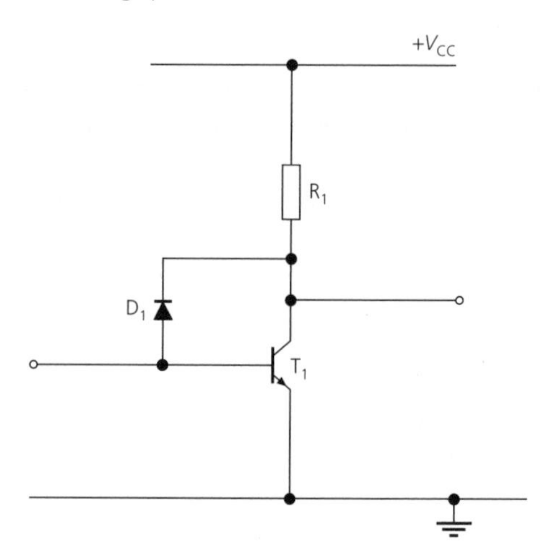

t_s is the time delay that occurs between the removal of the base voltage and the collector/emitter voltage rising to 10% of its final value of V_{CC} volts.

t_r is the time taken for the collector/emitter voltage to rise from 0.1 V_{CC} to 0.9 V_{CC} volts.

Typically, the ON and OFF times are about 6 and 10 ns, respectively, and to increase the switching speed the transistor must be prevented from saturating. This can be achieved by the connection of a diode between the base and the collector terminals of the transistor as shown in Fig. 7.7. When the transistor is turned ON, its collector/emitter voltage falls, and when the collector potential becomes less positive than the base potential the diode D_1 conducts and prevents an excess current from entering the base. As a result the transistor is not driven into saturation and there is no storage of charge in the base region. The best results are obtained if a *Schottky diode* is employed since these devices have zero charge storage and so are very fast switches. A Schottky

Fig. 7.8 *(a) Schottky transistor and (b) symbol*

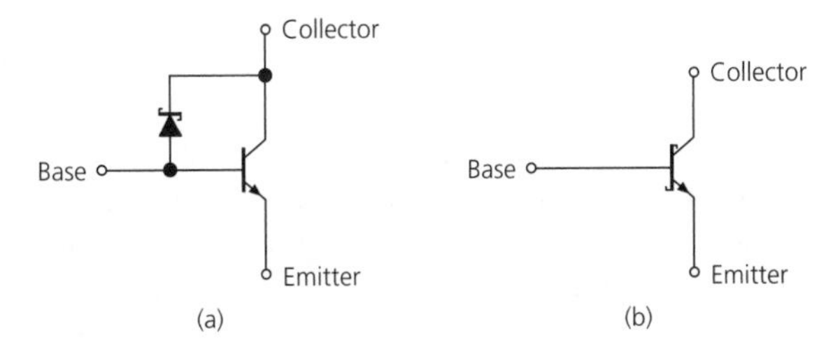

Fig. 7.9 *Enhancement MOSFET switch; T_1 acts as an active load for T_2*

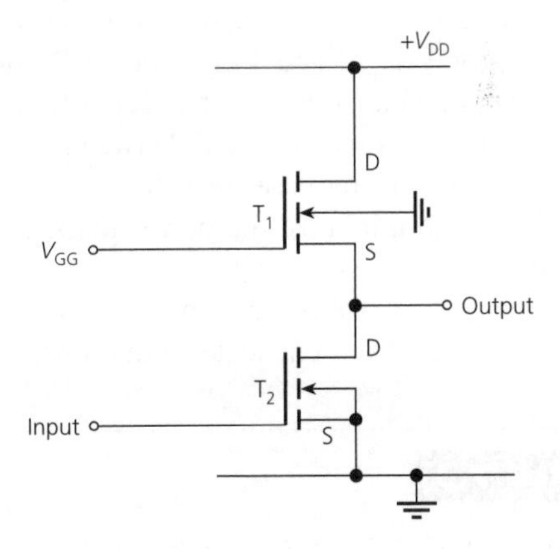

transistor is a bipolar transistor that has a Schottky diode internally connected between its base and collector terminals (see Fig. 7.8(a)). The symbol for a Schottky transistor is given in Fig. 7.8(b). The voltage drop across a Schottky diode, or p–n junction, is smaller than for an ordinary diode, typically 0.3 V when conducting and 0.5 V when full ON.

The enhancement-type MOSFET

An enhancement-type MOSFET can also be employed as an electronic switch since its drain current can be turned ON and OFF by the application of a suitable gate-source voltage. When the MOSFET is ON, the gate-source voltage will have moved the operating point to the top of the load line (similar to the bipolar transistor switch), and the voltage across the MOSFET, known as the saturation voltage $V_{DS(SAT)}$, is small, typically 0.2–1.0 V.

When OFF the MOSFET passes a very small current, typically 1 nA for a junction MOSFET and 50 pA for a MOSFET. The drain load resistance needed for a MOSFET switch is often of the order of some tens of thousands of ohms and such high values are not conveniently fabricated within a monolithic integrated circuit. For this reason the drain load is often provided by another MOSFET as shown in Fig. 7.9. The bottom transistor T_2 is operated as the switch while the upper transistor T_1 is biased to act as an *active load* by the voltage V_{GG}, where $V_{GG} \geqslant V_{DD}$. Very often $V_{GG} = V_{DD}$ and a separate bias voltage supply is then not necessary. The active load transistor T_1 is always conducting current and often has its substrate connected to its source instead of to earth.

The switching speed of a MOSFET is determined by the stray and transistor capacitances which are unavoidably present in the circuit. Because of the very high input impedance of a MOSFET, the time constant of its input circuit is a speed-limiting factor. The charge storage effects encountered with the bipolar transistor circuit are now insignificant because, in a MOSFET, current is carried by the majority charge carriers.

When a transistor is used as a switch a voltage is applied to its base, or gate, of sufficient magnitude to drive it into either saturation or cut-off depending upon the polarity of the voltage. When the transistor is ON the collector, or drain, current is at its maximum value and the voltage across it is the collector, or drain, saturation voltage. Typical values for $V_{CE(SAT)}$ and $V_{DS(SAT)}$ are 0.2 and 0.2–1 V, respectively. When the transistor is OFF the voltage across it is equal to the power supply voltage V_{CC} or V_{DD}. The transistor can be switched rapidly between its ON and OFF states by the application of a pulse waveform to the base, or gate, terminal. Rapid switching between the two states is necessary to minimize the power dissipated within the transistor, since either the current flowing in the device, or the voltage across it, is small, the power dissipation when either ON or OFF is also small. Most of the power dissipation occurs as the transistor is switching from one state to the other.

EXAMPLE 7.3

The circuit in Fig. 7.5 has $R_L = 1\ \text{k}\Omega$ and the output terminal is connected to a circuit with an input resistance of 20 kΩ. The transistor has $V_{CE(SAT)} = 0.2$ V and is driven alternately into saturation and cut-off by the applied base voltage. Calculate the voltages between which the output pulse waveform varies.

Solution

When T_1 is ON, $V_{out} = V_{CE} = 0.2$ V
When T_1 is OFF, $V_{out} = 5 \times 20/(1 + 20) = 4.76$ V
The output waveform varies between 0.2 and 4.76 V (*Ans.*)

Parameters of the logic technologies

The various logic technologies possess different characteristics, which means that any one of them may be the best suited for a particular application. For example, for one application the most important consideration might be the highest possible speed of operation, while for another application it might be the minimum possible power dissipation. The characteristics of the various logic technologies can be classified under the following headings: logic levels, speed of operation, fan-in and fan-out, noise margin or noise immunity, and power dissipation.

Fig. 7.10 *Logic levels*

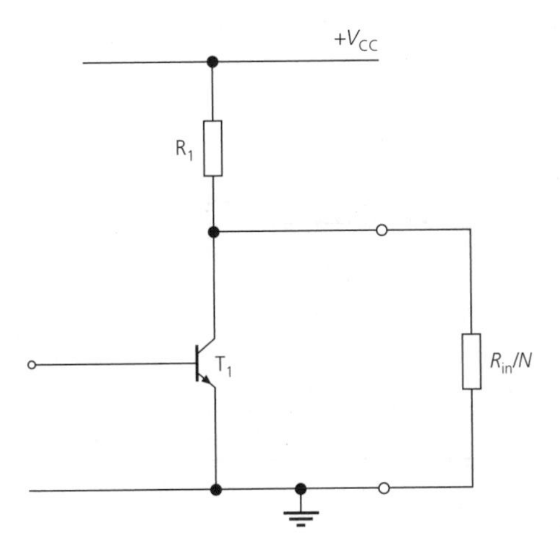

Logic levels

The logic levels of a gate are the input and output voltages that represent the binary values 0 and 1. For any practical gate the logic levels are nominally 0 V and $+V_{CC}$ volts but may be anywhere in a range of values above 0 V or below V_{CC} because of voltage drops.

Consider Fig. 7.10 which shows a transistor that represents the output stage of a gate. When the transistor is ON the voltage across it – the logic 0 voltage level – is $V_{CE(SAT)}$ or approximately 0.2 V. The logical 1 voltage level will depend upon the number N of similar gates (the fan-out) that are connected in parallel across the output terminals. The total resistance across the output terminals is R_{IN}/N, where R_{IN} is the input resistance of each of the N identical gates.

The output 1 voltage level is

$$(V_{CC}R_{IN}/N)/(R_1 + R_{IN}/N)$$

and it is clear that the greater the fan-out, the lower will be the logical 1 voltage level. The maximum fan-out is limited by the allowable minimum voltage specified for the logical 1 state.

For a TTL gate having a power supply voltage of +5 V the nominal level for binary 1 is +5 V. The minimum input voltage that is guaranteed to be taken as binary 1 is 2 V while the maximum input voltage that can represent binary 0 is 0.8 V. At the output the minimum voltage for binary 1 is 2.4 V and the maximum 0 voltage is 0.4 V. The TTL logic levels are shown in Fig. 7.11(a).

The levels for the other main logic family, the CMOS family, depend upon the supply voltage used.

Logic 1 is the in the range V_{DD}–$2V_{DD}/3$, and logic 0 is in the range 0–$V_{DD}/3$; this is shown in Fig. 7.11(b). The supply voltage V_{DD} is in the range 3–18 V except for some low-voltage families where the voltage may be as low as 2 V.

Manufacturers of digital ICs specify the maximum voltage that a gate will interpret as LOW and the minimum voltage that it will interpret as HIGH. The resistive loading of

Fig. 7.11 *Logic levels for TTL and 5 V CMOS families*

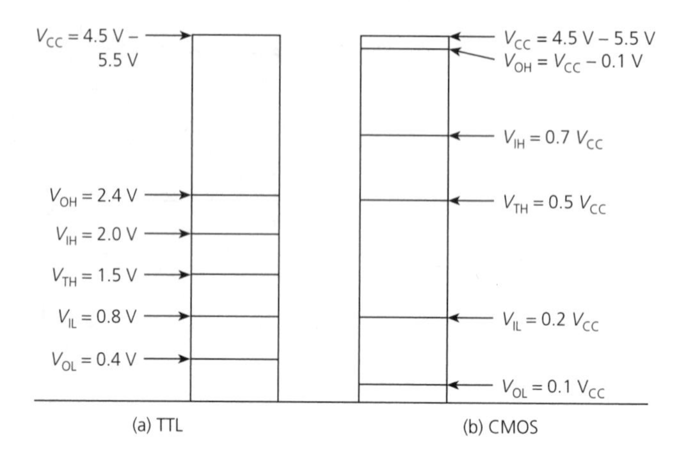

a gate is therefore significant whenever the output of a gate is the input to one, or more, other gates.

PRACTICAL EXERCISE 7.1

To investigate the logic levels of a TTL IC.
Components and equipment: one 74LS00 (or 74HC00) NAND gate IC. Power supply. d.c. voltmeter. Breadboard.
Procedure:

(a) Connect up the circuit shown in Fig. 7.12. State the maximum voltage for binary 0 and the minimum voltage for binary 1 for the NAND gates.

Fig. 7.12

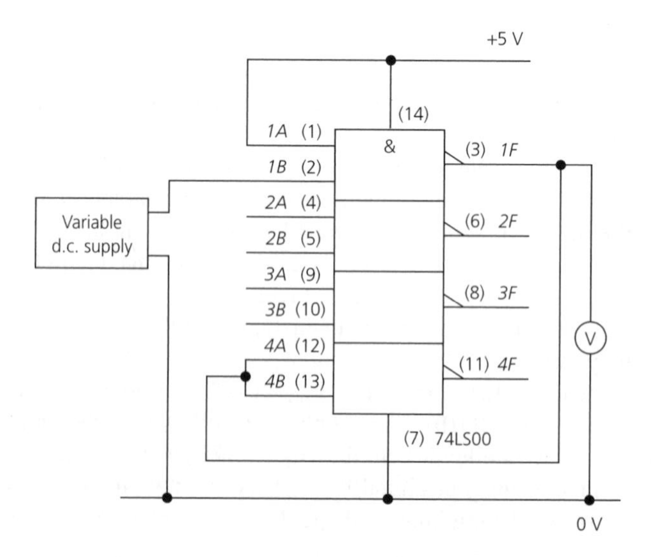

(b) With the positive input voltage provided by the variable d.c. power supply set to 5 V, note the output voltage measured by the d.c. voltmeter.
(c) Decrease the output of the d.c. supply in a number of discrete steps down to 5 V. At each step note the output voltage at pin 4.
(d) Use the results to state: (i) The range of input voltages that give a LOW (binary 0) output voltage at pin 3. (ii) The input voltage at which the output voltage suddenly increases to a HIGH (binary 1) value. What is the value of this HIGH voltage? (iii) The voltage at pin 3 when the input voltage is 5 V. (iv) Calculate the noise margin of the gate.

Fan-in and fan-out

The *fan-in* of a gate is the number of inputs connected to the gate. The *fan-out* of a gate is the maximum number of standard loads that can be connected to its output terminals without the output voltage falling outside the limits at which the logic levels 0 and 1 are specified.

A *standard load* is the load provided by a single simple input stage; for TTL circuits this is 1.6 mA. Some of the more complex circuits may be equivalent to two, or more, standard loads.

Fan-out is normally limited by the number of package pins available, but also, to some extent, by the switching speed required since each gate contributes extra capacitance.

Current sinking and current sourcing

When the output of a driving gate is LOW, current will flow into that gate from the driven gate, or gates. The driving gate is then said to be *sinking* current. Conversely, when the output of the driving gate is HIGH, current flows out of that gate into the driven gate or gates. The driving gate is then said to be *sourcing* current. The two actions are illustrated for TTL gates in Figs 7.13(a) and (b), respectively.

When calculating the fan-out of a gate the parameters that are taken into account are the input and output currents for the logic 1 and logic 0 levels. Each gate has a maximum LOW current specification I_{OL} that sets the maximum current that the gate is able to sink. Also, each gate has a maximum HIGH current specification I_{OH} that determines the maximum current that the gate may source. The fan-out of a gate is calculated using equations (7.2) and (7.3), i.e.

$$\text{Fan-out (LOW)} = I_{OL}/I_{IL} \tag{7.2}$$

$$\text{Fan-out (HIGH)} = I_{OH}/I_{IH} \tag{7.3}$$

where I_{IL} and I_{IH} are, respectively, the input LOW, and HIGH, current specifications of the driven gate.

The actual fan-out in any particular case will be the lower of the two figures calculated using these equations.

Fig. 7.13 *(a) Sinking and (b) sourcing current*

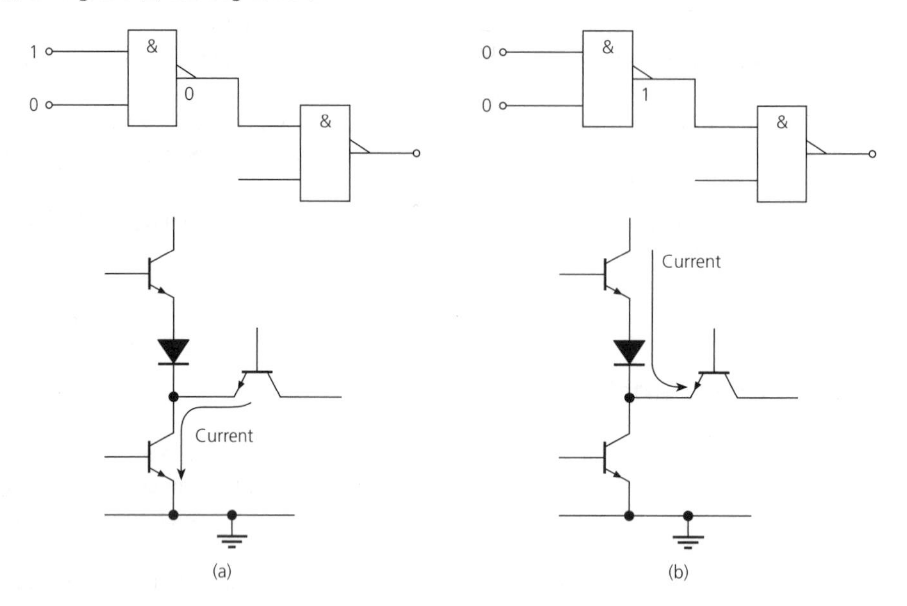

EXAMPLE 7.4

Calculate the fan-out of a low-power Schottky (LS) TTL gate that is driving other LS TTL gates.

Solution

An LS gate has $I_{IL} = -0.4$ mA, $I_{IH} = 20$ μA, $I_{OL} = 8$ mA, and $I_{OH} = -0.4$ mA.

[The minus signs indicate the direction of current flow and are ignored in the calculation.]

Fan-out (LOW) = 8/0.4 = 20
Fan-out (HIGH) = $0.4/(20 \times 10^{-3}) = 20$
Therefore, fan-out = 20 (*Ans.*)

PRACTICAL EXERCISE 7.2

To investigate the fan-out of a gate.
Components and equipment: two 74LS00 (or 74HC00) NAND gate ICs. Power supply, d.c. voltmeter, breadboard.
Procedure:

(a) Connect up the circuit shown in Fig. 7.14(a) and then measure the voltage at the output of gate 1.

Fig. 7.14

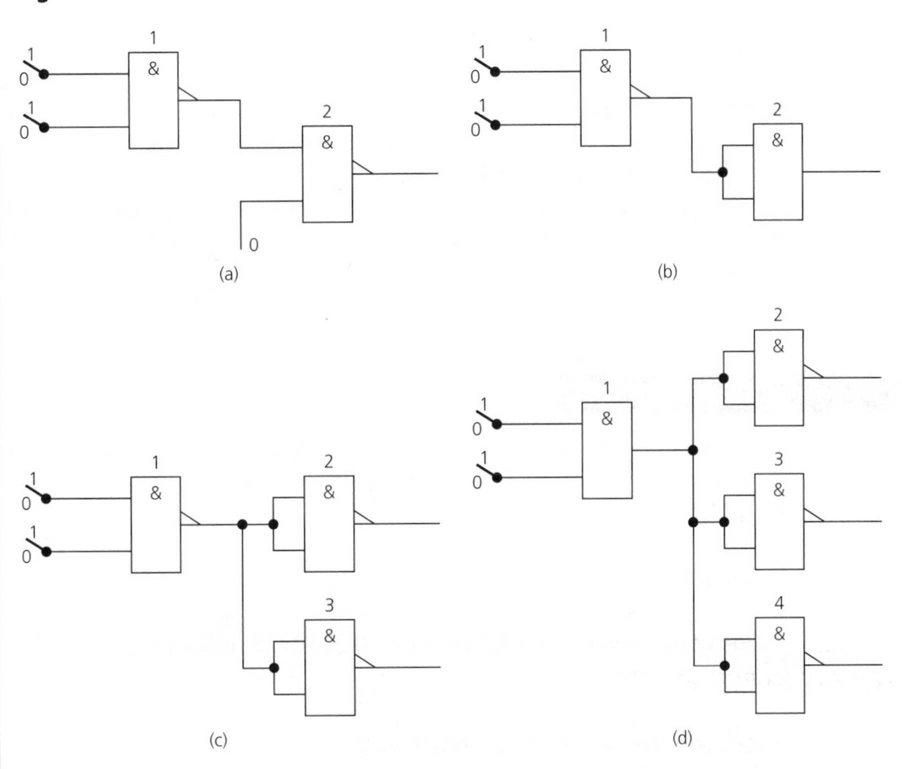

(b) Alter the logic 1 voltage level at one input of gate 1 to logic 0 so that the output of gate 1 is HIGH. Measure the HIGH voltage.

(c) Connect the second input of gate 2 to the output of gate 1 as shown in Fig. 7.14(b) to obtain a fan-out of 2. Measure both the HIGH and the LOW voltages at the output of gate 1.

(d) Connect a third NAND gate with both inputs linked to the output of gate 1 as shown in Fig. 7.14(c). Measure the HIGH and LOW output voltages of gate 1 with a fan-out of 4.

(e) Repeat procedure (d) with the fourth NAND gate connected (see Fig. 7.14(d)), to give a fan-out of 6.

(f) Now connect, in turn, NAND gates 5, 6, 7 and 8 from the other 74LS00 IC to obtain successive fan-outs of 8, 10, 12 and 14. Each time measure both the HIGH and the LOW output voltages of gate 1.

(g) Copy the results of your measurements into a table and draw graphs of output voltage plotted against fan-out and on it mark the minimum voltage that a TTL gate will interpret as binary 1.

(h) Draw a graph of LOW output voltage against fan-out and on it mark the maximum voltage that is taken as logic 0.

(i) Use your results to predict the maximum fan-out for the gate under test.

Power dissipation

Power is dissipated within a digital IC as it switches from one state to the other, known as the *dynamic dissipation*. The *static dissipation* is equal to the product of the d.c. supply voltage and the mean current taken from the supply. For TTL devices the power dissipation is not greatly influenced by the switching frequency and the power dissipated at the maximum operating frequency (usually about 30 MHz) is no more than about 20% greater than the static dissipation. The static power dissipation of a CMOS device is very small. The dynamic dissipation increases linearly with increase in clock frequency and may at some frequency result in a CMOS device dissipating more power than its TTL equivalent.

EXAMPLE 7.5

The total supply current for a 74LS02 NOR gate IC is 14 mA when the output is LOW, and 8 mA when the output is HIGH. Calculate the static power dissipated in the IC.

Solution

Power dissipated = $5 \times [(14 + 8) \times 10^{-3}]/2 = 55$ mW (*Ans.*)

Noise immunity or margin

Noise is the general term for any unwanted voltages that appear at the input to a gate. If the noise voltage has a sufficiently large amplitude, it may cause the gate to change its output state even though the input signal voltage has remained constant. Such false operation of a gate will lead to errors in the circuit performance. The *noise immunity* or *noise margin* of a gate is the maximum noise voltage that can appear at its input terminals without producing a change in the output state. Usually, manufacturers of integrated circuit gates quote d.c. values of noise margin, giving both typical and worst-case values.

The *threshold value* of a gate is the input voltage at which a change of the output state of the circuit is just triggered. A reasonable approximation to this value is the voltage midway between the two logic levels. For the TTL logic family the threshold voltage is 1.4 V but the maximum input voltage that will definitely be read as logic 0 is 800 mV, while the minimum input voltage giving a definite logic 1 is 2.0 V.

The noise margin of a gate is equal to the difference between the logic level at the output of the gate and the threshold value of the gate(s) to which its output is connected. This is shown in Fig. 7.15 which refers to TTL NAND gates. In Fig. 7.15 it is supposed that binary 0 is applied to the inputs of the first NAND gate so that its output is at binary 1. This means that the output voltage lies within the limits of 2.4 and 3.3 V. The threshold voltage is taken as 1.4 V and so the noise margin varies between 1.0 V at the worst and 1.9 V at best. The maximum value of the threshold voltage is 2.0 V and, should this exist, the worst-case noise margin will be only 400 mV.

Fig. 7.15 *Noise margin of TTL gates*

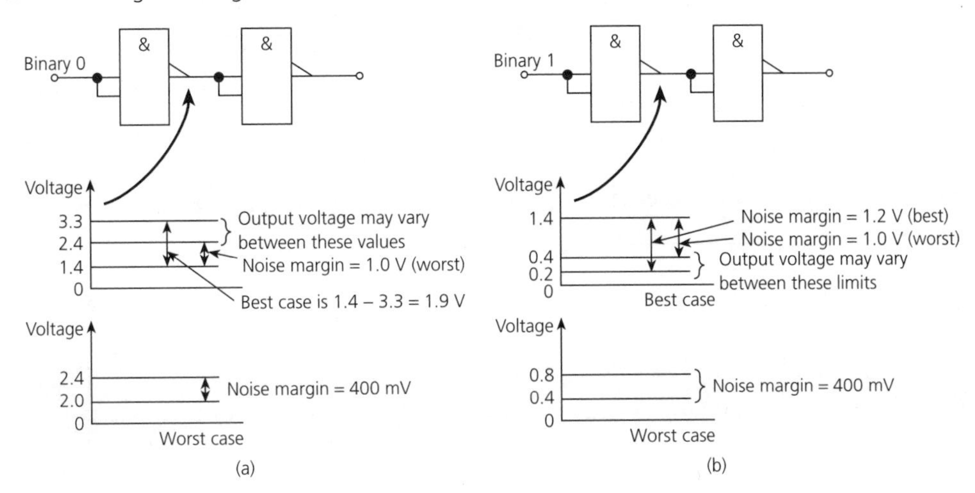

When the output of the first gate is at binary 0 its voltage will be within the range 0.2–0.4 V. The threshold voltage is 1.4 V and so the noise margin varies from 1.0 V at worst to 1.2 V at best. The minimum threshold voltage is only 0.8 V and hence the worst-case noise margin is 400 mV. Thus, for TTL gates the noise margin is taken as 0.4 V even though it is the worst possible case. In practical cases the noise margin is usually much better.

Gates in the CMOS family can be operated within a wide range of supply voltages (3–15 V for the 4000 series or 2–6 V for the 74HC series). This makes the noise margin rather vague, but it is approximately equal to $V_{DD}/2$.

Propagation delay

The propagation delay of a logic gate is the time that elapses between the application of a signal to an input terminal and the resulting change in logical state at the output terminal. The delay arises because the output voltage of a switching transistor is unable to change instantaneously from one logic value to another when its input voltage is changed. This was discussed on pages 149–150.

Manufacturer's data sheets usually specify the propagation delay between the 50% amplitude points on the input and output pulse waveforms. Two propagation delays are usually quoted:

t_{pLH}: the propagation delay when the output voltage switches from LOW to HIGH

t_{pHL}: the propagation delay as the output voltage switches from HIGH to LOW

Fig. 7.16 shows the two propagation delays for an inverter. When the input voltage changes from LOW to HIGH the output voltage must change from HIGH to LOW. Because of the inherent transistor switching delays the change in the output voltage

Fig. 7.16 *Switching a bipolar transistor circuit: (a) basic circuit; (b) input voltage and (c) output voltage*

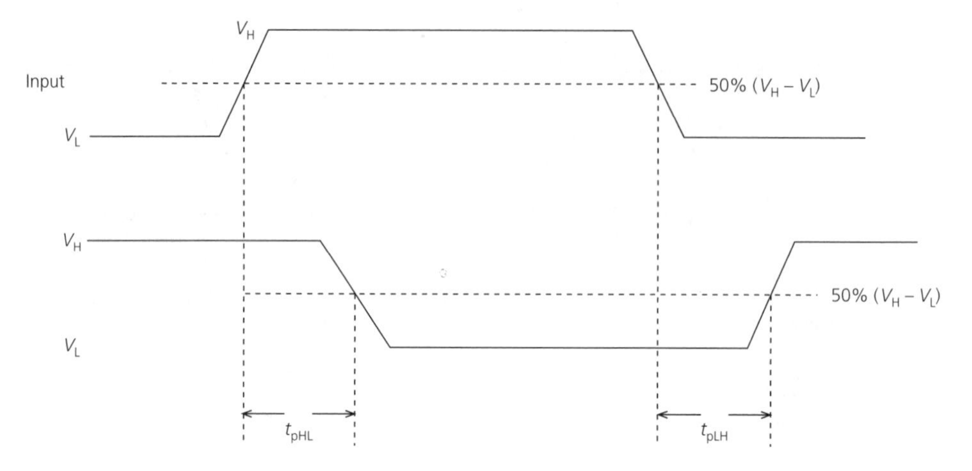

occurs some time after the occurrence of the input transition. In general, t_{pLH} and t_{pHL} are not of equal value but they are both functions of the load capacitance C_L. Data sheets always state what value of C_L was used in measuring the quoted values for t_{pLH} and t_{pHL}.

Note that in some cases t_{pLH} and t_{pHL} have the same value; when this is the case the common value is often labelled as t_{pd}. This is the label given to the average value of t_{pLH} and t_{pHL} and sometimes it is quoted even when t_{pLH} and t_{pHL} are different. The propagation delay depends upon the load capacitance as already stated, but also upon the supply voltage V_{CC}. Some of the CMOS logic technologies allow a device to be operated at either 5 V or at 3.3 V, but operation at the lower voltage considerably increases the propagation delay of the device. Some examples of this are:

74AC: t_{pLH} = 7 ns (typ) at 3.3 V, = 6 ns (typ) at 5 V
74AHC: t_{pLH} = 5.5 ns (typ) at 3.3 V, = 5.5 ns at 5 V

EXAMPLE 7.6

The propagation delay times for the 74AHC02 quad 2-input NOR gate IC are:

$$t_{pLH} = t_{pHL} \text{ from } A \text{ or } B \text{ to } Y = 3.6 \text{ ns when } C_L = 15 \text{ pF}$$
$$= 5.1 \text{ ns when } C_L = 50 \text{ pF}$$

Draw the input and output waveforms if input *B* is held LOW and a positive pulse is applied to input *A*. C_L = 50 pF.

Solution

The input and output waveforms are shown in Fig. 7.17 (*Ans.*)

Fig. 7.17 *Use of a diode to increase switching speed*

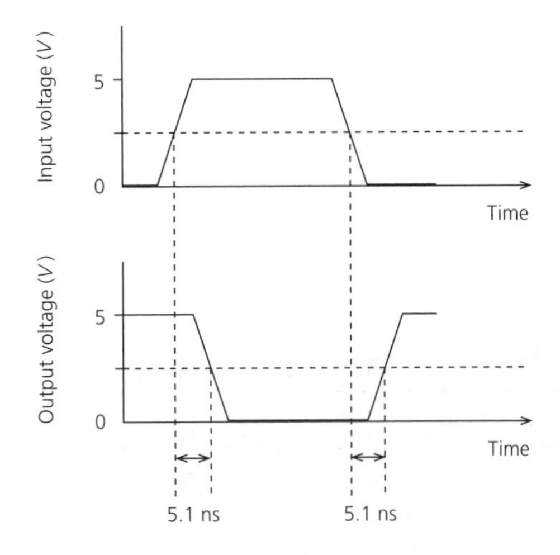

EXAMPLE 7.7

A gate has a quoted propagation delay of $t_{pLH} = t_{pHL} = 10$ ns when the load capacitance is 50 pF. If the output capacitance of the gate is 5 pF, estimate the values of t_{pLH} and t_{pHL} when the load capacitance is 25 pF.

Solution

The propagation delay is proportional to load capacitance. Hence

$t_{pLH} = t_{pHL} = 10 \times 30/55 = 5.45$ ns (*Ans.*)

The propagation delays of a gate limit the highest frequency at which the gate can be operated. The lower the t_{pLH}/t_{pHL} figures the higher the maximum frequency of operation.

Speed-power Product

The *speed-power product* gives a basis for the comparison of logic circuits when both propagation delay and power dissipation are of importance. The lower the speed-power product the better the device. 74CMOS devices are better than 74TTL devices in this respect. Typically, at a frequency of 100 kHz, a 74HC00 has a speed-product of 1.4 pJ and a 74LS00 has a speed-power product of 20 pJ.

Flip-flops

In a flip-flop (p. 208) the propagation delays t_{pLH} and t_{pHL} are quoted from all of the inputs to the Q output. The inputs always include the clock terminal and may, depending upon the device, also include a clear and/or a preset terminal. Fig. 7.18 shows the waveforms for the clock input.

Fig. 7.18 *(a) Schottky transistor and (b) symbol*

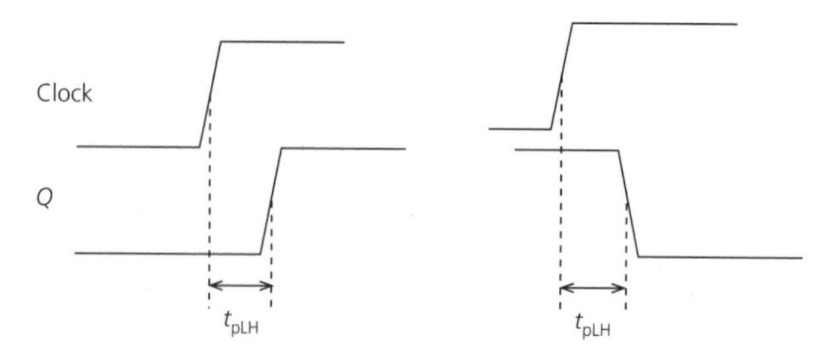

Transistor–transistor logic

Transistor–transistor logic (TTL) has been used for several years because it offers a good performance, it is readily available from several manufacturers, and it is easily inter-connected, or *interfaced*, with other digital circuitry – all at relatively low cost. The original version of TTL, known as *standard TTL*, the 54/74 series, is no longer used. It has been superseded by the *low-power Schottky TTL* (74LS family), by *advanced Schottky TTL* (AS), and by *advanced low-power Schottky TTL* (ALS) logic families. The 54 series devices are intended for use in military equipment and can operate at temperatures of up to 125 °C, whereas series 74 devices have a maximum temperature of just 90 °C.

NAND gate

The basic circuit of a TTL NAND gate is given in Fig. 7.19. The input transistor has a number of emitters equal to the desired fan-in of the circuit; in the figure, a fan-in of 2 has been assumed. In the 54/74 series, fan-ins of 2, 3, 4 and 8 are available. When both input terminals are at +5 V the emitter/base junction of T_1 is reverse biased but its collector/base junction is forward biased. Current then flows from the collector power supply, through R_1, into the base of T_2. Transistor T_2 turns full ON and the output voltage of the circuit falls to the saturation voltage of the transistor. The output of the circuit is then at logic 0. When either or both of the input terminals is at approximately zero volts, logic 0, the associated emitter/base junction will be forward biased (its base

Fig. 7.19 *Basic TTL NAND gate*

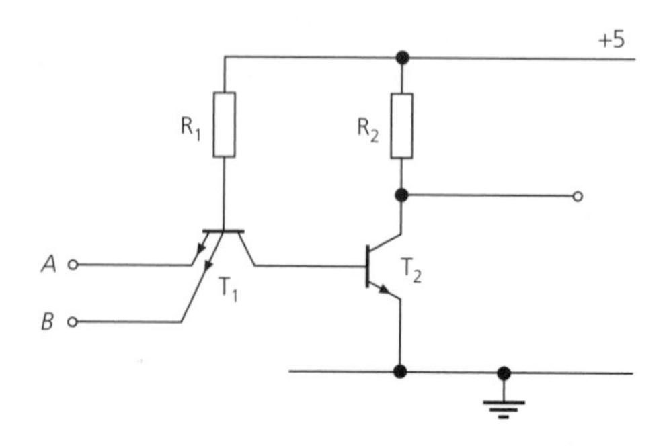

Fig. 7.20 *Standard TTL NAND gate*

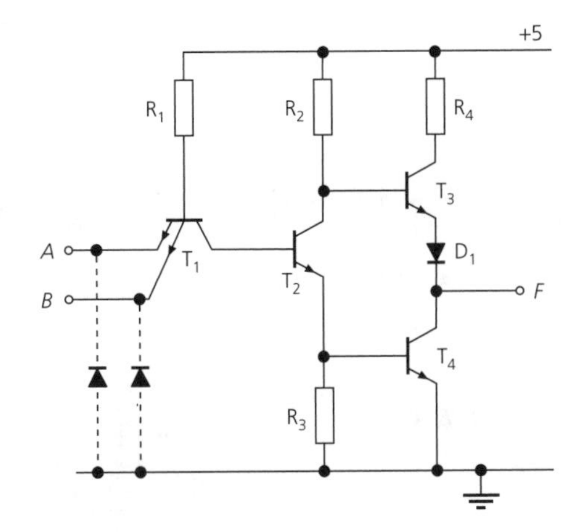

is more positive than its emitter). The value of resistor R_1 is selected to ensure that T_1 is then full ON and so the base voltage of T_1 is only V_{BEI} volts (approximately 0.7 V) above earth potential. This potential is insufficient to keep T_2 ON and so T_2 turns OFF. The output voltage of the circuit then rises to +5 V, i.e. becomes logic 1. Thus, transistor T_1 performs the AND function and T_2 acts as an inverter to give an overall circuit rendering of the NAND function.

The standard TTL gate adds an output stage to the basic circuit shown in Fig. 7.19 in order to increase both the operating speed and the fan-out, the complete circuit being given in Fig. 7.20. The output stage, consisting of transistors T_3 and T_4 and diode D_1, is often known as a *totem-pole* stage. When T_2 has turned ON the base/emitter potential of T_3 is approximately zero and so T_3 does not conduct. At the same time T_4 is turned ON by the voltage developed across resistor R_3. Thus, when T_2 is ON transistor T_3 is OFF and T_4 is ON; this means that the potential at the output terminal of the circuit is LOW and so the output state is logic 0. The fan-out can be up to about 10 without the saturation voltage of T_4 rising above the 0 level. Similarly, when T_2 is OFF its collector voltage is +5 V and its emitter voltage is 0 V. Now T_4 is turned OFF and T_3 conducts to an extent that is determined by the external load connected to the output terminals of the circuit. The output voltage is then equal to 5 V minus the voltage dropped across T_3 and D_1, i.e. logic 1. T_3 acts as an active pull-up resistor.

T_2, with resistors R_2 and R_3, acts as a phase splitter to give opposite polarity logic levels at the collector and emitter. Because of the low output impedance of the totem-pole stage the pin-out is up to 10. The function of the diode D_1 is to ensure that T_3 and T_4 never conduct current simultaneously. Standard TTL is rarely, if ever, used in new systems.

Low-power Schottky TTL

Low-power Schottky TTL, 54LS/74LS, combines the advantages of fast operation, low power dissipation, and high-frequency capability.

The circuit of the low-power Schottky NAND gate is shown in Fig. 7.21. When both inputs are at logical 1, diodes D_1 and D_2 are OFF. T_1 and T_3 turn ON and then the

Fig. 7.21 *Low-power Schottky TTL NAND gate*

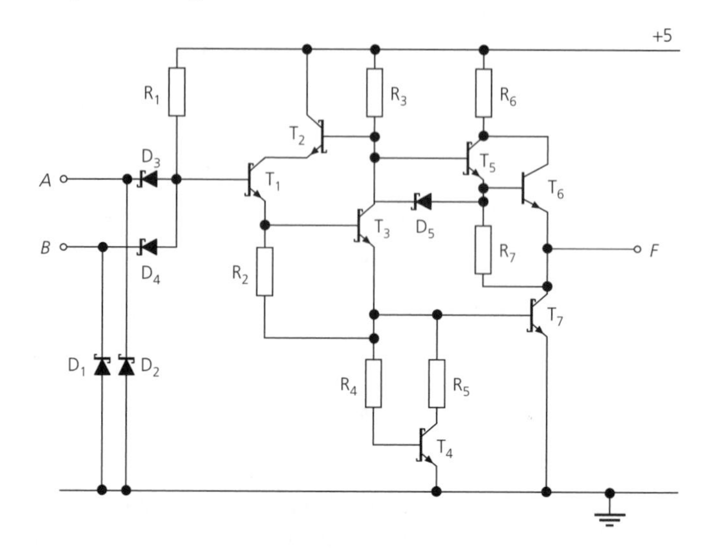

voltage at the base of T_1 is equal to 0.5 + 0.5 + 0.5 = 1.5 V. The collector potential of T_3 is low and so the Darlington pair T_5/T_6 turns OFF. When the base/emitter voltage of T_3 is very nearly equal to $V_{BE(SAT)}$, then T_4 will be conducting and will take most of the emitter current of T_3. Pull-down transistor T_7 will not receive sufficient base current to turn it ON until the base potential of T_1 is high enough for T_4 to supply sufficient current. T_7 then rapidly turns ON. The speed of switching is increased by T_2 and D_5, whereby T_2 supplies a current surge during the turn-on of T_1 while D_5 provides a discharge path for pull-up transistor T_6.

When one or more inputs are at logical 0, the associated input diode conducts and the base potential of T_1 falls below the value needed to keep T_1 conducting. Therefore, T_1 turns OFF. This makes transistors T_3 and T_7 turn OFF as well. The collector potential of T_3 rises and turns T_5/T_6 ON. The base of the output transistor T_6 is returned via R_7 to the output terminal since this allows the output voltage to pull up to $V_{CC}-V_{BE}$ volts.

The maximum and minimum input and output voltages for logical 0 and 1 are, in the main, the same as those for the standard TTL gates. Differences are as follows: $V_{OH(min)}$ = 2.7 V and $V_{OL(max)}$ = 0.5 V. I_{IL} = 0.4 mA, I_{IH} = 20 μA, I_{OL} = 8 mA and I_{OH} = –0.4 mA.

Advanced Schottky TTL

Further development of the TTL logic family has brought about the introduction of two further versions of Schottky TTL. One of these, known as *advanced Schottky* (AS) provides the designer with a speed performance which is comparable with that given by the ECL family of devices. The other version, known as *advanced low-power Schottky* (ALS), is an improved version of the low-power Schottky logic family giving both increased speed and a reduction in power dissipation. The speed of operation of ALS is, however, less than that of AS.

Table 7.1 *TTL gates*

Type of gate		*Standard*	*Low-power Schottky*	*Advanced low-power Schottky*	*Advanced Schottky*
Propagation delay (ns)		9	7	4	1.5
Guaranteed	1	0.4	0.7	0.7	0.4
noise margin (V)	0	0.4	0.3	0.4	0.3
Power dissipation (mW)		10	2	1	2
Fan-out		40	20	10	10

A large number of SSI and MSI circuits are presently available in both versions of advanced Schottky, and further circuits are continually being added to the catalogues. In general, the circuits offered correspond to those that are readily available in the standard and LS branches of the TTL family.

A comparison between the various types of TTL gates is given in Table 7.1, which shows typical values for the parameters quoted.

Open-collector gates

Very often it is desirable to be able to connect together the outputs of several gates to increase the fan-out, or to perform a particular logical function, or perhaps to connect several gates to a common output line or *bus*.

If two gates with totem-pole output stages have their output terminals connected together, one of the gates is very likely to pass an excessive current which may well damage it. If, for example, the output of one gate is at logic 0, while the output of the other gate is at logic 1, the first gate would have a low resistance to earth and a current of high enough magnitude to cause damage could flow.

An *open-collector gate* is one in which the pull-up transistor T_6 of the output stage has been omitted. The output of the gate is the collector of the pull-down transistor T_7 which may be connected to an external *pull-up resistor*. An input to the gate that causes the output to be LOW also allows the pull-down transistor to be ON and this allows current to flow through the external resistor and into the transistor. When the output is HIGH the transistor has turned OFF.

Wired-AND logic function

Two open-collector NAND gates have their output terminals paralleled and then connected to an external *pull-up resistor* R_1 to produce what is generally called the *wired-AND gate*. The logical function performed by wired-AND NAND gates is $F = (\overline{AB})(\overline{CD}) = \overline{AB + CD}$ (see Fig. 7.22).

If the same function is produced using totem-pole output stages, an extra gate is necessary, thus increasing the propagation delay (also the power dissipation is larger).

Fig. 7.22 *Direct connection of open-collector gates*

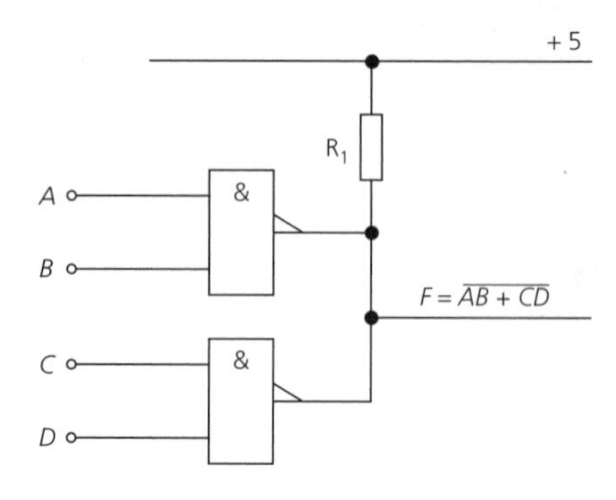

Fig. 7.23 *Exclusive-OR gate using open-collector gates*

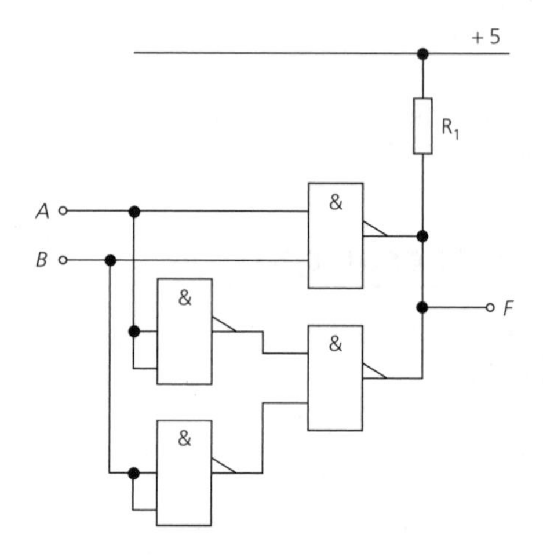

Disadvantages of open-collector gates are: an external pull-up resistor is needed, and the noise margin is poorer.

The connection is known as the wired-AND because, if the output of either gate goes to 0 V, the output of the paralleled gates must also become 0 V. Only if both outputs are at logic 1 can the combined output be 1. The method of connection can be extended to more than two open-collector gates.

If two open-collector NOR gates are connected in the wired-AND circuit, the output F is

$$F = (\overline{A + B})(\overline{C + D}) = (\bar{A}\bar{B})(\bar{C}\bar{D}) = \overline{A + B + C + D}$$

This means that the wired-AND circuit expands the NOR logic operation.

The wired-AND principle can be applied to form an exclusive-OR gate (see Fig. 7.23). For this circuit

$$F = \overline{AB + \bar{A}\bar{B}} = (\overline{AB})(\overline{\bar{A}\bar{B}}) = (\bar{A} + \bar{B})(A + B)$$
$$= \bar{A}A + \bar{A}B + A\bar{B} + B\bar{B} = \bar{A}B + A\bar{B}, \text{ the exclusive-OR function.}$$

Fig. 7.24 *Logic symbol for 74LS09 quad 2-input open-collector AND gate IC*

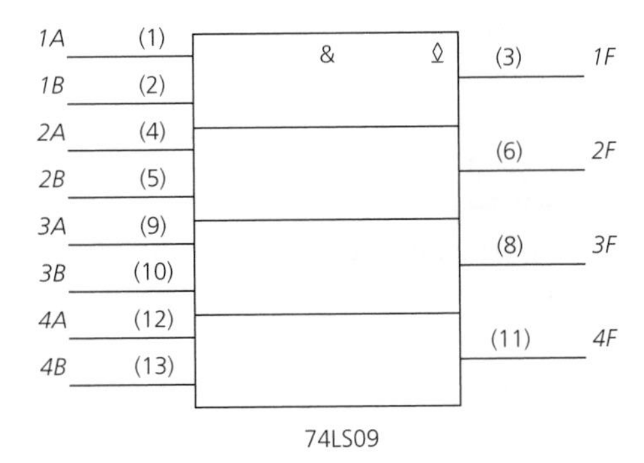

The main use of open-collector gates occurs when the outputs from two, or more, gates are to be connected together.

EXAMPLE 7.8

Determine the logic function performed by each of the wired-OR circuits (a) two 74LS01 NAND, (b) three 74LS05 inverters, and (c) two 74LS33 NOR open-collector gates.

Solution

(a) $F = (\overline{AB})(\overline{CD}) = (\bar{A} + \bar{B})(\bar{C} + \bar{D})$ (*Ans.*)
(b) $F = \bar{A}\bar{B}\bar{C} = \overline{A + B + C}$ (*Ans.*)
(c) $F = (\overline{A + B})(\overline{C + D}) = \bar{A}\bar{B}\bar{C}\bar{D} = \overline{A + B + C + D}$ (*Ans.*)

The sign ◊ is placed once on the symbol block of an open-collector device. This is shown for the 74LS09 quad 2-input open-collector AND gate IC in Fig. 7.24.

Pull-up resistor

The value of the pull-up resistor R_1 depends upon the number N of open-collector gates that are connected together and the required fan-out n. The *minimum* value that R_1 can have is set by the wanted fan-out since the total current taken from a gate must not exceed 8 mA (74LS). Therefore,

$$R_{1(\min)} = \frac{V_{CC} - \text{maximum 0 state output voltage}}{8\ \text{mA} - n \times \text{input current for the 0 state}}$$
$$= (5 - 0.4)/(8 - 1.6n)\ \text{k}\Omega \qquad (7.4)$$

The maximum value for the pull-up resistor is found by considering the output of a gate when at the logic 1 level. The current through R_1 will then be the sum of the currents supplied to the OFF outputs (a maximum of 250 μA in 74LS) and the input current of the next stage (a maximum of 40 μA per gate for 74LS). Therefore,

$$R_{1(max)} = \frac{V_{CC} - \text{minimum 1 state output voltage}}{N \times 250\ \mu A + n \times 40\ \mu A}$$
$$= (5 - 2.4)/(0.25N + 0.04n)\ k\Omega \quad (7.5)$$

EXAMPLE 7.9

Calculate the maximum and minimum values for the pull-up resistor in a wired-OR circuit if three NAND gates are connected and the fan-out is one.

Solution

$$R_{1(min)} = (5 - 0.4)/(8 - 1.6) = 719\ \Omega$$
$$R_{1(max)} = (5 - 2.4)/(0.75 + 0.04) = 3291\ \Omega$$

In most cases a 1000 Ω resistor is deemed to be a suitable value (*Ans.*)

CMOS technology

4000 series

The *complementary metal-oxide semiconductor* or *CMOS 4000* logic family offers some significant advantages over bipolar logic such as very low power dissipation, good noise immunity, and the ability to operate from a wide range of power supply voltages (2–15 V). Its main disadvantages are its relatively long propagation delay and its low output current capability. The long propagation time arises from the input time constant of the enhancement-mode MOSFETs that are employed because they turn OFF when their gate-source voltage is at zero volts.

The circuit of a CMOS NAND gate is given in Fig. 7.25. It can be seen that the n-channel MOSFETs are connected in series and the p-channel devices are connected in

Fig. 7.25 *CMOS NAND gate*

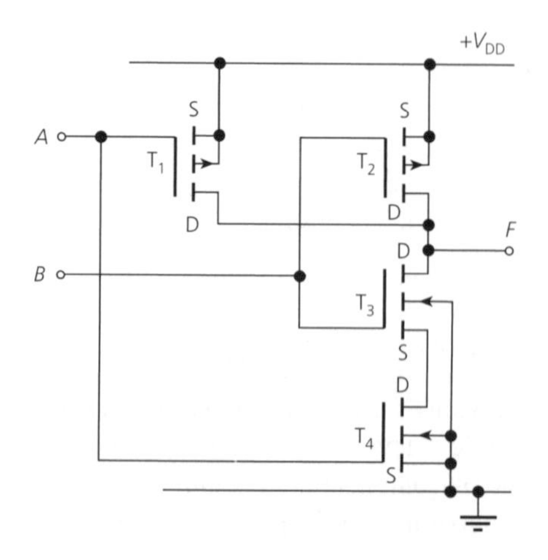

parallel. The operation of the circuit is as follows. If either, or both, of the inputs is at logical 0 ($\simeq$ 0 V), then the associated p-channel MOSFETs, T_1 and/or T_2, are turned ON, while the associated n-channel MOSFETs, T_3 and/or T_4, are turned OFF. The output terminal of the circuit is then at +5 V minus the saturation voltage of an ON MOSFET. Conversely, if both inputs are at logical 1 ($\simeq$ 5 V), T_1 and/or T_2 are turned OFF, and T_3 and/or T_4 are turned ON. The output of the circuit is then at approximately 0 V or logical 0.

The only path between the $+V_{DD}$ line and earth is via the series connection of an n-channel and a p-channel MOSFET and so zero power is dissipated when the gate is in either logic state.

CMOS devices are available in both the A and B series, the B series operating from drain supply voltages of 3–20 V as opposed to 3–15 V for the A series. Also, some of the electrical characteristics are slightly different. Some typical figures for CMOS 4000 series devices are

Low output voltage $V_{OL} = 0.05$ V max
High output voltage $V_{OH} = V_{DD} - 0.05$ V min
Noise margin 1 V min
Propagation delay 30 ns
Sink current $I_{OL}(V_{DD} = 5\text{ V}) = 1$ mA
Sink current $(V_{DD} = 15\text{ V}) = 6.8$ mA
Source current $(V_{DD} = 5\text{ V}) = 1$ mA
Source current $(V_{DD} = 15\text{ V}) = 6.8$ mA

CMOS 4000 series devices are no longer used in new designs. Many CMOS gates are available with open-drain outputs, the equivalent of TTL open-collectors.

74HC/HCT/AC/ACT/AHC/AHCT series

To overcome the low-speed disadvantage of the 4000 series, further CMOS logic families have been introduced that combine the high speed of AS and ALS TTL devices with the low power dissipation of CMOS.

- High-speed CMOS (HC). This family of devices has two sub-groups; (i) 74HC, which has CMOS compatible inputs and outputs, and (ii) 74HCT, which has TTL compatible input and output terminals and is used for interfacing with TTL circuitry. HC series devices have the same numbers and pinouts as the TTL device that performs the same function, e.g. 74LS00 and 74HC00 are both quad 2-input NAND gate, and 74LS27 and 74HC27 are both triple 3-input NOR gate ICs. The 74 HCT devices are a direct replacement for 74LS devices. The power supply voltage for HC devices is 2–6 V and for HCT devices is 4.5–5.5 V.
- 74AC/ACT series. Devices in this family have the same pinout as 74HC and 74LS devices and have a better performance than either. AC/ACT devices have shorter propagation delays, a higher maximum frequency, and an increased drive capability. 74ACT devices have TTL compatible inputs. The family also includes devices with a different pinout: these have the V_{CC} and earth pins) in the centre of the IC package (pins 11 and 4 instead of 14 and 7). The new pinout reduces noise caused by high-speed switching. Such devices are identified by adding 11000 to the standard device numbering, e.g. the 74AC11000 is a quad 2-input NAND gate.

- 74AHC/AHCT. This latest logic family provides even greater speed of operation and reduced noise. It includes single-gate devices, e.g. the 74AHC1G00 contains one 2-input NAND gate.

AHC/AHCT devices are three times faster than their HC/HCT equivalents and take approximately 50% of the static current from the power supply. The supply voltage is 1.5–5.5 V for AHC and 4.5–5.5. V for AHCT devices. The noise immunity is the same as for HC/HCT. The number of devices available in the family is still limited but includes commonly employed gates such as NAND, AND and exclusive-OR. AHC devices have the same pinouts as the corresponding 4000 series devices, and AHCT devices have 74LS device pinouts.

CMOS hazards

Many CMOS gates have protective diodes connected between their input terminals and earth to reduce the possibility of damage to the device caused by handling or soldering. The gate terminal of a MOSFET is insulated from the channel by a very thin layer of insulation which effectively forms the dielectric of a capacitance. Any electric charge which accumulates on the gate terminal may easily produce a voltage that is large enough to cause the dielectric to break down. Once this happens the device has been destroyed. The charge necessary to damage a MOSFET need not be very large since the capacitance between the gate and the channel is very small and $V = Q/C$. This means that a dangerously high voltage can easily be produced by merely touching the gate lead with a finger or a tool. Great care must be taken, therefore, when a CMOS circuit is fitted into, or is removed from, a circuit. The leads of a CMOS device in store are usually short-circuited together by springy-wire clips or by conductive jelly or grease, and the short circuit should be maintained while the device is being fitted into circuit, particularly during the soldering process. The solderer should stand on a non-conducting surface such as a rubber mat and should use a non-earthed soldering iron. The wearing of a wrist strap that is connected to earth via a high value resistance will also help.

The IC leads should be protected from heat by the use of a heat shunt (pliers) and the device should be allowed to cool after each connection is soldered before tackling another one. When an IC is removed from a circuit a de-soldering tool should be used to remove all the solder from the connections and then the IC can be lifted off from its tags. The problem is, of course, simplified if IC holders are used.

A CMOS device should never be removed from, or inserted into, a circuit while the power is applied (even if it has been switched off).

Interfacing TTL and CMOS

Very often the need may arise for gates in the TTL and CMOS families to be interconnected or *interfaced*. The output of a TTL gate cannot be directly connected to the input of a CMOS gate without adversely affecting the noise immunity of the circuit.

There is no problem when the TTL output is LOW. The TTL minimum LOW voltage is 0.4 V and CMOS takes any voltage up to $V_{CC}/3$ as being LOW. The TTL HIGH output level may be as low as 2.4 V, while a CMOS device expects at least 3.33 V for a HIGH

Fig. 7.26 *Interfacing TTL and CMOS gates*

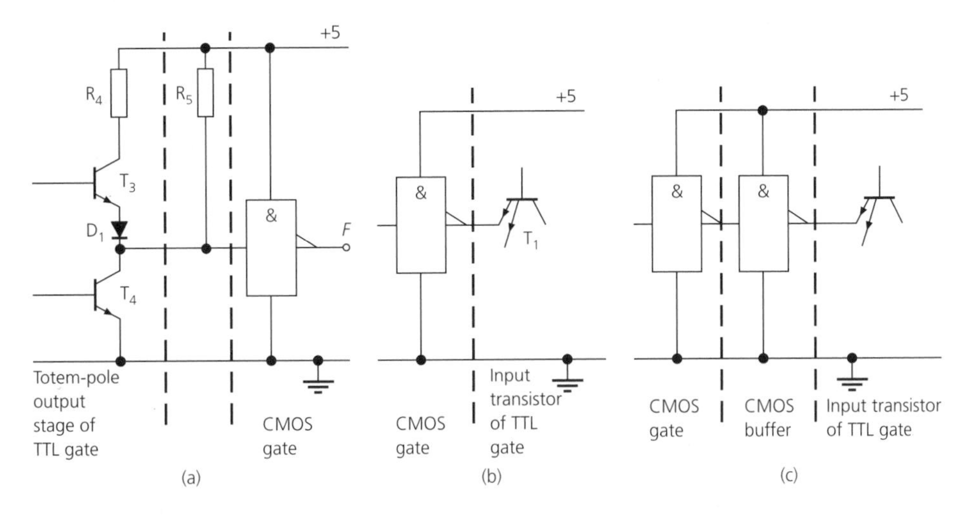

signal. To overcome this problem, a pull-up resistor can be connected between the junction of the two gates and the + 5 V supply line (Fig. 7.26(a)).

Typically, $R_5 = 10\ k\Omega$. The output of a CMOS gate can be connected directly to the input of a TTL gate as shown in Fig. 7.26(b), but current problems exist with a CMOS device not being able to sink enough current. To overcome this problem, a CMOS buffer stage, such as the 4010, is used (see Fig. 7.26(c)). There are several combinations of interfaces and, in general, when connecting TTL to CMOS a pull-up resistor is necessary, the only exception being 74HCT/ACT/AHCT devices which have been designed to be TTL compatible.

Tri-state outputs

In some equipments, such as microprocessors, information is transferred from one point to another via buses. A number of different circuits may require to be connected to a bus, yet must not interact with one another. There is thus a need for circuits whose outputs can be connected directly in parallel with one another without affecting their operation.

Three-state (often called tri-state) circuits have *three* possible output states. Two of these are the usual logic 0 and logic 1 states, while the third state is one in which the output circuit is of high impedance. A *select* or *inhibit* terminal is provided to allow the output to be switched into, or out of, its high-impedance condition; see Fig. 7.27 which shows a tri-state inverter. The truth table for a tri-state inverter is given in Table 7.2.

Comparison between logic technologies

The main characteristics of the various logic families which are available in integrated circuit form are listed in Table 7.3. Typical figures are quoted.

Fig. 7.27 *Tri-state inverter*

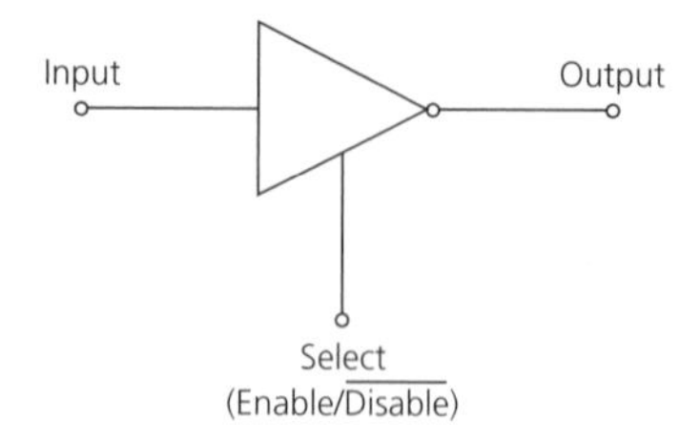

Table 7.2 *Tri-state inverter*

Input	*Select*	*Output*
0	1	High impedance
1	1	High impedance
0	0	1
1	0	0

EXERCISES

7.1 A bipolar transistor has a collector resistor of 1 kΩ, a base resistance of 15 kΩ and 5 V collector supply voltage. The output terminal of the circuit is connected to a circuit whose input resistance is R Ω. Calculate the lowest value of R for the output voltage when the transistor is OFF to be greater than 4 V.

7.2 The propagation delay times for a TTL gate are $t_d = 5$ ns, $t_f = 10$ ns, $t_s = 15$ ns and $t_r = 7$ ns. Draw the output waveform when the device is turned OFF and ON.

7.3 Figure 7.28 shows how the collector/emitter voltage of a transistor varies when a rectangular pulse of +5 V is applied to its base terminal. Calculate
(a) $V_{CE(SAT)}$.
(b) The fall-time t_f.
(c) The rise-time t_r of the transistor.

7.4 (a) Figure 7.29 shows the output circuit of a gate. If the logic 1 level is defined by the voltage limits 3.5–5 V calculate the fan-out of the circuit. Assume that each of the driven gates takes a current of 1.6 mA.
(b) The gate shown in Fig. 7.29 is connected to four similar gates. Each gate has an input resistance of 4000 Ω. Calculate the logic 1 output voltage level.
(c) A TTL gate has a minimum logic 1 voltage level of 2.4 V and a maximum logic 0 voltage level of 400 mV. Calculate the worst-case noise margin of the gate.

7.5 (a) Referring to Fig. 7.30. What are (i) the required fan-in for each gate, and (ii) the required fan-out for each gate?
(b) The logical function $F = \overline{ABC + DEF}$ is to be implemented. Draw possible arrangements using (i) NAND gates only, and (ii) open-collector gates only.

Table 7.3 ***Integrated circuit logic technologies***

Logic family	Propagation delay (ns)	Power dissipation (mW)	Noise margin (mV)		Fan-out	Supply voltage (V)	Maximum frequency (MHz)	Logic levels (V)				Source current (mA)	Sink current (mA)
			1	0				V_{IH} (min)	V_{IL} (max)	V_{OH} (min)	V_{OL} (max)		
Standard TTL	9	10	400	400	40	5	35	2	0.8	2.4	0.4	16	1.6
Low-power Schottky TTL	7	2	700	400	20	5	40	2	0.8	2.7	0.4	8	0.4
Advanced Schottky TTL	1.5	2	400	300	10	5	200	2	0.8	2.4	0.5	8	0.1
Advanced low-power Schottky TTL	4	1	700	400	10	5	70	2	0.8	2.7	0.4	4 8	0.01 0.1
74HC	10	2.5×10^{-6}	28% V_{DD}	18% V_{DD}	10	2–6	40	3.15	1.35	3.84	0.33	4	0.001
74AHC	4.8	2.5×10^{-6}	28% V_{DD}	18% V_{DD}	50	2–6	160	3.85	1.65	2.48	0.44	24	0.001
74AC	7.8	2.1×10^{-6}	28% V_{DD}	18% V_{DD}	50	2–6	100	3.85	1.65	2.46	0.44	24	0.001

[Note: the power dissipation of CMOS devices increases with increase in the clock frequency. At 1 MHz, for example, 4000 series devices dissipate 1 mW power and 74HC series devices dissipate 1.5 mW power.]

Fig. 7.28

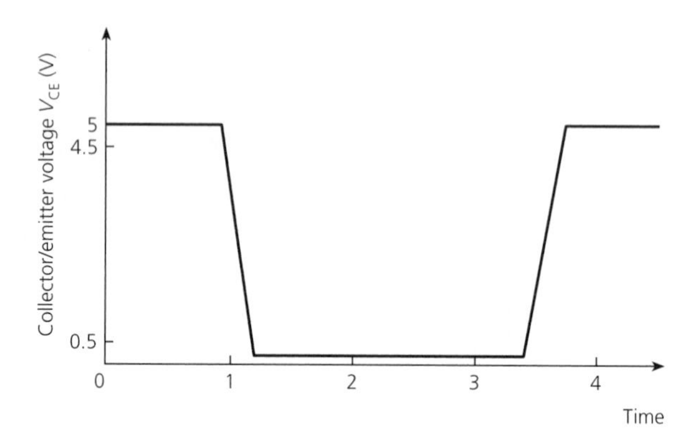

Fig. 7.29

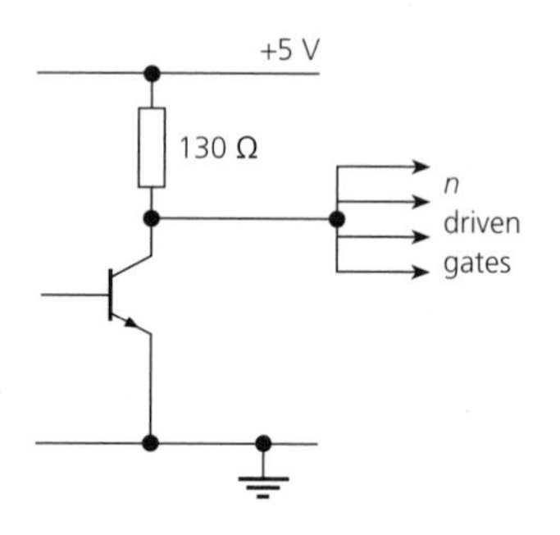

Fig. 7.30

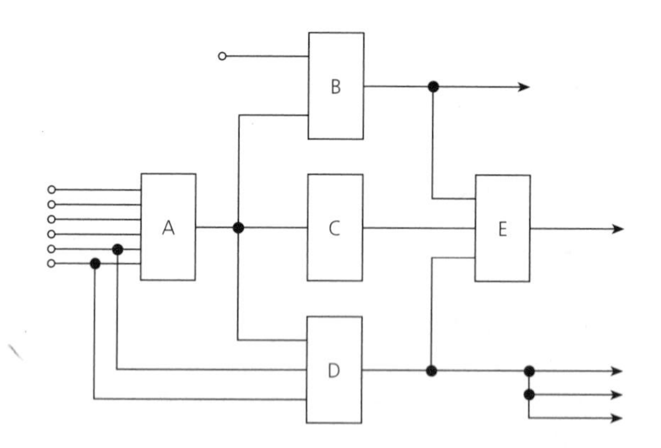

7.6 (a) Explain when and why a tri-state device may be required.
(b) Explain the principle of operation of such a device.

7.7 For the circuit shown in Fig. 7.31 calculate
(a) The collector current when the transistor is saturated.
(b) V_{BE} when the transistor is just saturated.

Fig. 7.31

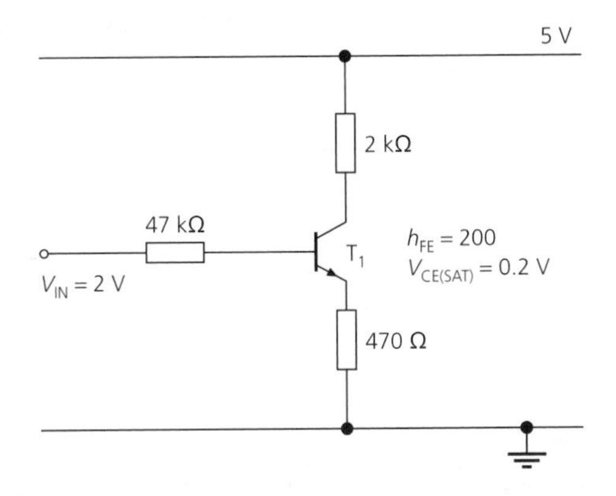

7.8 The rising output voltage of a low-power Schottky TTL gate is given by $V(t) = V_{OH} + 3.7(1 - e^{t/(8 \times 10^{-9})})$. State the meaning of the symbol V_{OH} and quote a typical value. Use this value to determine the time taken for the output voltage to rise to 90% of $V_{CC} = 5$ V.

7.9 What is meant by saturation in a bipolar transistor circuit? A transistor has a collector resistance of 2.2 kΩ and a supply voltage of 5 V. Calculate the saturated collector current if $V_{CE(SAT)} = 0.2$ V. If $V_{BE(SAT)} = 0.75$ V determine the forward bias voltage of the collector–base junction.

7.10 A switching transistor has the following data:
$V_{CB(max)} = 20$ V, $V_{CE(max)} = 20$ V
$I_{C(max)} = 450$ mA
$P_{Diss(max)} = 520$ mW
$V_{CE(SAT)} = 250$ mV
$V_{BE(SAT)} = 800$ mV
Rise-time = 22 ns
Storage time = 100 ns
Delta time = 6 ns
Fall-time = 28 ns

Draw typical characteristics and, hence, show how the given ratings limit the choice of power supply voltage and collector load resistance. Explain the meanings of each of the four switching times.

7.11 A TTL gate and its CMOS equivalent both operate from a 5 V power supply.
(a) State the worst case noise margin of the TTL device.
(b) Calculate the noise margin of the CMOS device.
(c) If the power supply voltage is increased to 12 V what would then be the noise margin of the CMOS gate and what would happen to the TTL gate?

7.12 (a) The outputs of the 74LS05 open-collector hex inverter are connected via a pull-up resistor to 5 V. What logic function is implemented?
(b) One of the gates in each one of two open-collector ICs have their outputs connected together and to 5 V via a pull-up resistor. The ICs are (i) 74LS33 NOR gate and (ii) 74LS01 NAND gate. Determine the logic function that is performed.

8 MSI combinational logic circuits

After reading this chapter you should be able to:

(a) Explain the operation of a full-adder, and combine two 4-bit full-adders to give an 8-bit full-adder.
(b) Describe the function of a demultiplexer/decoder.
(c) Describe the function of an encoder.
(d) Describe the function of a multiplexer.
(e) Use a multiplexer to implement a Boolean equation.
(f) Understand the use of a code converter.

A combinational logic circuit is one whose output, or outputs, are determined by the logical states of its existing input, or inputs. This is to be contrasted with *sequential logic* circuits whose output is set by *both* the present input state *and* the previous output state of the circuit.

Many different combinational logic circuits are possible and they find a wide variety of applications. In most cases an MSI circuit can be employed.

One of the devices considered is the full-adder which is able to perform the addition of two binary numbers. In a microprocessor, or the processing unit of a main-frame computer, arithmetic operations on data are performed by a circuit known as the *arithmetic logic unit* (ALU). These circuits may operate on 16-, 32-, or even 64-bit numbers and employ complex circuitry and hence a very large number of transistors.

Multiplexers

A *multiplexer*, or *data selector*, is a circuit whose basic function is to select any one out of n input lines and to transmit the data present on that line to a single output line. The basic concept is illustrated in Fig. 8.1. At any instant in time only one of the switches is closed, connecting the associated data input to the output.

A multiplexer has 2^n data inputs, n control or select inputs, and an output terminal, where $n = 1, 2, 3$ or 4. The selection of one of the 2^n data inputs is made by the signals applied to the select lines. The block diagram of a multiplexer with eight data inputs is shown in Fig. 8.2. The address on the select inputs determines which data input is switched to the output. Each data input has its own unique address, e.g. data input D_1

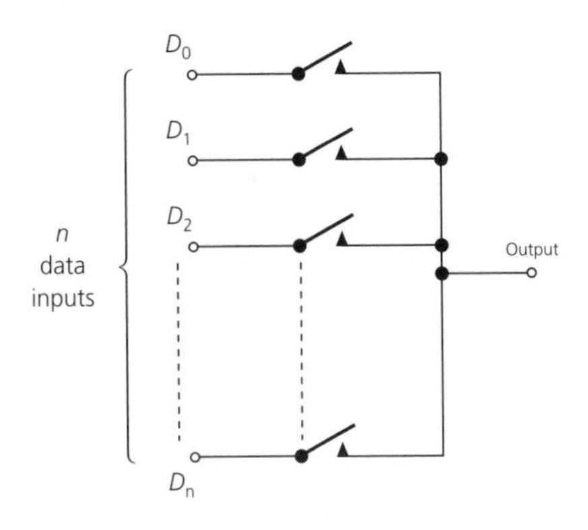

Fig. 8.1 *Multiplexer principle*

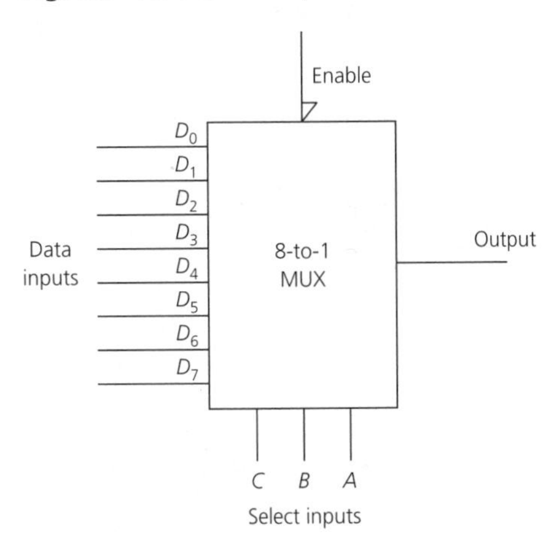

Fig. 8.2 *An 8-to-1 multiplexer*

Fig. 8.3 *(a) Pin connections and (b) logic symbol of the 74153 dual 4-to-1 multiplexer*

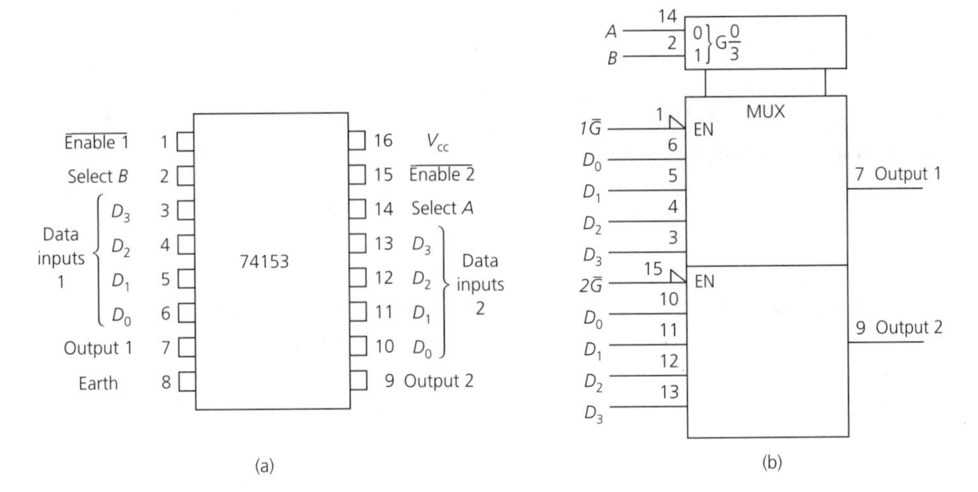

has the address 001 for a 3-input MUX and 0001 for a 4-input MUX. Several different multiplexers are included in the TTL and CMOS logic families: some examples are:

Quad 2-to-1:	74LS157,	74HC157,	74HCT157
Dual 4-to-1:	74LS153,	74HC153	
Single 8-to-1:	74LS151,	74HC151	

Some multiplexers have three-state outputs and others have an open-collector output. The output of some multiplexers is active-HIGH while others have an active-LOW output, while the 74LS151 has both true and inverted outputs.

74LS/HC153 dual 4-to-1 multiplexer

The pinout and the logic symbol of the 74153 circuit are shown in Fig. 8.3, and the truth table for each multiplexer is listed in Table 8.1.

Table 8.1 *The 4-to-1 multiplexer*

Inputs							*Output*
Select		*Data*				*Enable*	
B	A	D_3	D_2	D_1	D_0		
×	×	×	×	×	×	1	0
0	0	×	×	×	0	0	0
0	0	×	×	×	1	0	1
0	1	×	×	0	×	0	0
0	1	×	×	1	×	0	1
1	0	×	0	×	×	0	0
1	0	×	1	×	×	0	1
1	1	0	×	×	×	0	0
1	1	1	×	×	×	0	1

As before, × indicates don't care.

Note: manufacturers' data sheets for multiplexers employ different symbols for the input and output signals, e.g. the 74LS153 uses C for the data inputs and Y for the output.

Fig. 8.4 *Parallel-to-serial conversion*

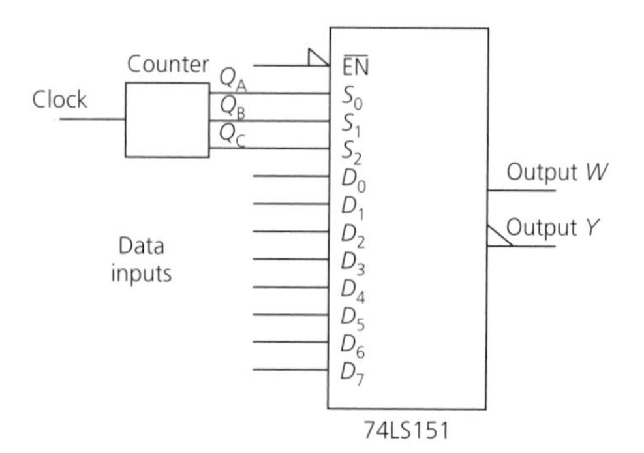

When the enable input is HIGH the output will be LOW whatever the state of the other inputs. When the enable input is LOW the control of the circuit operation is passed to the other inputs. The select inputs A and B control *both* circuits and are employed to select one of the data inputs D_0, D_1, D_2 and D_3 and cause its logical state to become the output state of the multiplexer. Consider, for example, row 5 of the truth table: in this $A = 1$, $B = 0$ so that input D_1 is selected; the logical state of D_1 is 1 so that the output state becomes 1 also.

A multiplexer can be used as a parallel-to-serial converter, and Fig. 8.4 shows one application. Data in parallel form is applied to the data inputs of a 74LS151 and is fed out of the circuit in serial form at a speed that is determined by the frequency of the clock applied to the counter.

EXAMPLE 8.1

A 74HC151 multiplexer has binary 0 applied to its data input pins D_0, D_3, and D_6, and binary 1 applied to the remaining data inputs. The waveforms shown in Fig. 8.5(a), (b) and (c) are applied to the three select pins, and the enable, respectively. Determine the waveform at the output of the multiplexer. The select inputs are derived from a clocked counter.

Solution

The output waveform is shown in Fig. 8.5(e) (*Ans.*)

Fig. 8.5

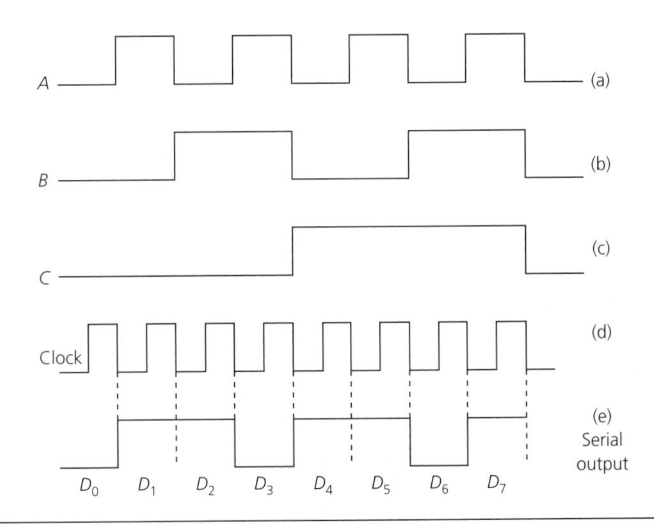

Logic function generation

An n-input multiplexer can be employed to generate any n-input truth table and, hence, to implement a Boolean equation. The logic function is formed by setting the data inputs D_0, D_1, etc. to either logic 0 or 1. The select inputs become the inputs for the variables $ABCD$ in the equation to be implemented. If the logic function has more variables than there are select inputs, then one, or more, variables are also applied to the data inputs. The advantages gained by the use of a multiplexer for this purpose are (a) one multiplexer can replace several gate ICs, and (b) it is relatively easy to change the implemented logic function.

The Boolean expression for a 4-to-1 multiplexer can be determined from Table 8.1. Thus,

$$F = \bar{A}\bar{B}D_0 + A\bar{B}D_1 + \bar{A}BD_2 + ABD_3 \tag{8.1}$$

The truth table of an 8-to-1 multiplexer is shown in Table 8.2.

Table 8.2 ***The 8-to-1 multiplexer***

A	×	0	1	0	1	0	1	0	1
B	×	0	0	1	1	0	0	1	1
C	×	0	0	0	0	1	1	1	1
Enable	1	0	0	0	0	0	0	0	0
Output	1	D_0	D_1	D_2	D_3	D_4	D_5	D_6	D_7

Table 8.3

A	0	1	0	1	0	1	0	1
B	0	0	1	1	0	0	1	1
C	0	0	0	0	1	1	1	1
$F = AB + \bar{A}C + \bar{B}\bar{C}$	1	1	0	1	1	0	1	1
Data input	D_0	D_1	D_2	D_3	D_4	D_5	D_6	D_7

Fig. 8.6

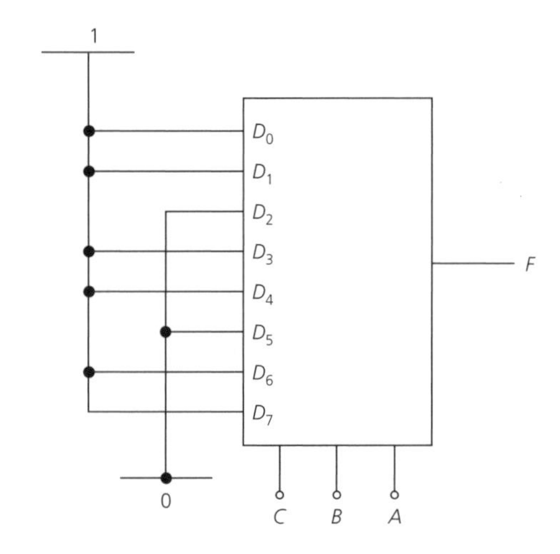

From this the Boolean equation describing its operation is

$$F = \bar{A}\bar{B}\bar{C}D_0 + A\bar{B}\bar{C}D_1 + \bar{A}B\bar{C}D_2 + AB\bar{C}D_3 + \bar{A}\bar{B}CD_4 + A\bar{B}CD_5 + \bar{A}BCD_6 + ABCD_7 \quad (8.2)$$

Select inputs equal to number of variables

The first step in implementing a Boolean function with a multiplexer is to write down the truth table of the function. Then, for every 1 in the output (F) column of the table there must be a HIGH voltage connected to the corresponding data inputs. A LOW voltage is connected to all the other data terminals. The variables are then connected to the data select terminals of the multiplexer. Consider the truth table given in Table 8.3.

The implementation of this function by an 8-to-1 multiplexer is shown in Fig. 8.6.

EXAMPLE 8.2

Use the 74HC151 8-to-1 multiplexer to implement the logic function $F = AB + \bar{B}C + \bar{A}BC$.

Solution

The truth table of the function is given in Table 8.4.

Connecting the data inputs in those columns where $F = 1$ to logic 1, and all other data inputs to logic 0, gives the circuit implementation shown in Fig. 8.7 (*Ans.*)

Table 8.4

A	0	1	0	1	0	1	0	1
B	0	0	1	1	0	0	1	1
C	0	0	0	0	1	1	1	1
AB	0	0	0	1	0	0	0	1
$\bar{B}C$	0	0	0	0	1	1	0	0
$\bar{A}BC$	0	0	0	0	0	0	1	0
F	0	0	0	1	1	1	1	1
Data	D_0	D_1	D_2	D_3	D_4	D_5	D_6	D_7

Fig. 8.7

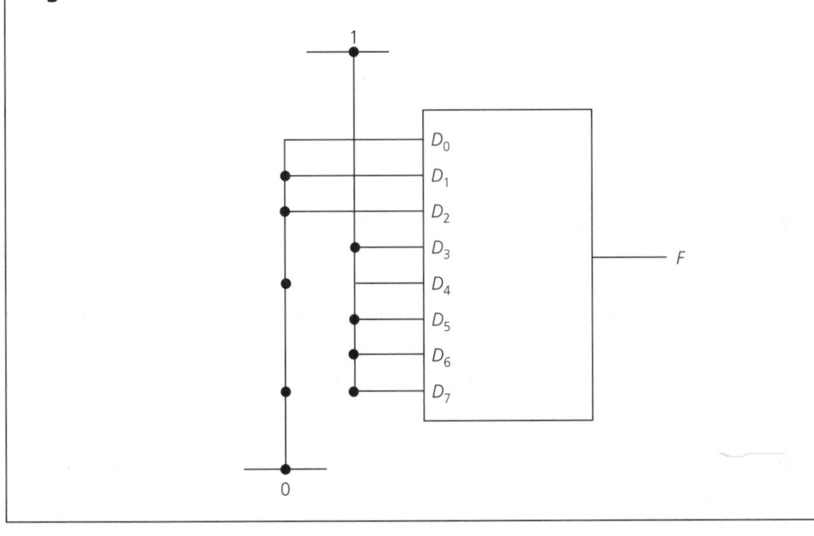

It is not always necessary to write down the truth table of a function before it is implemented. If the number of variables is equal to the number of select inputs the equation to be implemented can be compared term-by-term with the Boolean expression for the multiplexer.

EXAMPLE 8.3

Implement the logic function $F = ABC + ABC + ABC + ABC$ using an 8-to-1 multiplexer.

Solution

Comparing the function with equation (8.2), it can be seen that data inputs D_0, D_3, D_4 and D_5 must be connected to logic 1 voltage and all other data inputs to logic 0. The required circuit is shown in Fig. 8.8 using the 74LS151 (*Ans.*)

Fig. 8.8

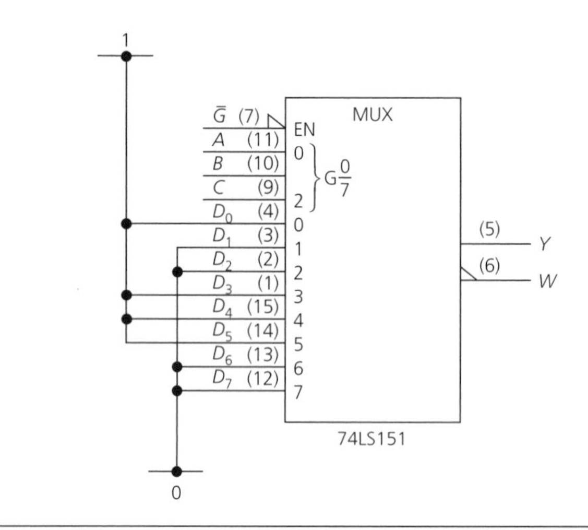

Select inputs one less than number of variables

A multiplexer with n select inputs can also be employed to implement a Boolean equation that has $n + 1$ variables. The most-significant input variable and its complement are then applied to some of the data inputs. Suppose that the logic function $F(A, B, C, D)$ is to be implemented, where D is the most-significant bit. The design procedure is as follows: write down the truth table of the logic function. Each combination of A, B and C must occur for both $D = 0$ and for $D = 1$. If

- $F = 0$ both times an ABC combination occurs, then the logic 0 voltage must be connected to the data input selected by that combination.
- $F = 1$ both times a particular ABC combination occurs, connect the logic 1 voltage to the selected data inputs.
- F is different for the two particular combinations of A, B and C and $F = D$ each time, then connect D to the selected data input.
- F is different for the two combinations of A, B and C and $F = \overline{D}$, then $\overline{D}$ must be connected to the selected data input.

EXAMPLE 8.4

Use a 74HC151 8-to-1 multiplexer to implement the logic function $F = \bar{A}\bar{B}\bar{C}\bar{D} + \bar{A}B\bar{C}\bar{D} + A\bar{B}C\bar{D} + \bar{A}BC\bar{D} + \bar{A}\bar{B}\bar{C}D + AB\bar{C}D + A\bar{B}CD + \bar{A}BCD + ABCD$.

Solution

The truth table is given in Table 8.5.

Table 8.5

A	0	1	0	1	0	1	0	1	0	1	0	1	0	1	0	1
B	0	0	1	1	0	0	1	1	0	0	1	1	0	0	1	1
C	0	0	0	0	1	1	1	1	0	0	0	0	1	1	1	1
D	0	0	0	0	0	0	0	0	1	1	1	1	1	1	1	1
F	1	0	1	0	0	1	1	0	1	0	0	1	0	1	1	1
	D_0	D_1	D_2	D_3	D_4	D_5	D_6	D_7								
	1	0	$\bar{D}$	D	0	1	1	D								

When $CBA = 000$, $F = 1$ whether $D = 0$ or 1 and so input D_0 is connected to logic 1. When $CBA = 001$, $F = 0$ for both $D = 0$ and $D = 1$ and so input D_1 is connected to logic 0. When $CBA = 010$, $F = 1$ when $D = 0$ and 0 when $D = 1$; in both cases $F = \bar{D}$ and, hence, input D_2 is connected to $\bar{D}$. When $CBA = 011$, $F = 0$ when $D = 0$ and 1 when $D = 1$. In both cases $F = D$ and, hence, input D_3 is connected to D. Carrying on in this way gives the connections shown in Fig. 8.9 (*Ans.*)

Fig. 8.9

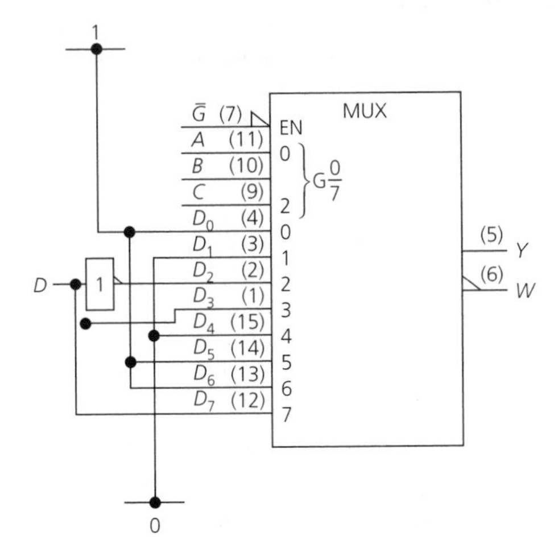

EXAMPLE 8.5

Use a multiplexer to implement the logical function $F = \bar{A}\bar{B}C + \bar{A}\bar{B}\bar{C} + A\bar{B}C + AB\bar{C}$.

Solution

Comparing the function to be implemented with equation (8.1)

$\bar{A}\bar{B}$: $D_0 = C + \bar{C} = 1$
$A\bar{B}$: $D_1 = C$
$\bar{A}B$: $D_2 = 0$ (no $\bar{A}B$ term)
$\bar{A}B$: $D_3 = \bar{C}$

Figure 8.10 shows the required implementation (*Ans.*)

Fig. 8.10

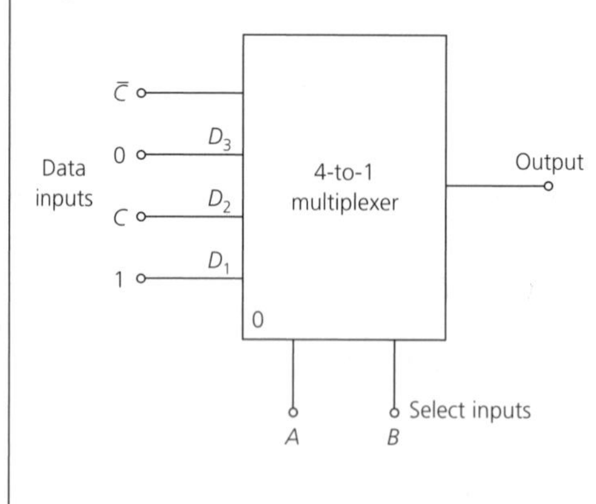

Fig. 8.11

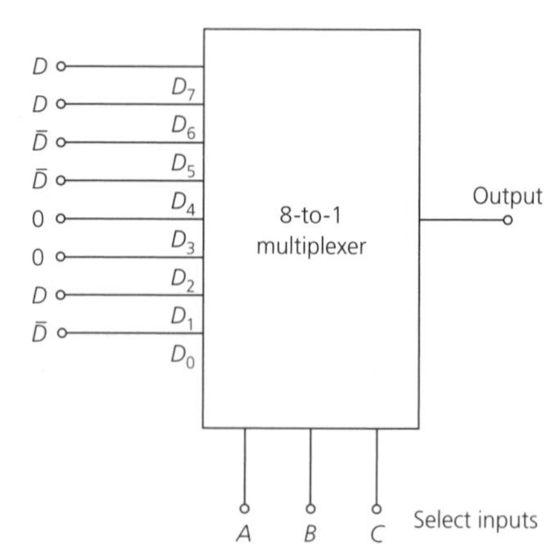

EXAMPLE 8.6

Use a multiplexer to implement the logical function $F = \bar{A}\bar{B}\bar{C}\bar{D} + A\bar{B}\bar{C}D + \bar{A}BCD + \bar{A}\bar{B}C\bar{D} + ABCD + A\bar{B}C\bar{D}$.

Solution

There are four input variables so a multiplexer with three select terminals is necessary, i.e. an 8-to-1 multiplexer. Comparing the equation to be implemented with equation (8.2),

$\bar{A}\bar{B}\bar{C}$: $D_0 = \bar{D}$, $A\bar{B}\bar{C}$: $D_1 = D$, $\bar{A}B\bar{C}$: $D_2 = 0$
$AB\bar{C}$: $D_3 = 0$, $\bar{A}\bar{B}C$: $D_4 = D$, $A\bar{B}C$: $D_5 = \bar{D}$
$\bar{A}BC$: $D_6 = D$, ABC: $D_7 = D$

The multiplexer implementation of the function is shown in Fig. 8.11 (*Ans.*)

It may sometimes be convenient to use a dual 4-to-1 multiplexer IC rather than an 8-to-1 multiplexer. All the terms containing variable C can be implemented by one multiplexer and all the terms containing $\bar{C}$ can be implemented by the other multiplexer. The outputs of the two multiplexers can then be combined by means of an OR gate.

EXAMPLE 8.7

Repeat example (8.6) using a dual 4-to-1 multiplexer such as the 74LS153 or 74HC153.

Solution

Rearranging the expression slightly, $F = \bar{C}(\bar{A}\bar{B}\bar{D} + A\bar{B}D) + C(\bar{A}BD + \bar{A}\bar{B}\bar{D} + ABD + A\bar{B}\bar{D})$. Comparing each product term with equation (8.1) gives:

(a) $\bar{A}\bar{B}$: $D_0 = \bar{D}$; $A\bar{B}$: $D_1 = D$. There are no terms in either $\bar{A}B$ or AB, so $D_2 = D_3 = 0$.

(b) $\bar{A}\bar{B}$: $D_0 = \bar{D}$; $A\bar{B}$: $D_1 = \bar{D}$; $\bar{A}B$: $D_2 = D$; AB: $D_3 = D$.

The required connections are shown in Fig. 8.12 (*Ans.*)

Fig. 8.12

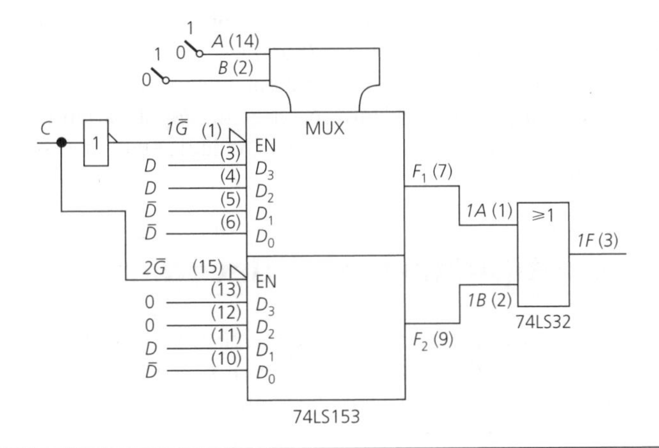

PRACTICAL EXERCISE 8.1

To investigate the use of the 74LS153 (or 74HC153) 4-to-1 multiplexer for parallel-to-serial conversion.

Components and equipment: one 74LS153 (or 74HC153) dual 4-to-1 multiplexer IC. One 270 Ω resistor. One LED. Power supply. Breadboard.

Procedure:

(a) Connect the earth and V_{CC} pins of the IC to 0 V and +5 V, respectively.
(b) Enter a 4-bit parallel data word on to pins 3, 4, 5 and 6 with the most significant bit on pin 3. Connect output pin 7 to 0 V via the resistor and the LED. Connect $1\overline{G}$ to 0 V.
(c) Now connect, in turn, the four possible combinations of 0 V (logic 0) and +5 V (logic 1) to the select A and B pins, 14 and 2, respectively. For each combination of 0 and 1 note the logic state of the LED.
(d) Change the input data word and repeat procedure (b).
(e) Confirm that the truth table given in Table 8.1 has been followed.

PRACTICAL EXERCISE 8.2

Components and equipment: one 74LS153 (or 74HC153) quad 4-to-1 multiplexer IC. One 270 Ω resistor. One LED. Power supply. Breadboard.
Procedure: the logic function given in exercise (8.7) is to be implemented.

(a) Connect the 74LS153 (or 74HC153) 4-to-1 multiplexer to implement the circuit shown in Fig. 8.12. Connect the 270 Ω resistor and the LED in series with one another, then connect the resistor to pin 7 and the LED to 0 V. Connect pin 16 to +5 V and pin 8 to 0 V. Connect the enable $\overline{G}$ input, pin 1, to 0 V to enable the circuit.
(b) (i) Take the select inputs A and B LOW by connecting 0 V to pins 2 and 14. Note whether the LED is ON or OFF. (ii) Put $A = 1$, $B = C = 0$ by applying +5 V to pin 14 and again note the logic state of the LED. (iii) Repeat for two other combinations of A, B and C.
(c) The logic function to be implemented now is $F = \bar{A}\bar{B}C + \bar{A}\bar{B}\bar{C} + A\bar{B}C + AB\bar{C}$. Repeat (b). Put, in turn, each of the input variable combinations used in (b) into the equation and compare the logic state of F with the observed states of the LED.

Demultiplexers and decoders

A demultiplexer performs the inverse function to a multiplexer in that it has a single input terminal and effectively switches it to any one of a number of possible outputs. The required output is selected by an input address. Non-selected outputs are either non-active or are open-circuit.

A decoder performs a similar function but the *only* input is the address of the output to be selected.

The basic idea of a demultiplexer can best be understood by reference to Fig. 8.13. Clearly, any one of the four outputs can be connected to the input.

The truth table of a demultiplexer has one data input, n select address inputs, and 2^n data outputs. Table 8.6 gives the truth table for a 1-to-4 demultiplexer.

The Boolean expressions describing the logical function of the circuit are: $Y_0 = \bar{A}\bar{B}D$, $Y_1 = A\bar{B}D$, $Y_2 = \bar{A}B$ and $Y_3 = ABD$.

Some examples of demultiplexers in the TTL and CMOS logic families are:

Table 8.6

Data input	Select B	inputs A	Enable G	Outputs Y_3	Y_2	Y_1	Y_0
×	×	×	1	1	1	1	1
D	0	0	0	1	1	1	*D*
D	0	1	0	1	1	*D*	1
D	1	0	0	1	*D*	1	1
D	1	1	0	*D*	1	1	1

Fig. 8.13 *Demultiplexer principle*

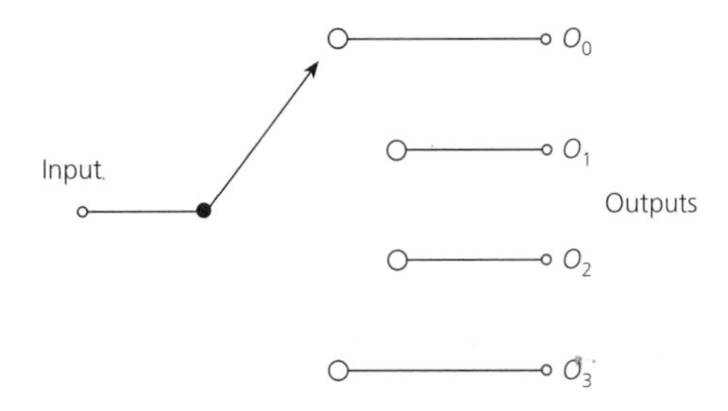

Fig. 8.14 *(a) Pin connections and (b) logic symbol of the 74139 dual 1-to-4 demultiplexer/decoder*

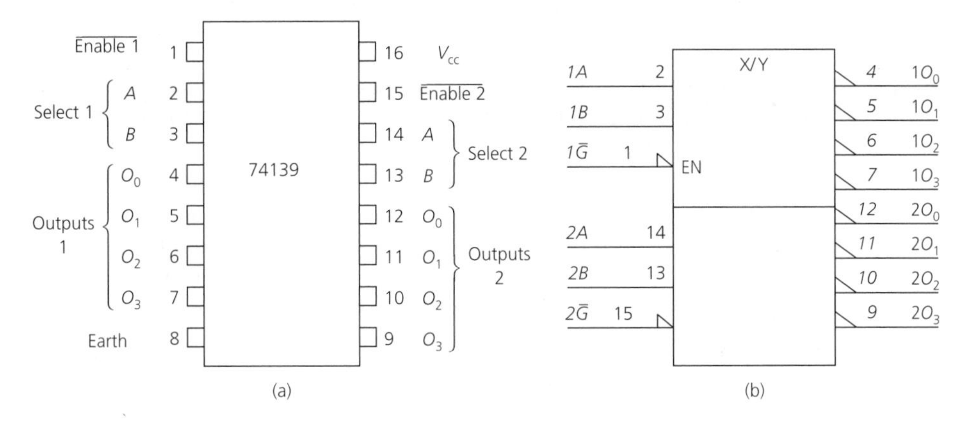

Dual 2-to-4: 74LS139, 74HC139 and 74HCT139
3-to-8: 74LS138, 74HC138 and 74HCT138.

Consider the 74LS or 74HC139 1-of-4 demultiplexer/decoder, Figure 8.14 shows the pinout and the logic symbol, and Table 8.7 the truth table, of the device.

Each of the two circuits in the IC package has one enable, and two select input pins and four output pins. When the enable pin is HIGH all four outputs are also HIGH, regardless of the logical state of the select inputs. When the enable pin is LOW *only* the output selected by the input address pins *A* and *B* will be LOW.

Table 8.7 The 1-of-4 demultiplexer/decoder

Inputs			*Outputs*			
Enable	*Select*		O_3	O_2	O_1	O_0
	B	*A*				
1	×	×	1	1	1	1
0	0	0	1	1	1	0
0	0	1	1	1	0	1
0	1	0	1	0	1	1
0	1	1	0	1	1	1

Fig. 8.15 *Use of a demultiplexer/decoder as (a) a demultiplexer; (b) and (c) as a decoder*

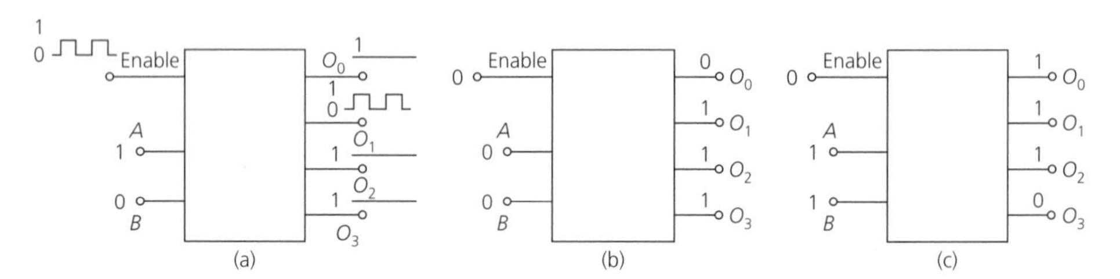

When the circuit is employed as a demultiplexer the enable pin is used as the data input terminal. The output that is selected by the input address (*AB*) will change its logical state to follow the state of the input data. (see Fig. 8.15(a)).

If the enable pin is held permanently LOW the circuit will then operate as a decoder. The signals applied to the input address pins will ensure that only the required output line is decoded. Two examples are given in Fig. 8.15(b) and (c).

The 74LS138/74HC138 has three enable inputs, any one of which can act as the data input when the device is used as a demultiplexer. G_1 is active-HIGH while $\overline{G}_{2A}$ and $\overline{G}_{2B}$ are both active-LOW. When G_1 is used the output data is inverted; if a G_2 input is used the output is true. The two G_2 inputs can be connected together to act as a data input terminal.

Code converters

A code converter is a circuit that takes an input in one code and produces an output in some other code. It is usual to refer to code converters as decoders; some examples follow

TTL: 74LS42 BCD-to-decimal decoder. 74LS47 BCD-to-7 segment decoder.
CMOS: 74HC42 BCD-to-decimal decoder.

Table 8.8 ***BCD-to-decimal decoder***

Decimal	*BCD input*				*Decimal output*									
	D	*C*	*B*	*A*	*9*	*8*	*7*	*6*	*5*	*4*	*3*	*2*	*1*	*0*
0	0	0	0	0	1	1	1	1	1	1	1	1	1	0
1	0	0	0	1	1	1	1	1	1	1	1	1	0	1
2	0	0	1	0	1	1	1	1	1	1	1	0	1	1
3	0	0	1	1	1	1	1	1	1	1	0	1	1	1
9	1	0	0	1	0	1	1	1	1	1	1	1	1	1

10–15 are don't cares

Fig. 8.16 *A 74LS42 BCD-to-decimal decoder: (a) pinout and (b) logic symbol*

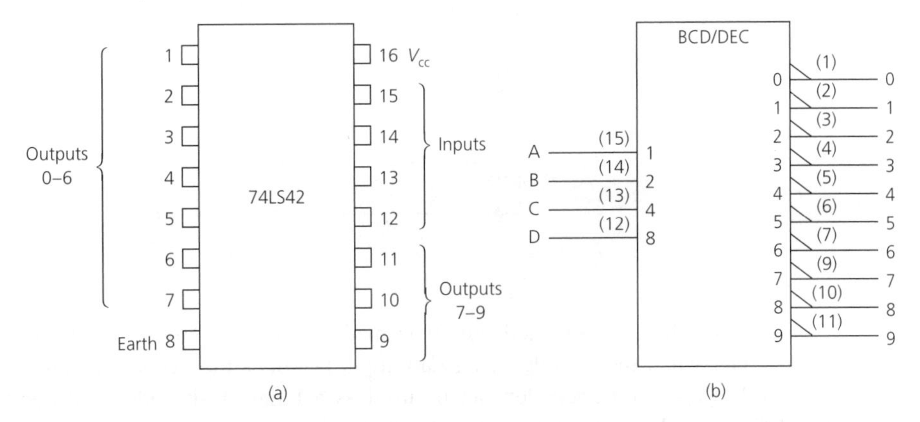

The operation of each of these devices is explained fully by its truth table and, as an example, Table 8.8 gives the truth table for the 74LS42 or 74HC42 BCD-to-decimal decoder.

A BCD-to-decimal decoder is also known as a 4-to-10 line decoder since it has a 4-bit input and its output is one of the 10 decimal digits. Figure 8.16(a) gives the pinout of the 74LS42 BCD-to-decimal decoder and Fig. 8.16(b) gives its logic symbol. The qualifying symbol for a decoder, encoder or code converter is *X*/*Y*, where *X* is the input code and *Y* is the output code. *X* and *Y* can be replaced by letters that indicate the type of code, e.g. BCD/DEC.

Decoders

A decoder can detect a code at its input and deliver a single output that indicates the presence of that code. The code is the address of the output to be made active. The applications of decoders include: alerting a system that a specified input has arrived, selecting memory chips in a microprocessor system, or a number comparator.

The operation of a decoder is similar to that of a demultiplexer in that there is only one active output at a time. However, no data is routed to the output and the address is

Fig. 8.17 *A 74LS154 4-to-16 line decoder used as a binary-to-hex converter*

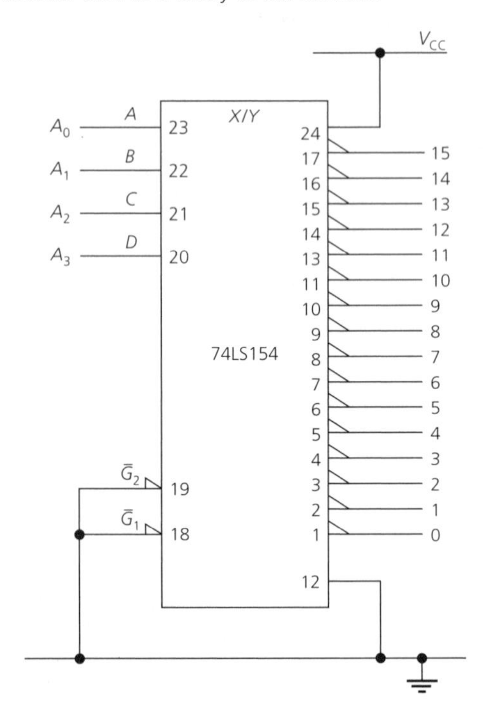

decoded when the selected output is made active. A demultiplexer can be used as a decoder when the demultiplexer data input becomes the decoder enable input.

A 4-to-16 line decoder can be used as a binary-to-hexadecimal converter. Figure 8.17 shows how the 74LS154 is connected for this purpose. Each output, 0 through to 15, is normally HIGH but goes LOW when it is selected by the digital word applied to the inputs A_0, A_1, A_2 and A_3. The two enable pins $\overline{G}_1$ and $\overline{G}_2$ can be connected to earth as shown, or they may have the clock connected to them to prevent glitches occurring.

EXAMPLE 8.8

The logic symbol of an IC is shown in Fig. 8.18. (a) What kind of device is it? (b) When do all outputs go HIGH? (c) What input combination will take output (i) 4 and (ii) 15 LOW?

Solution

(a) A 4-to-16 decoder with active-LOW outputs.
(b) Either $\overline{G}_1$ or $\overline{G}_2$ or both are HIGH.
(c) $\overline{G}_1$ and $\overline{G}_2$ both LOW and (i) $A = \overline{B} = D = 0$, $C = 1$; (ii) $A = B = C = D = 1$ (*Ans.*)

Fig. 8.18

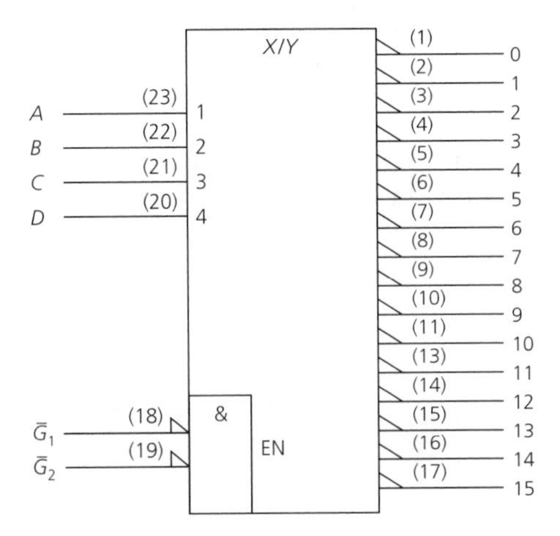

Fig. 8.19 *Number detector*

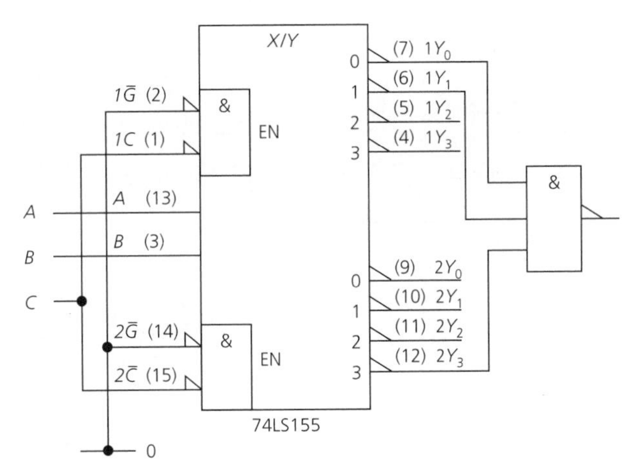

Number detector

A decoder can be employed to determine whether a 3-bit input number is within a predetermined range. The logic symbol of the 74LS155 dual 2-to-4 decoder/demultiplexer is shown in Fig. 8.19 with connections made to give a number detector. The circuit has inputs A, B which are common to both parts of the device and individual inputs $1C$ and $2C$. Data applied to $1C$ is inverted at its outputs and data applied to $2\bar{C}$ is not inverted. If the $1\bar{G}$ and $2\bar{G}$ inputs are both connected to 0 V and inputs $1C$ and $2C$ are linked together, a 3-bit binary input number will take one of the outputs LOW. When C is 0 (and, hence, the number is less than 4), the upper half of the circuit is active, when C is 1 (and the number is between 4 and 7) the lower half of the circuit becomes active. To use the decoder as a number detector the numbers whose presence is to be signalled can be connected to the inputs of a NAND gate. The circuit shown in Fig. 8.19 will signal the presence of input numbers 3, 6 and 7.

PRACTICAL EXERCISE 8.3

To investigate the action of the 74LS139 (or 74HC139) dual 2-to-4 decoder/demultiplexer (a) as a demultiplexer and (b) as a decoder.
Components and equipment: One 74LS139 (or 74HC139) dual 2-to-4 demultiplexer/decoder. Four 270 Ω resistors. Four LEDs. Power supply. Breadboard.
Procedure:

Fig. 8.20

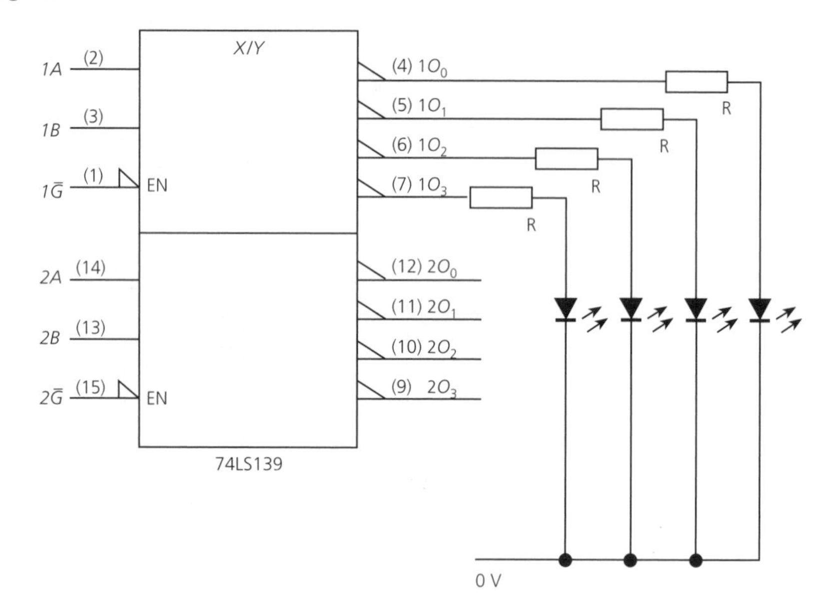

(a) Connect the circuit shown in Fig. 8.20. When the device is used as a decoder the enable input $1\overline{G}$ is held LOW.

(b) Connect +5 V to the enable pin $1\overline{G}$. Apply, in turn, all four combinations of HIGH and LOW voltage to inputs *1A* and *1B* and note that all four LEDs glow continuously. What does this indicate?

(c) Apply 0 V to the enable pin $1\overline{G}$. Then apply, in turn, the four possible combinations of 0 V and +5 V to inputs *A* and *B*. Each time a different LED should turn OFF. Each time note which LED is not lit.

(d) Modify the circuit so that the LEDs glow visibly when their associated output is LOW and turn OFF when that output is HIGH.

(e) To use the circuit as a demultiplexer the input data must be applied to the enable pin $1\overline{G}$. The input data will then be routed to the output that has been specified by the address applied to inputs *A* and *B*.

(f) Select the output circuit $1O_2$ by applying 0 V to input *1A* and +5 V to input *1B*. Now apply a low-frequency pulse waveform to the enable pin $1\overline{G}$ and observe that only one LED flashes ON and OFF.

PRACTICAL EXERCISE 8.4

To build and then test a decoder constructed using NAND gates only.

Components and equipment: two 74LS00 (or 74HC00) quad 2-input NAND gate ICs. One 270 Ω resistor. One LED. Power supply.

Procedure:

Fig. 8.21

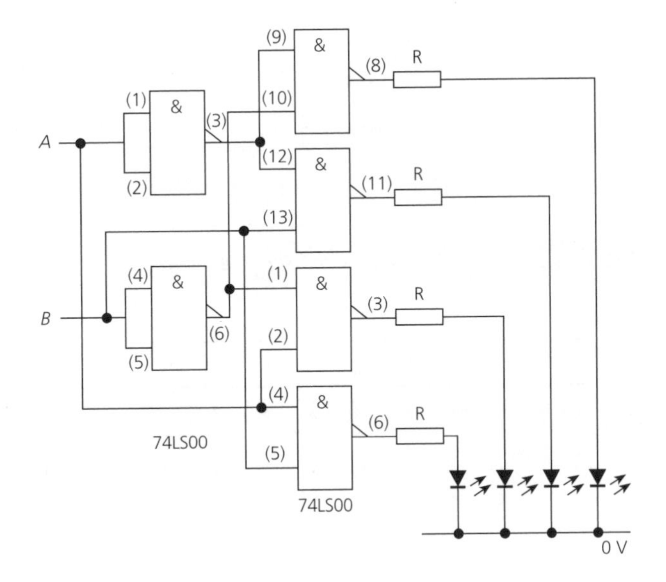

(a) Build the circuit shown in Fig. 8.21. Look up the pinout of the '00 in Fig. 3.24(a) on p. 61.
(b) Apply each of the possible combinations of 1 and 0 to the inputs A and B. Each time note the logical state of the LED.
(c) Write down the truth table of the circuit and then use it to deduce the function of the circuit.

Encoders

A code converter that converts from either decimal or hexadecimal into some other code is usually known as an encoder. An encoder performs the opposite function to a decoder. Only one of its inputs is active at a time and this input produces a specified output data word. The main application of encoders is in conjunction with keyboards, where a single key operation must produce a unique binary code. If two, or more, keys are pressed simultaneously two inputs become active together and will give an incorrect output. To avoid this happening, a *priority encoder* is often employed. If more than one input is applied to a priority encoder it encodes the input with the highest magnitude and ignores all other lower magnitude inputs.

Fig. 8.22 *Pinout of the 74LS148 8-to-3 line priority encoder*

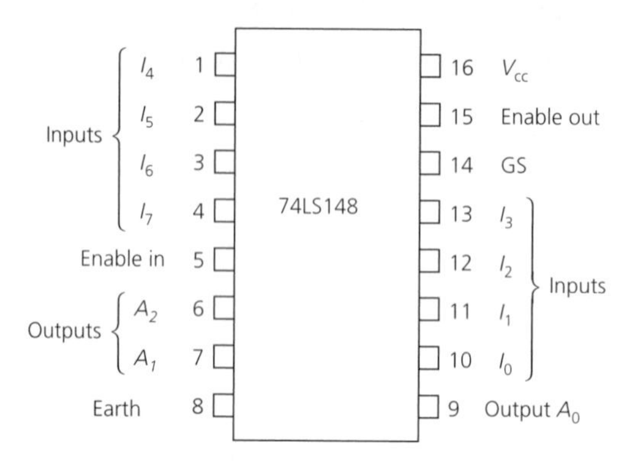

Table 8.9 *The 8-to-3 line priority encoder*

Enable in	*Inputs*								*Outputs*				*Enable out*
	I_7	I_6	I_5	I_4	I_3	I_2	I_1	I_0	GS	A_2	A_1	A_0	
1	×	×	×	×	×	×	×	×	1	1	1	1	1
0	1	1	1	1	1	1	1	1	1	1	1	1	0
0	×	×	×	×	×	×	×	0	0	0	0	0	1
0	×	×	×	×	×	×	0	1	0	1	0	0	1
0	×	×	×	×	×	0	1	1	0	0	1	0	1
0	×	×	×	×	0	1	1	1	0	1	1	0	1
0	×	×	×	0	1	1	1	1	0	0	0	1	1
0	×	×	0	1	1	1	1	1	0	1	0	1	1
0	×	0	1	1	1	1	1	1	0	0	1	1	1
0	0	1	1	1	1	1	1	1	0	1	1	1	1

The 74LS148 and 74HC148 are 8-to-3 line encoders. Figure 8.22 gives its pinout and Table 8.9 its truth table. It has eight active-LOW inputs and three active-LOW outputs and it performs octal-to-binary conversions.

Binary adders

A binary adder is a circuit which is able to add together two binary numbers. The *half-adder* adds two inputs *A* and *B* to produce a *sum* and a *carry* but it is unable to take into account any carry from a previous stage. The full-adder also adds together two binary numbers but it can take account of any input carry as well. Its block diagram is shown in Fig. 8.23. The truth table of a full-adder is given in Table 8.10.

Fig. 8.23 *Full-adder*

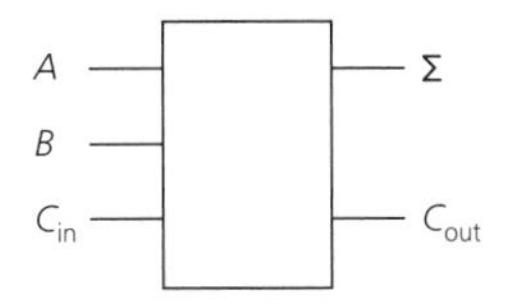

Table 8.10 *Full-adder*

Carry-in (C_{in})	Inputs		Sum	Carry-out (C_{out})
	B	A		
0	0	0	0	0
0	0	1	1	0
0	1	0	1	0
0	1	1	0	1
1	0	0	1	0
1	0	1	0	1
1	1	0	0	1
1	1	1	1	1

The Boolean equation describing the logical operation of the full-adder can be derived from the truth table. It is

$$\begin{aligned} S &= A\bar{B}\bar{C}_{in} + \bar{A}B\bar{C}_{in} + \bar{A}\bar{B}C_{in} + ABC_{in} \\ &= (A\bar{B} + \bar{A}B)\bar{C}_{in} + (AB + \bar{A}\bar{B})C_{in} \\ &= (A \oplus B)\bar{C}_{in} + (\overline{A \oplus B})C_{in} \end{aligned} \quad (8.3)$$

$$= A \oplus B \oplus C_{in} \quad (8.4)$$

$$\begin{aligned} C_{out} &= AB\bar{C}_{in} + A\bar{B}C_{in} + \bar{A}BC_{in} + ABC_{in} \\ &= AB + (A \oplus B)C_{in} \end{aligned} \quad (8.5)$$

The implementation of equations (8.4) and (8.5) can be achieved using gates (see Fig. 8.24). Usually, however, two 4-bit numbers, or bigger, are to be added and then an MSI circuit is employed.

To build a 4-bit adder, four full-adders are required connected in the way shown in Fig. 8.25. The least-significant input bits A_1 and B_1 as well as the carry-in C_{in0} come into the full-adder FA1. The carry-out from full-adder FA1 is connected to the carry-in terminal of full-adder FA2 where it is added to inputs A_2 and B_2. Similarly, the carry-out from full-adders FA2 and FA3 are connected to the carry-in terminals of the next more-significant full-adder. The sum Σ produced by each full-adder is available as well as the carry-out of the most-significant full-adder.

It can take some time for the carries to ripple through the circuit and MSI versions of 4-bit full-adders incorporate *fast look-ahead carry* circuitry to speed up the process. The 74LS283 includes fast look-ahead carry.

Fig. 8.24 *Full-adder using NAND and exclusive-OR gates*

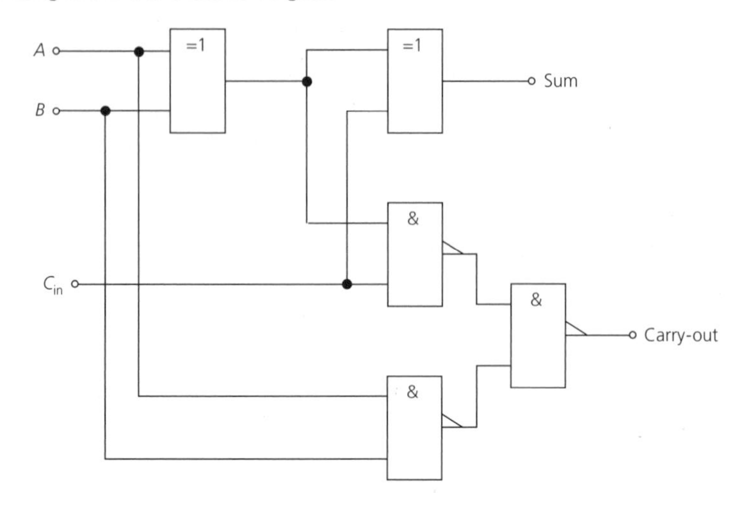

Fig. 8.25 *A 4-bit full-adder*

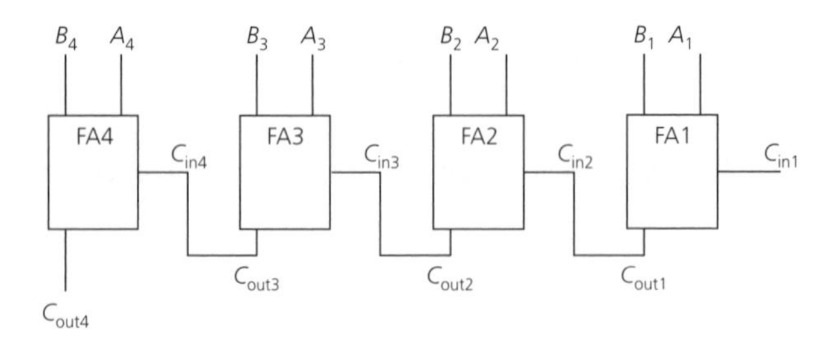

Fig. 8.26 *Pin connections of the 74LS283 binary adder*

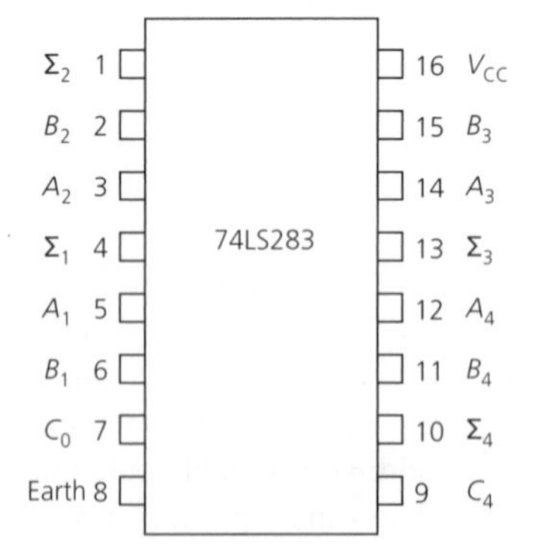

The pin connections of the 74LS83 is shown in Fig. 8.26. The 74LS83 is able to accept two 4-bit numbers A and B *and* a carry C_0 from a previous stage. The sum of the two

Table 8.11

Input									*Output*				
A				*B*				C_0	Σ_4	Σ_3	Σ_2	Σ_1	C_4
1	0	0	1	0	0	1	1	0	1	1	0	0	0
1	0	1	0	0	1	1	1	0	0	0	0	1	1
1	0	1	0	0	1	1	1	1	0	0	1	0	1

Fig. 8.27 *Two 74LS83 4-bit binary adders connected to give 8-bit addition*

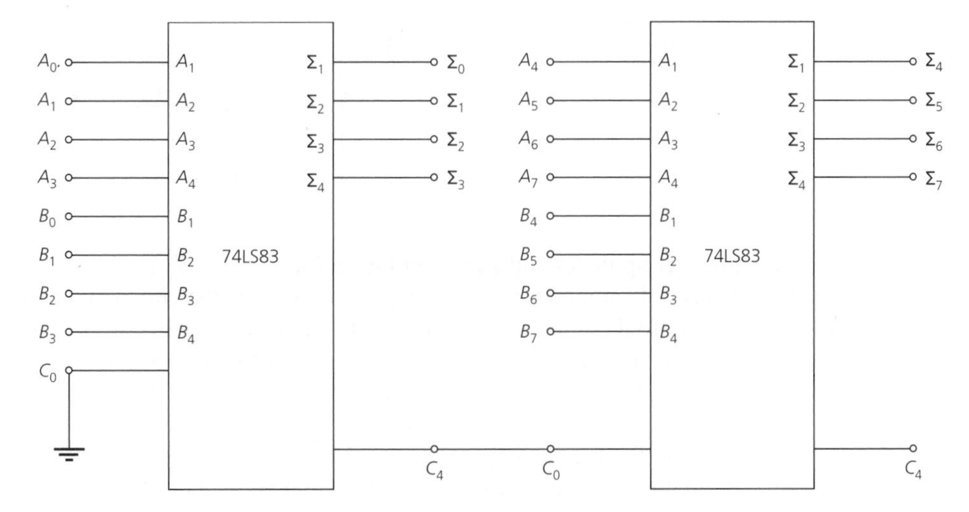

numbers appears at the terminals labelled as Σ_1, Σ_2, Σ_3 and Σ_4 with any carry-out appearing at terminal C_4. If (a) the sum of A and B and C_0 is anywhere between 0 and 15 the output carry C_4 will be 0; and if (b) the sum of A and B and C_0 is anywhere between 16 and 31 the output carry is 1 and the Σ outputs indicate a number equal to the sum of A and B *minus* 16. Table 8.11 gives some examples.

Should the two numbers to be added have more than 4 bits two 74LS83s can be cascaded in the manner shown in Fig. 8.27 for 8-bit arithmetic. The carry-out C_4 of the least-significant stage is connected to the carry-in C_0 of the next most-significant stage.

PRACTICAL EXERCISE 8.5

To investigate the action of a full-adder.

Components and equipment: one 74LS283 full-adder IC, Two 270 Ω resistors. Five LEDs. Power supply. Breadboard.

Procedure:

Fig. 8.28

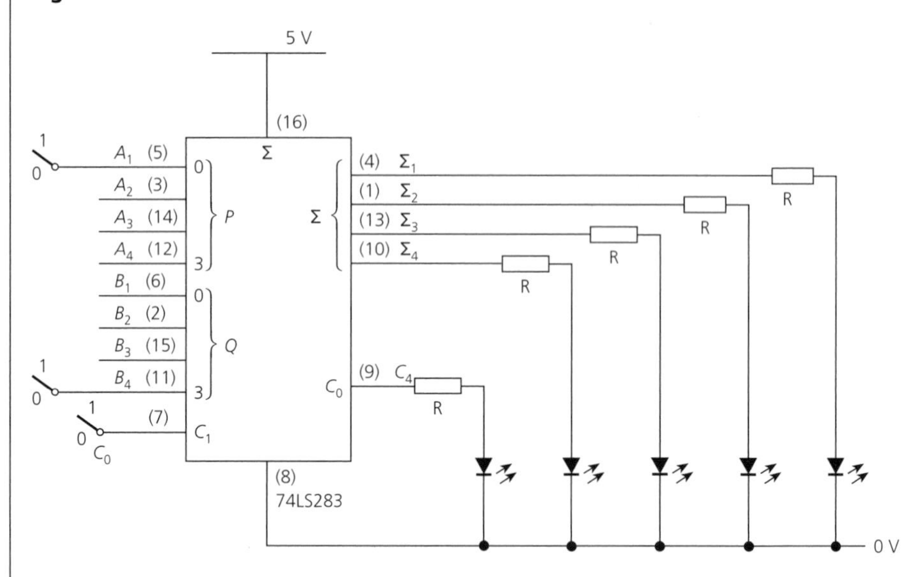

(a) Connect up the circuit shown in Fig. 8.28.

(b) Apply a 4-bit number to the *A* input terminals of the circuit and another 4-bit number to the *B* input terminals. Do not apply a signal to the carry-in terminal. Note the logical state of the LEDs connected to the output of the 4-bit adder. Write down the two input binary numbers and the logical states of the sum and carry-out LEDs. Now apply a binary 1 voltage to the carry-in terminal (C_0) and note the logical state of the two LEDs.

(c) Write down a table that shows the two input numbers and their sum both with, and without, the carry-in. List also any carry-out that occurs.

(d) Repeat the procedure for three other pairs of binary numbers of your choosing.

(e) Explain how (i) two 8-bit numbers can be added together, and (ii) how one 4-bit number can be subtracted from another making clear what circuit modifications would be required.

Twos complement subtraction

Subtraction using 2s complement arithmetic is performed by changing the number to be subtracted into a negative number by complementing each bit and then adding 1. The complementing of the 4-bit number can be achieved by the circuit shown in Fig. 8.29. When the control signal is at logic 1 the output bits are complemented. The top exclusive-OR gate, for example, has an output *F* of either $F = D_0 0 + \bar{D}_0 1$, or $F = D_0 1 + \bar{D}_0 0$.

To obtain the 2s complement the circuit shown in Fig. 8.29 is employed and 1 is added to its output. Figure 8.30 shows the circuit for a 4-bit binary subtractor.

Fig. 8.29 *Complementing B inputs when control signal = 1*

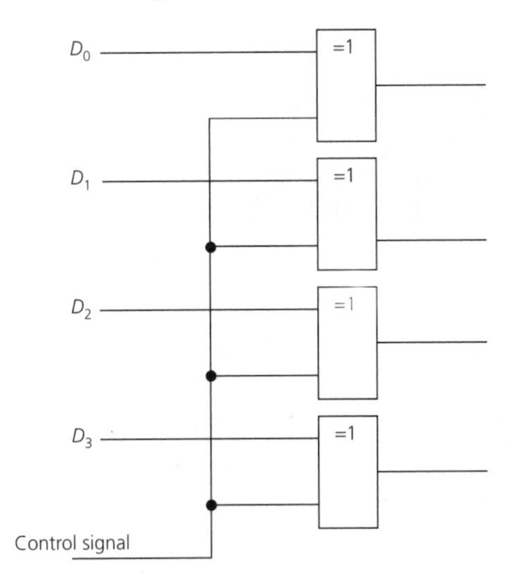

Fig. 8.30 *Four-bit binary subtraction*

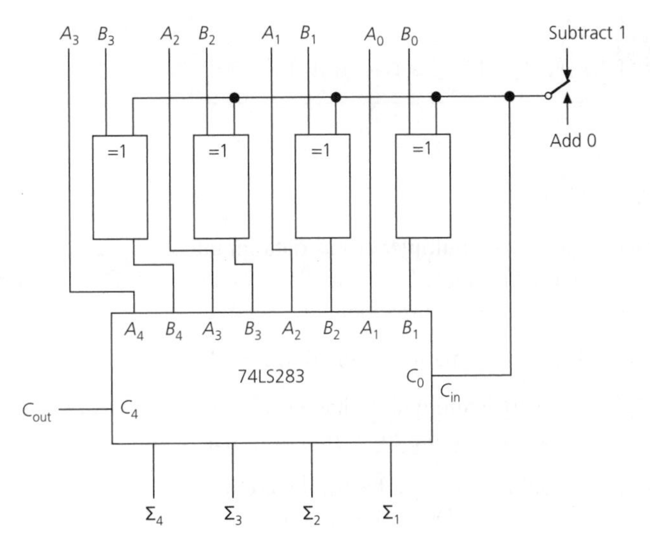

EXAMPLE 8.9

Apply the binary signals $A = 1011$ and $B = 0101$ to the circuit shown in Fig. 8.30 and confirm that the output is correct (a) when adding, and (b) when subtracting.

Solution

(a) Figure 8.31(a) shows the logic levels at the inputs and outputs of the exclusive-OR gates and the signals at the inputs A_1, B_1 etc. to the full-adder. The output of the circuit is then 10000 (*Ans.*)

Fig. 8.31

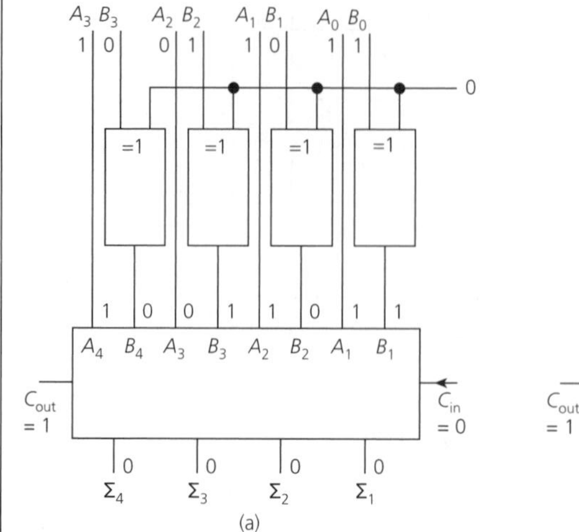

(a)

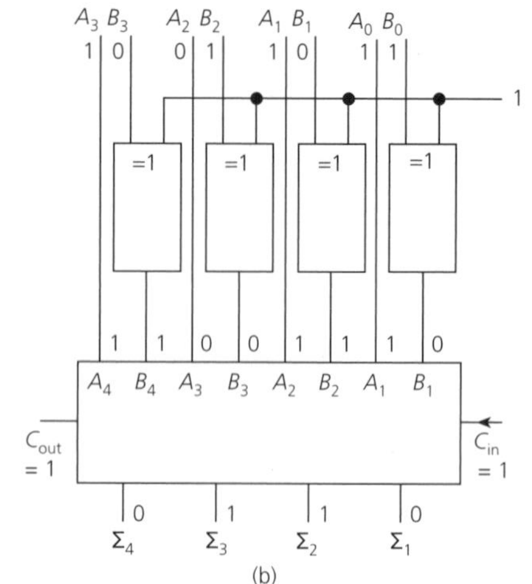

(b)

To check: 1011 = 11 and 0101 = 5 and their sum = 16 or 10000.

(b) Figure 8.31(b) shows the signals in the circuit. The output is now 10110 (*Ans.*)

[To check: 11 − 5 = 6 = 0110.]

EXERCISES

8.1 An 8-to-1 multiplexer has data inputs D_0, D_1, D_3, D_4, and D_5 connected to logic 1 and data inputs D_2, D_6 and D_7 connected to logic 0. Determine the logic function that has been implemented.

8.2 Implement the logic function $F = AB + \bar{A}\bar{B}$ on a 4-to-1 multiplexer.

8.3 An 8-to-1 multiplexer has $D_0 = D_1 = D_2 = D_3 = D_4 = D_5 = 1$ and $D_6 = D$ and $D_7 = 0$. Determine the logic function that is implemented.

8.4 A multiplexer is to be employed to allow any one of four digital computers access to a printer. Draw the circuit.

8.5 Implement the logic function $F = \bar{A}\bar{B}\bar{C} + \bar{A}B\bar{C} + AB\bar{C} + A\bar{B}C$ on a 8-to-1 multiplexer.

8.6 Implement $F = \bar{A}\bar{B}\bar{C} + \bar{A}B\bar{C} + AB\bar{C} + A\bar{B}C$ using one 4-to-1 multiplexer.

8.7 Draw the circuit of a 74LS138 or 74HC138 demultiplexer used to connect a data source to any one of four different destinations, *A*, *B*, *C* or *D*.

8.8 Use a multiplexer to implement the function $F = AB + \bar{B}C + \bar{A}BC$

8.9 Use a multiplexer to implement the function $F = \bar{A}\bar{B}\bar{C}\bar{D} + \bar{A}B\bar{C}\bar{D} + A\bar{B}C\bar{D} + \bar{A}BC\bar{D} + \bar{A}\bar{B}\bar{C}D + AB\bar{C}D + A\bar{B}CD + \bar{A}BCD + ABCD$.

8.10 Implement the function $F = \bar{A}\bar{B}\bar{C} + BC + AC$ using a multiplexer.

8.11 Write down the truth table of a BCD-to-decimal decoder. Explain how the 74HC42 4-to-10 decoder can perform this conversion.

8.12 Show how four 74LS154 ICs can be connected to decode 64 addresses from 00H to 40H.

8.13 (a) Show that $C(\overline{A \oplus B}) + \bar{C}(A \oplus B) = A \oplus B \oplus C$.

(b) The truth table for a full-adder is shown in Table 8.10. From it obtain the Boolean expression describing the operation of the circuit. Show that this expression can be written as sum $= A \oplus B \oplus C$ and carry $= C(A \oplus B) + AB$.

8.14 The inputs applied to a 74LS283 full-adder are (a) $A_4A_3A_2A_1 = 0101$, $B_4B_3B_2B_1 = 1001$, $C_{in} = 1$, (b) $A_4A_3A_2A_1 = 0111$, $B_4B_3B_2B_1 = 1011$, $C_{in} = 0$, and (c) $A_4A_3A_2A_1 = 1001$, $B_4B_3B_2B_1 = 0110$, $C_{in} = 1$. Determine the sum outputs and the output carry in each case.

8.15 The adder/subtractor circuit shown in Fig. 8.30 is to be used to subtract 6 from 11.

(a) In which position should the switch be?

(b) What is the carry-in?

(c) What are the inputs applied to $A_4A_3A_2A_1$ and $B_4B_3B_2B_1$?

(d) What is the output of the circuit (i) in binary and (ii) in decimal?

(e) Is there a carry-out and what happens to it?

8.16 Show how four 74LS151 8-to-1 multiplexers can be connected to give a 32-to-1 multiplexer.

8.17 Draw the block diagram of a 3-to-8 line decoder and, with the aid of a truth table, explain its operation.

(a) If the input signal is 101 which output line is selected?

(b) If output line 7 is to be selected what should be the input signal?

9 Latches and flip-flops

After reading this chapter you should be able to:

(a) Explain the difference between a latch and a flip-flop.
(b) Explain the operation of an S-R latch.
(c) Understand how to use a J-K flip-flop.
(d) Explain the operation of a master–slave flip-flop and say why modern devices are nearly all edge-triggered.
(e) Understand the operation, and the use, of a D latch and a D flip-flop.
(f) Understand the operation of a T flip-flop.

The *latch* and the *flip-flop* are circuits that have two stable states and are able to remain in either one for an indefinite length of time. The circuit will change state only when a switching operation is initiated by a trigger pulse applied to the appropriate terminal. Once switched the flip-flop will remain in its other stable state until another trigger pulse is received that will force it to revert to its original state. A latch or flip-flop has two output terminals that are always labelled as Q and $\bar{Q}$ since the logical state of one output terminal is *always* the complement of the logical state of the other terminal. When the circuit is in the state $Q = 1$, $\bar{Q} = 0$ it is said to be set; conversely, when $Q = 0$ and $\bar{Q} = 1$ the circuit is said to be reset.

Two types of latch are available, the S-R (or $\bar{\text{S}}$-$\bar{\text{R}}$) and the D; and three types of flip-flop, the J-K, the D and the T. A latch is *level triggered*, i.e. its operation is initiated by the voltage level applied to its input and not by any change in the clock. The operation of a flip-flop is triggered by the transition of an input signal, either the leading edge (0 to 1), or the trailing edge (1 to 0), of the clock waveform. The *clock* is a rectangular voltage pulse waveform of constant frequency. The symbol of a latch or flip-flop indicates the triggering method.

The majority of modern IC flip-flops are edge-triggered because the much smaller time in which changes in state can occur reduces the possibility of false operation. Figure 9.1 shows the basic symbols for the triggering methods, the active-HIGH level triggered device (Fig. 9.1(a)), operates when the clock is HIGH, and the active-LOW circuit operates when the clock is LOW. The other parts of this figure show the symbols for edge-triggered devices, a leading-edge-triggered circuit is indicated by a small wedge (Fig. 9.1(c)) and a trailing-edge-triggered circuit has both the small wedge and a triangle at the clock input (see Fig. 9.1(d)).

Fig. 9.1 *(a) Active-HIGH; (b) active-LOW; (c) leading-edge-triggered and (d) trailing-edge-triggered flip-flops*

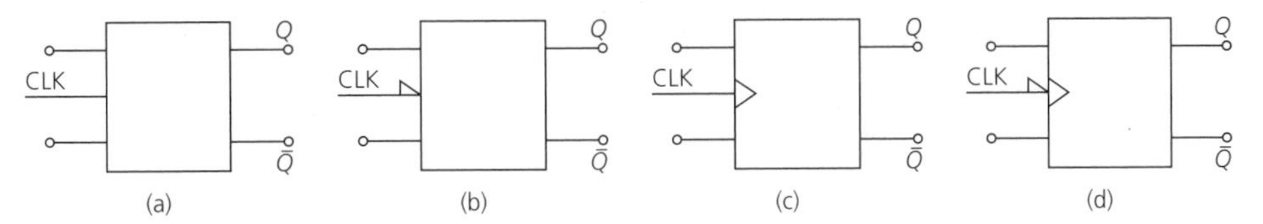

Fig. 9.2 *Symbol for an S-R latch*

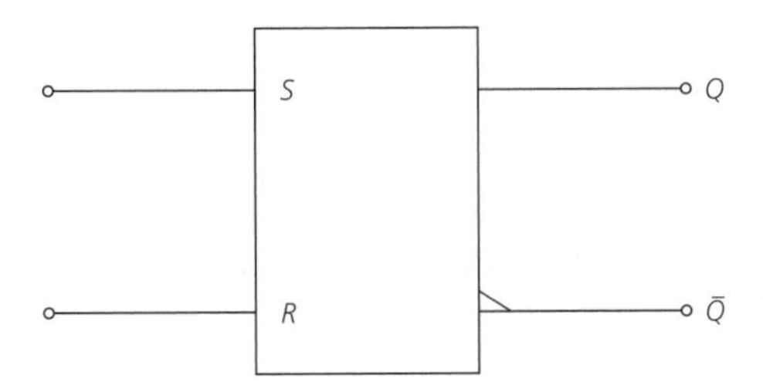

Table 9.1 ***S-R latch***

S	R	Q	Q^+	
0	0	0	0	no change in state
0	0	1	1	
1	0	0	1	set operation
1	0	1	1	
0	1	0	0	reset operation
0	1	1	0	
1	1	0	×	indeterminate operation
1	1	1	×	

The S-R latch

The *S-R latch* has two input terminals labelled as S (for set) and R (for reset) and two output terminals labelled as Q and $\bar{Q}$.

The symbol for an S-R latch is given in Fig. 9.2 and its truth table in Table 9.1. In the table the symbol Q represents the *present* state of the Q output terminal, and Q^+ represents the *next* state of the terminal *after* a set (S) or reset (R) pulse has been applied to the appropriate input terminal.

If the inputs S and R are both at logical 0, the output state of the circuit will not change and the latch can be said to have stored one bit of information.

When $S = 1$ and $R = 0$, the latch will change state if $Q = 0$ to $Q^+ = 1$ but it will not change state if $Q = 1$. Conversely, if $Q = 1$, a reset pulse, i.e. $R = 1$, $S = 0$, will cause the circuit to switch to $Q^+ = 0$, but the $R = 1$, $S = 0$ input condition will have no effect on the circuit if $Q = 0$ and $\bar{Q} = 1$.

Fig. 9.3 *(a) S-R flip-flop using two NOR gates and (b) timing diagram*

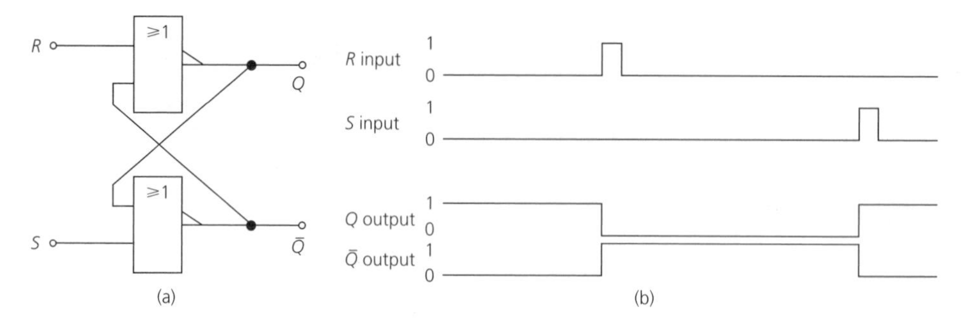

Thus, $S = 1$, $R = 0$ will always produce the state $Q^+ = 1$; and $S = 0$, $R = 1$ will always give $Q^+ = 0$, regardless of the original state of Q.

Lastly, if $R = S = 1$ the latch may or may not change state and the operation of the circuit is said to be *indeterminate*. This condition is denoted in the table by ×.

NOR gate S-R latch

Figure 9.3(a) shows how an S-R latch can be made by connecting together two NOR gates. The output of a 2-input NOR gate is 1 only when both of its inputs are at 0; if either or both of its inputs is at logic 1, the output state will be 0.

Suppose the circuit is initially set, i.e. $Q = 1$, $\bar{Q} = 0$. If $S = R = 0$ the upper gate will then have both of its inputs at 0 and so Q remains at 1. The lower gate has one input at 0 and the other at 1, hence, its output $\bar{Q} = 0$. If now a pulse is applied to the input (reset) terminal only, giving $S = 0$, $R = 1$, the upper gate will have one input (R) at 1 and the other ($\bar{Q}$) at 0; the output (Q) of this gate will then become equal to logic 0. Both inputs to the lower gate are now at 0 and so the output of this gate, the $\bar{Q}$ terminal, becomes logic 1. Thus the circuit has switched states from $Q = 1$, $\bar{Q} = 0$ to $Q = 0$, $\bar{Q} = 1$, i.e. the latch has been *reset*.

With the latch in the reset condition, suppose that the input conditions change to $S = 1$, $R = 0$. The lower gate will now have one input (S) at 1 and the other (Q) at 0 and so its output ($\bar{Q}$) will be 0. This means that the upper gate now has both of its inputs at 0 and so its output (Q) becomes 1. The timing diagram is shown in Fig. 9.3(b).

Whether the latch is set or reset the application of a pulse to both the S and R inputs simultaneously may or may not switch the circuit; in other words the operation of the circuit is unpredictable.

NAND gate S-R latch

The S-R latch can also be constructed using four NAND gates interconnected in the manner shown in Fig. 9.4. The output of a NAND gate is 0 only if both of its input terminals are at 1; if either or both inputs is at 0, the output will be 1. Suppose that, initially, the latch is set and $S = R = 0$. Both inputs to the lower gate are 1 and, hence,

Fig. 9.4 *S-R latch using four NAND gates*

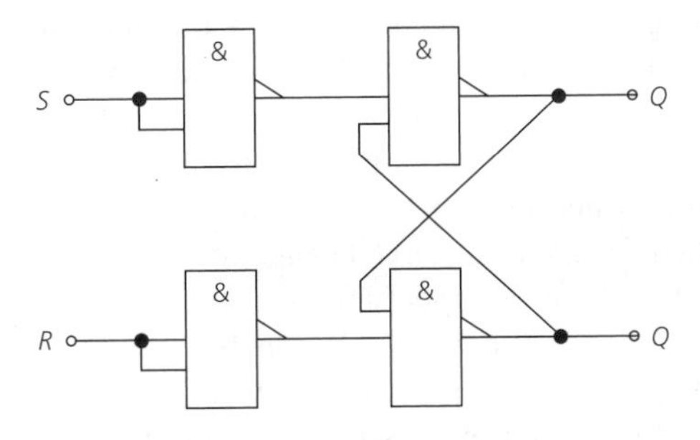

$\bar{Q}$ remains at 0. This means that the upper gate has one of its inputs ($\bar{S}$) at 1 and the ($\bar{Q}$) at 0 and so its output terminal remains at 1. Resetting the circuit requires the input condition $S = 0$, $R = 1$. Then, the lower gate will have one input ($\bar{R}$) at 0 and its other input (Q) at 1 and so its output ($\bar{Q}$) will become 1. This results in the upper gate having both its inputs ($\bar{S}$ and $\bar{Q}$) at 1 and its output becomes $Q = 0$. To set the circuit $S = 1$, $\bar{R} = 0$; then the upper gate has one input ($\bar{S}$) at 0 and the other ($\bar{Q}$) at 1 and, hence, its output is $Q = 1$. Finally, the lower gate has now both of its inputs at 1 and it switches to give $\bar{Q} = 0$. Once again the input state $S = R = 1$ is indeterminate and may result in the latch either switching or its output state remaining unchanged.

$\bar{S}$-$\bar{R}$ latch

If the 2-input NAND gate inverters are not included in the circuit an $\bar{S}$-$\bar{R}$ latch is obtained. This circuit is active-LOW. When the $\bar{S}$ input goes LOW, the Q output becomes HIGH, and if $\bar{R}$ goes LOW, Q becomes LOW. The two inputs must not be taken LOW simultaneously because the output will then be unpredictable. The no-change state occurs when $\bar{S} = \bar{R} = 1$.

Clocked S-R latch

Very often it is desirable for the set and reset operations to occur at particular instants in time determined by the *clock*. A *clocked* latch will change state only when a clock pulse is received. The truth table of a clocked S-R latch is given in Table 9.2 and its symbol in Fig. 9.5(a).

A method of clocking the NAND gate type of S-R latch is given in Fig. 9.5(b). Whenever the clock is 0, the outputs of both gates must be 1 whatever the logical states

Table 9.2 ***Clocked S-R latch***

S	0	0	1	1	0	0	1	1	0	0	1	1	0	0	1	1
R	0	0	0	0	1	1	1	1	0	0	0	0	1	1	1	1
Clock	0	0	0	0	0	0	0	0	1	1	1	1	1	1	1	1
Q	0	1	0	1	0	1	0	1	0	1	0	1	0	1	0	1
Q^+	0	1	0	1	0	1	0	1	0	1	1	1	0	0	×	×

of the S and the R inputs. Suppose the latch is set; then $Q = 1$, $\bar{Q} = 0$ and so the upper right-hand gate has one input at 1 ($CS + \bar{C}\bar{S}$) and one at 0 ($\bar{Q}$) and, hence, the Q output remains at 1. The lower right-hand gate has both of its inputs at logic 1 and so the $\bar{Q}$ output remains unchanged at 0. Only when the clock input is 1 can the appropriate input gate (S or $R = 1$) have an output at 0 for a possible switching action to be initiated. In other words, when the clock input is at logic 1, the circuit follows the same sequence of operations as the basic NAND gate latch of Fig. 9.4. The clock determines the times at which the S and R input signals should be effective and this is illustrated by the waveforms given in Fig. 9.6. It can be seen that the latch does not set (or reset) immediately an S (or R) pulse is received but waits until the clock changes from 0 to 1.

Clocked operation of an S-R latch, and of the other types of latch and flip-flop discussed later in this chapter, ensures that all the circuits in a system operate in synchronism with one another. Synchronous operation of a digital system is generally advantageous since it (a) leads to faster operation, and (b) avoids the transient problems which may arise in a non-synchronous system.

The clocked S-R latch is often provided with clear and pre-set terminals (Fig. 9.5(a)) that allow the normal inputs to be overridden. These inputs are always non-synchronous even if the main operation is clocked.

Fig. 9.5 *Clocked S-R latch*

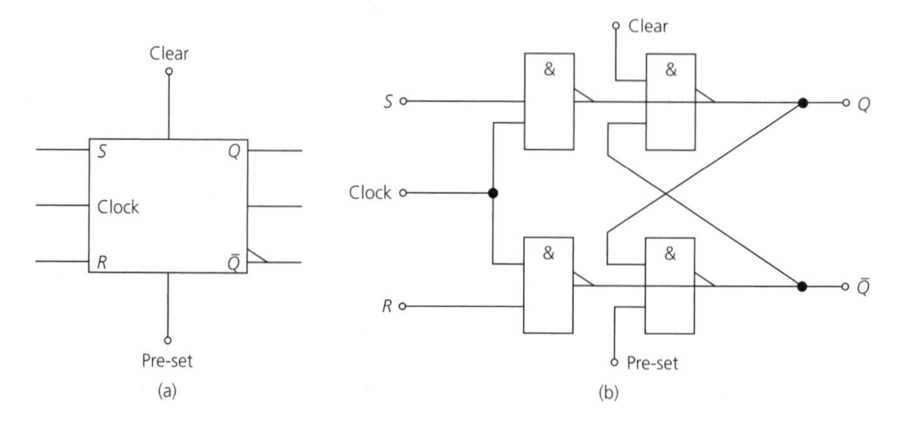

Fig. 9.6 *Waveforms in a clocked S-R latch*

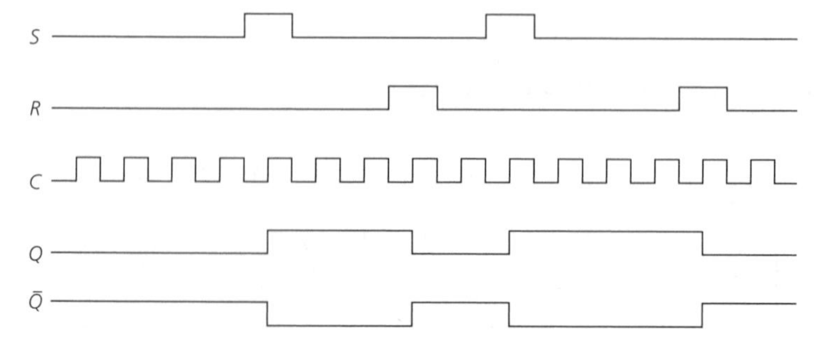

The TTL logic family includes the 74LS279A quad $\bar{S}$-$\bar{R}$ latches; the $\bar{S}$ and $\bar{R}$ inputs are normally held HIGH. When the $\bar{S}$ input goes LOW, the Q output becomes HIGH, and if $\bar{R}$ goes LOW, Q becomes LOW. Both inputs must not be taken LOW simultaneously because the output will be unpredictable.

Contact bounce

Many circuits in digital systems, such as counters and registers, are clock (pulse) operated. Only one pulse at a time should be applied to the circuit. When a mechanical switch is used some *contact bounce* often occurs. This term means that when the switch is operated to close the contacts they may open and shut several times in quick succession because of the inherent springiness of the contact arms. Contact bounce may result as a supposed single pulse actually being more like the waveform shown in Fig. 9.7. The digital circuit to which this waveform is applied may well interpret it as being several pulses instead of just one, and operate accordingly. To prevent false operation caused by contact bounce, a *de-bouncing circuit* can be employed. Figure 9.8(a) shows how an S-R latch can be used as a contact de-bouncing circuit. When the switch is in position B the latch is reset and $Q = 0$. When the switch is moved from B to position A it will bounce several times before it settles into its new position. However, immediately the first contact is made with A the latch will set and Q becomes 1. This action is shown in Fig. 9.8(b).

Fig. 9.7 *Contact bounce*

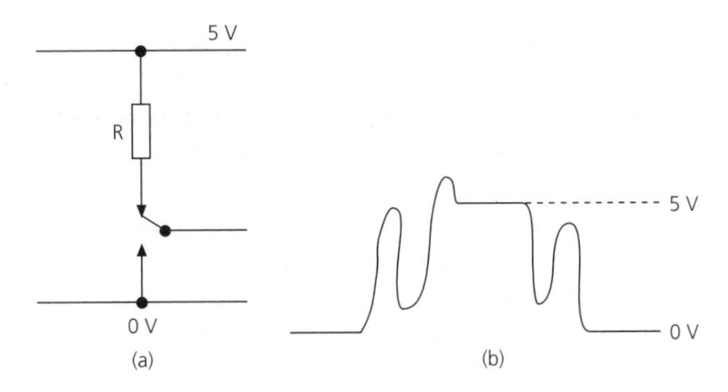

Fig. 9.8 *(a) De-bouncing circuit and (b) voltage at Q*

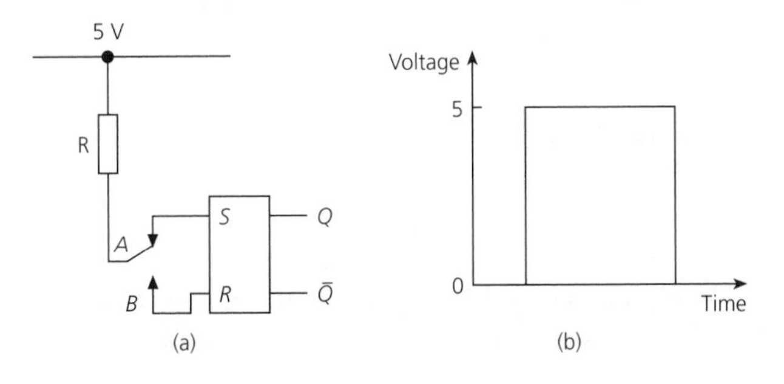

PRACTICAL EXERCISE 9.1

To investigate the operation of an S-R latch.
Components and equipment: one 74LS02 (or 74HC02) NOR gate IC. One 74LS00 (or 74HC00) NAND gate IC. Two 270 Ω resistors. Two LEDs. Power supply. Breadboard.
Procedure:

Fig. 9.9

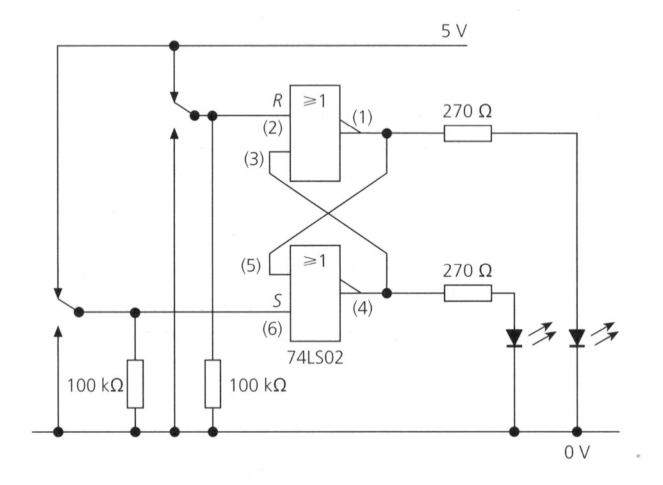

(a) Connect up the circuit shown in Fig. 9.9.
(b) With $S = 0$ connect the R input first to 0 V and then to +5 V. Note the logic state of the LEDs each time. The circuit is now reset.
(c) Apply, in turn, each of the logic states 1 and 0 to the S and R input terminals to determine whether the circuit follows the truth table shown in Table 9.1. For each combination of S and R input signals note the logic states of the two LEDs.
(d) Try the effect of applying $S = R = 1$ to the circuit, (i) when $Q = 0$, (ii) when $Q = 1$, and (iii) when $Q = 0$ again. Give the logic state of the LEDs each time. State what happened when $S = R = 1$. Did the same result occur when $S = R = 1$ and (i) $Q = 0$ and (ii) $Q = 1$?
(e) Give some reasons why condition (d) is undesirable in a practical digital system.
(f) Replace the NOR gate IC with the NAND gate IC leaving all other connections unchanged and repeat the above procedure.

The J-K flip-flop

Very often the indeterminate state $S = R = 1$ of an S-R latch cannot be permitted to occur and then a J-K flip-flop, is used. The operational difference between the S-R latch and J-K flip-flop lies in the response of the circuits to the input state $S = R = 1$. The truth table of a J-K flip-flop is shown in Table 9.3, and comparing this with the truth table of

Table 9.3 ***J-K flip-flop truth table***

J	K	Q	Q^+
0	0	0	0
0	0	1	1
1	0	0	1
1	0	1	1
0	1	0	0
0	1	1	0
1	1	0	1
1	1	1	0

Fig. 9.10 *Symbols for (a) a clocked J-K flip-flop; (b) a J-K flip-flop with clear and pre-set terminals; (c) a leading-edge-triggered J-K flip-flop and (d) a trailing-edge-triggered J-K flip-flop*

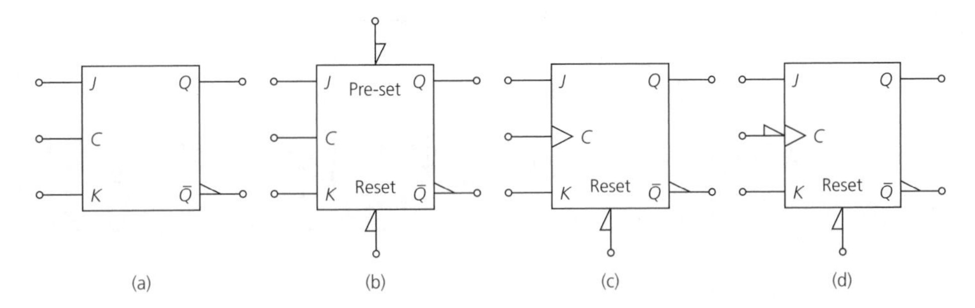

an S-R flip-flop (Table 9.1) makes it clear that the J-K flip-flop *always* changes state when $J = K = 1$. The J-K flip-flop *toggles* each time a clock pulse occurs.

The symbol for a master–slave J-K flip-flop is shown in Fig. 9.10(a). A clock input is shown since flip-flops are normally operated synchronously. Very often, clear (or reset) terminals and perhaps pre-set terminals are also provided (Fig. 9.10(b)); these are always non-synchronous. The small triangles indicate that the clear and pre-set terminals are active-LOW. A leading-edge-triggered J-K flip-flop is indicated by a small wedge on the clock input (see Fig. 9.10(c)).

Lastly, Fig. 9.10(d) shows the symbol for a trailing-edge-triggered J-K flip-flop.

Master–slave J-K flip-flops

The principle of operation of a master–slave J-K flip-flop is illustrated in the block diagram shown in Fig. 9.11. The switches S_1 and S_2 are both operated by the clock. At the leading edge of the clock waveform, switch S_1 is closed and S_2 is open so that the slave flip-flop is isolated from the master flip-flop. The input data is applied to the input (J-K) terminals of the master flip-flop and is then stored until the end of the clock pulse. When the trailing edge of the pulse occurs, the switch S_2 is closed; the logical state of the master flip-flop is transferred to the slave flip-flop and then appears at the output of the circuit.

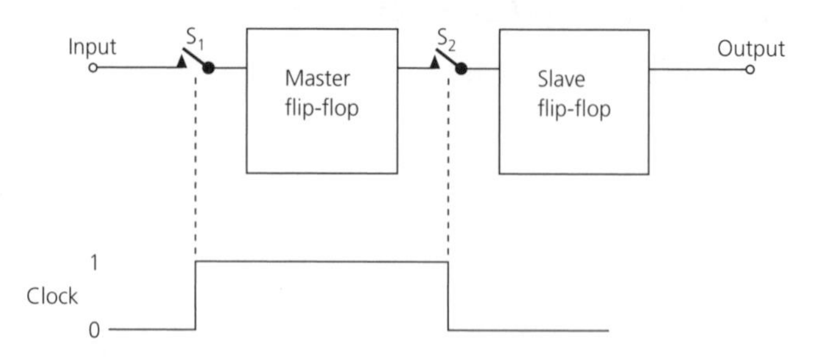

Fig. 9.11 *Master–slave principle*

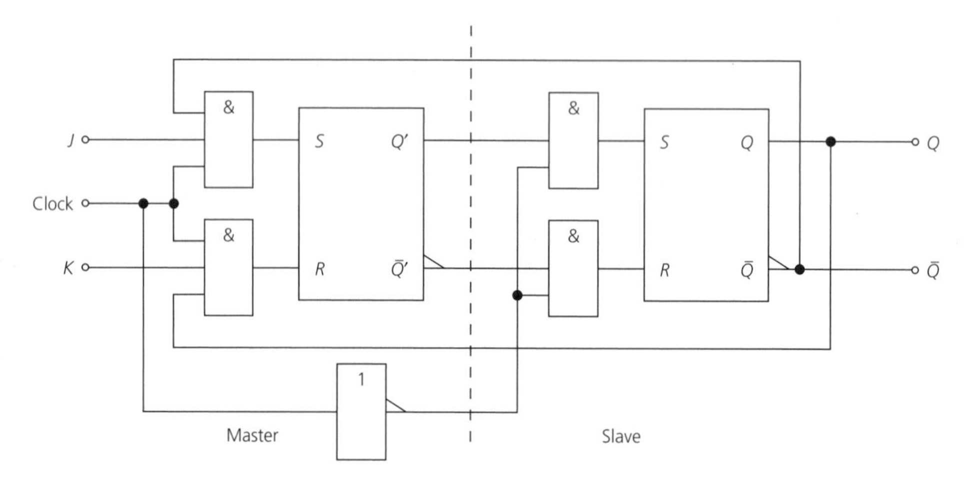

Fig. 9.12 *Master–slave J-K flip-flop*

The circuit of a master–slave J-K flip-flop is shown in Fig. 9.12. The left-hand flip-flop is the *master* and the right-hand flip-flop is the *slave*.

Suppose that initially the flip-flop is set so that $Q = 1$. When the clock is 0, the master will have $J = K = 0$ and, hence, its outputs Q' and $\bar{Q}'$ remain at 1 and 0, respectively. The clock input to the slave is inverted and so the slave has $J = 1$, $K = 0$. This condition sets the slave to retain Q and $\bar{Q}$ at 1 and 0, respectively.

When a clock pulse arrives, $C = 1$, either J or K or both could be at 1. If $J = 1$, $K = 0$, both inputs to the master are at 0 (since $\bar{Q} = 0$) and switching is not initiated. Suppose that now $J = 0$, $K = 1$; the three inputs to the lower input AND gate are all at 1 and so the master flip-flop has inputs $J = 0$, $K = 1$ applied. The master, therefore, changes state to have $Q' = 0$, $\bar{Q}' = 1$. However, since the clock is inverted, this change of state cannot be passed on to the slave, both of whose inputs remain at 0, so that at the output the condition $Q = 1$, $\bar{Q} = 0$ is retained. At the end of the clock pulse $C = 0$ and now the lower slave input AND gate has both its inputs at 1. Therefore, the slave has $J = 0$, $K = 1$ and switches states to give $Q = 0$, $\bar{Q} = 1$. Thus, the slave is reset by the trailing edge of the clock pulse. Lastly, consider that $J = K = 1$. Since the circuit is set $\bar{Q} = 0$ and one of the three inputs to the upper input AND gate is 0, the J input of the master is also 0. The lower input gate now has all three inputs at 1 when a clock pulse arrives and so the master has $K = 1$. This means that the operation when $J = K = 1$ is the same as for the condition $J = 0$, $K = 1$ just described.

Consider now the circuit operation when the flip-flop is initially reset, i.e. $Q = 0$, $\bar{Q} = 1$. When $J = 1$, $K = 0$, the J input of the master will become 1 once a clock pulse is present but the K input will be at 0. Hence, $Q' = 1$, $\bar{Q}' = 0$ but the state of Q' cannot be passed on to the slave's J input because the inverted clock pulse inhibits the slave input AND gates. When the clock changes from 1 to 0, the J input of the slave will become logic 1 and the slave will be set to have $Q = 1$, $\bar{Q} = 0$. The input state $J = 0$, $K = 1$ will not initiate switching since for this condition neither gate will have all its three inputs at 1.

The input signals must be kept constant for the duration of the clock pulse, although this requirement is not always easy to satisfy.

There are no master–slave J-K flip-flops in the 74LS family since these devices in the standard TTL family have been superseded by edge-triggered circuits.

Edge-triggered J-K flip-flops

The edge-triggered J-K flip-flop behaves in a similar manner to the master–slave circuit but its internal circuitry is different. The edge-triggered circuit reacts to the *J-K* inputs only at either the leading edge *or* the trailing edge, of the clock waveform. The edge-triggered flip-flop loads the input data at a clock edge but then the effects of any further input changes are locked out until the next corresponding edge of a clock pulse.

The circuit will only accept data on its J and K inputs that are present when the clock changes state; any change in the signals at the J and K inputs while the clock is steady (at 1 or 0) are ignored. Circuit operation may be initiated by either *leading-edge* or *trailing-edge* triggering depending on the flip-flop. Most 74LS flip-flops are trailing-edge-triggered. The input signals are not read until the start of the negative clock transition and the outputs change state (if they are going to) before the transition ends.

Figure 9.13 gives the logic symbol for the 74LS112A dual J-K trailing-edge-triggered flip-flop. The device is also available in most of the other logic families. The function table of the device is given in Table 9.4.

Fig. 9.13 *(a) Pinout and (b) logic symbol for a 74LS112A dual J-K trailing-edge-triggered flip-flop*

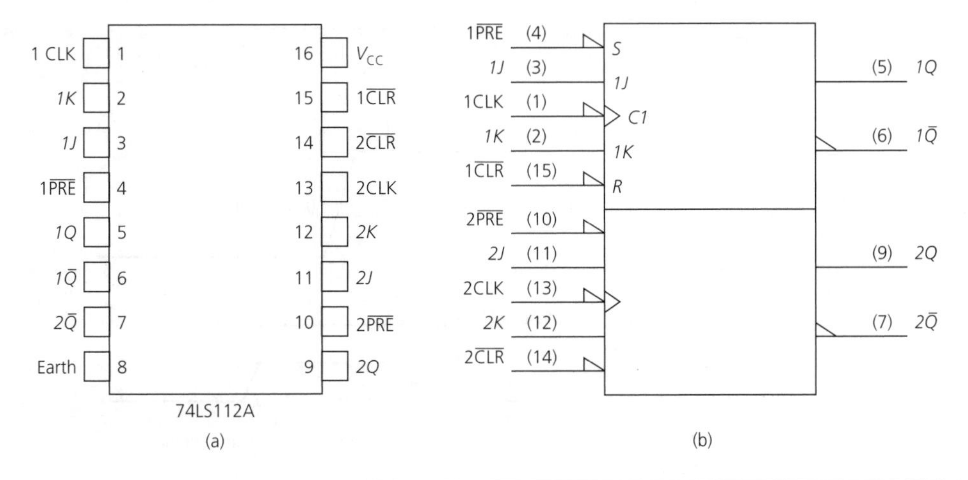

The non-synchronous inputs are active-LOW. Logic 0 on the pre-set terminal will set the flip-flop to $Q = 1$, $\bar{Q} = 0$, and logic 0 at the clear terminal will reset the flip-flop to $Q = 0$, $\bar{Q} = 1$. For synchronous operation the non-synchronous inputs are disabled by being held HIGH. The J and K inputs are read 20 ns before the clock changes from 1 to 0. When $J = K = 1$, the state of the flip-flop changes from $Q = 0$ to $Q = 1$, or from $Q = 1$ to $Q = 0$, at the trailing edge of the clock pulse.

The input data must be held constant for some time both before and after the clock transition takes place to ensure that it will be transferred to the output. The *set-up time* is the time that elapses between the leading edge of an input data pulse and the triggering edge of the clock pulse. Hence, it is the time for which the input data must be held before the triggering edge of the clock arrives. Set-up time is illustrated for leading-edge and trailing-edge flip-flops in Fig. 9.14.

Table 9.4 *74LS112/74HC112 function table*

	Inputs					Outputs	
Mode	*Pre-set*	*Clear*	*Clock*	*J*	*K*	*Q*	$\bar{Q}$
Non-synch set	0	1	×	×	×	1	0
Non-synch reset	1	0	×	×	×	0	1
Synch set	1	1	1 → 0	1	0	1	0
Synch reset	1	1	1 → 0	0	1	0	1
Synch toggle	1	1	1 → 0	1	1	change state	
Synch hold	1	1	1 → 0	0	0	no change	

Fig. 9.14 *Set-up time and hold-up time for an edge-triggered flip-flop: (a) leading edge and (b) trailing edge*

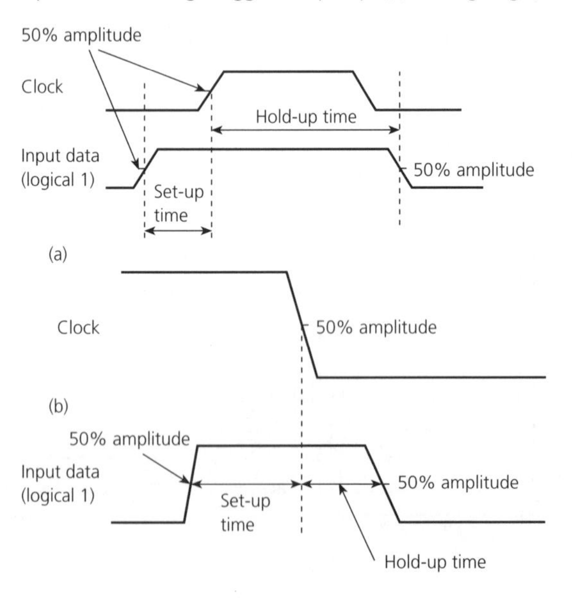

The *hold-up time* of an edge-triggered flip-flop is the interval of time between the clock pulse transition changing the state of the output and the end of the input data pulse. The hold-up time is illustrated in Fig. 9.14. After the hold-up time interval has passed, the J and K input data may be changed without affecting the output. The smaller the set-up and hold-up times, the better is the circuit's performance.

Typical figures for a 74LS flip-flop are a set-up time of 20 ns and a hold-up time of very nearly zero.

The *propagation time* is the time taken for the output to change state after a change in input(s). It is defined from the leading (or trailing) edge of the clock pulse to the 50% amplitude point of the output pulse. There are two delays; t_{pLH} is the time taken for the output to change from LOW to HIGH, and t_{pHL} is the time for a LOW to HIGH change to occur.

Divide-by-2 circuit

The J-K flip-flop can be used as a *divide-by-2* circuit by connecting it in the manner shown in Fig. 9.15. With the J and K input terminals connected to +V, both inputs are held in the logical 1 state and the circuit *toggles* with each clock pulse. The toggle action may, of course, occur at either the leading or the trailing edge of the clock pulse, depending upon the device concerned.

Fig. 9.15 *The J-K flip-flop as a divide-by-2 circuit*

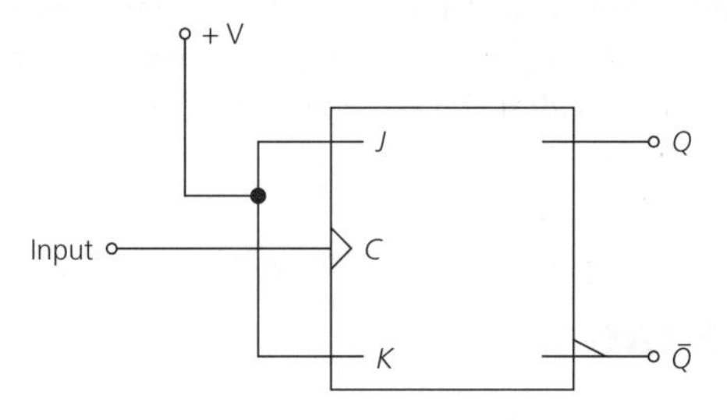

PRACTICAL EXERCISE 9.2

To investigate the operation of the 74LS112A (or 74HC112) dual J-K flip-flop.
Components and equipment: one 74LS112A (or 74HC112) dual J-K flip-flop IC. One 270 Ω resistor. One LED. Dual-beam CRO. Pulse generator. Power supply. Breadboard.
Procedure:

(a) Build the circuit shown in Fig. 9.16 on the breadboard with both the J and K terminals connected to the 5 V line.
(b) Set the pulse generator to output a square wave of about 100 Hz and connect it to the 1CLK terminal. Use the CRO to observe and draw the waveforms at both the clock input terminal and the Q output terminal. Now connect the K input to 0 V and draw the output waveform displayed on the CRO. Repeat

Fig. 9.16

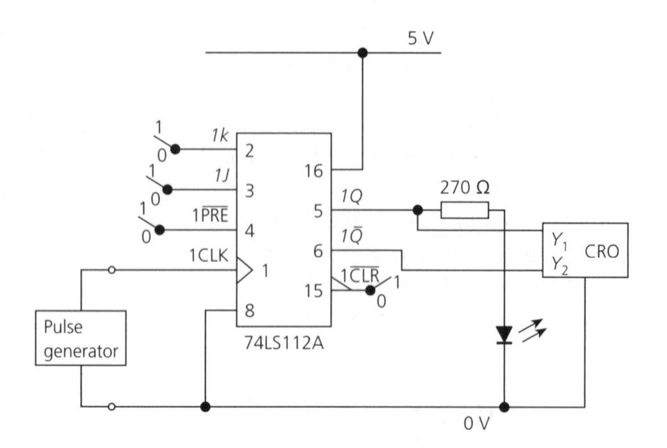

with the J terminal connected to 0 V and the K terminal connected to +5 V. Now connect the pre-set terminal to 0 V. What happens to the output of the circuit?

(c) Set the pulse generator to about 10 Hz and, by connecting the J and K terminals to logic 1 and logic 0, as required, confirm that the flip-flop follows the truth table shown in Table 9.4. What functions did the circuit perform for each connection in (b)? Give the function table obtained from the test in (c) and compare with Tables 9.3 and 9.4.

The D latch

A D latch has one data input, labelled D, one enable (clock) input, and one output Q. The circuit is employed for the temporary storage of a single bit. A bit, 0 or 1, present at the D input is transferred to the Q output whenever the enable (clock) input is HIGH. The logical state of Q will follow any changes in the logical state of the D input as long as the enable (clock) input remains HIGH. When the enable input goes LOW the bit present at the D input at that time will be retained at Q. Here it stays until such time as the enable input goes HIGH again. Then the Q output will again follow the logical state of the D input. A D latch is *transparent* when the clock is at logic level 1, i.e. any change at the D input appears immediately at the Q output. The D latch cannot be made to toggle.

An example of a D latch is the quad 74LS75 whose pinout is shown in Fig. 9.17(a) and whose logic symbol is shown in Fig. 9.17(b). The four flip-flops have just two enable pins so each controls two circuits; enable pin 13 enables flip-flops 1 and 2, while enable pin 4 enables flip-flops 3 and 4.

The function table of the 74LS75 D latch is shown in Table 9.5, where Q_0 is the logical state of Q before the clock changes from HIGH to LOW.

Fig. 9.17 *A 74LS75 quad D latch: (a) pinout and (b) logic symbol*

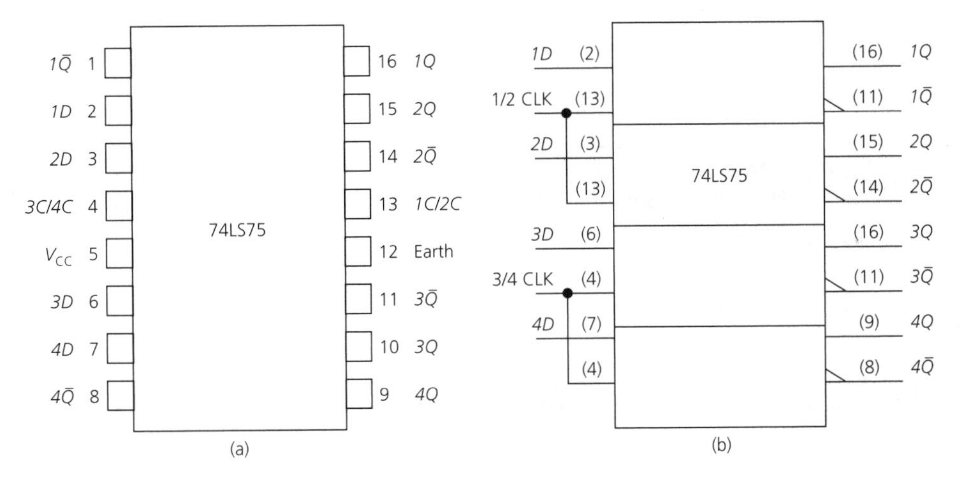

Table 9.5 D latch

D	CLK	Q	$\bar{Q}$
0	1	0	1
1	1	1	0
×	0	Q_0	Q_0

The 74HC logic family offers the 74HC373 octal three-state D latch.

The D flip-flop

With a D flip-flop, a bit present at the *D* input terminal at the end of the set-up time will be transferred to the *Q* output at either the leading edge or the trailing edge of the next clock pulse.

The D flip-flop is not transparent; the *Q* output will only take up the same state as the *D* input when there is a clock transition (either 0 to 1 or 1 to 0, depending on the circuit). The D flip-flop can be made to toggle by connecting its *D* output to the *D* input.

74LS/HC74 D flip-flop

The 74LS74 and the 74HC74D flip-flop has both synchronous and non-synchronous inputs. The *D* and clock inputs are synchronous, the state of the *D* input is transferred to the *Q* output at the leading edge of a clock pulse. When the clock is at either logic 0 or logic 1, any changes at the *D* input have no effect upon the *Q* output. Figure 9.18(a) shows the pinout and Fig. 9.18(b) shows the logic symbol.

Table 9.6

Mode	Inputs				Outputs	
	Pre-set	Clear	Clock	D	Q	Q
Non-synch set	0	1	×	×	1	0
Non-synch reset	1	0	×	×	0	1
Synch set	1	1	0 → 1	1	1	0
Synch reset	1	1	0 → 1	0	0	1

Fig. 9.18 *(a) Pinout and (b) logic symbol for the 74LS74A dual D leading-edge-triggered flip-flop*

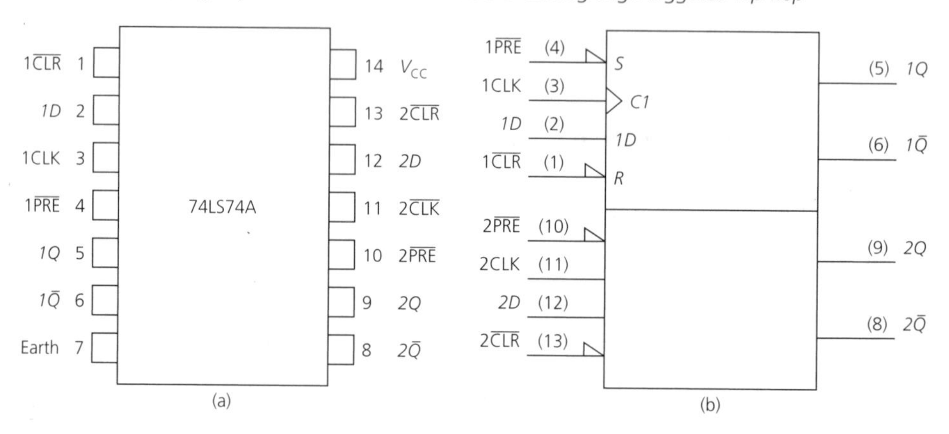

The pre-set and clear inputs (sometimes labelled as set and reset) are non-synchronous and operate independently from D and the clock. The Q output will respond immediately to any change in voltage level at the pre-set or clear terminals. Both non-synchronous inputs are active-LOW, so logic 0 on the pre-set pin will set $Q = 1$ and logic 0 on the clear pin will set $Q = 0$, whatever the states of the D and clock inputs. If only the pre-set and clear terminals are used the circuit will act as an S-R latch. The truth table of the 74LS74 is given in Table 9.6.

Divide-by-2 circuit

If a D flip-flop has its $\bar{Q}$ output connected to its D input, the circuit will divide by two the signal applied to its clock input. The necessary connection is shown in Fig. 9.19(a). Suppose that initially $Q = 1$, $\bar{Q} = D = 0$, then the first clock 1 pulse will cause the circuit to switch to $Q = 0$, $\bar{Q} = D = 1$. The second clock pulse will now switch the circuit back to $Q = 1$, $\bar{Q} = D = 0$, and so on. The circuit waveforms are given in Fig. 9.19(b) from which it is clear that the output pulse waveform has a frequency of one-half that of the input waveform.

Fig. 9.19 *Use of a D flip-flop to divide by 2*

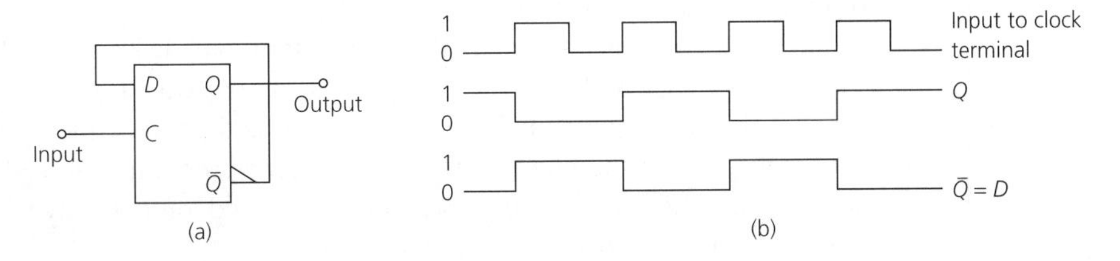

PRACTICAL EXERCISE 9.3

To investigate the action of a D flip-flop.

Components and equipment: one 74LS74 (or 74HC74) dual D-type flip-flop IC. One 74LS75 (or 74HC175) quad D latch IC. One LED. One 270 Ω resistor. Three-way switch. Low-frequency pulse generator. Dual-beam CRO. Power supply. Breadboard.

Procedure:

Fig. 9.20

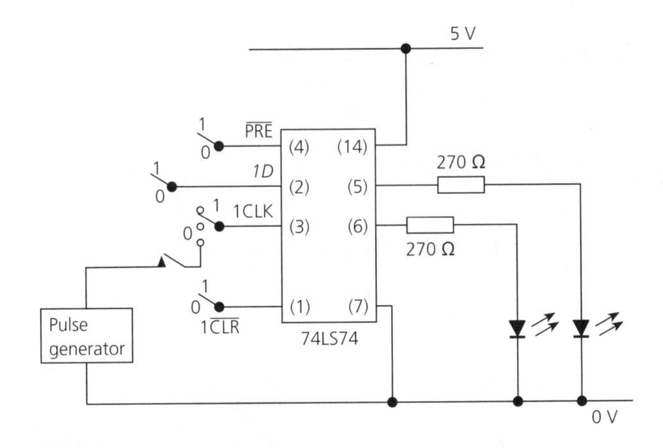

(a) Connect up the circuit shown in Fig. 9.20.

(b) Connect the pre-set terminal to logical 1 and then clear the circuit by connecting the clear terminal to logic 0. Now switch the clear terminal to logic 1.

(c) Connect the clock input to logic 0 and the 1*D* input to logic 1. Note what happens to the LED. Now change the clock input to logic 1 and note the logical state of the LED. See what happens when (i) the clock input is changed to 0, and (ii) the *D* input is changed to 0. The *Q* output should take up the logical state of the *D* input at the leading edge of a clock pulse. Did it? If not, suggest why.

(d) Change the clock input back to 1 and note the logic state of the LED. Now connect the pre-set terminal to 0. What happens to the LED? Try altering both the 1*D* and the clock inputs and note the effect on the LED.

(e) Put the pre-set terminal connection back to 1 and connect the clear terminal to 0. What happens to the LED?
(f) Connect the pulse generator to the 1CLK terminal. Disconnect the 1 and 0 connection to 1CLK. Observe the input and output waveforms on the CRO.
(g) Connect the $\bar{Q}_1$ output (pin 5) to the D_1 input and connect the low-frequency pulse generator to the clock input. Set the generator to about 100 Hz at 5 V and use the CRO to observe both the input and output waveforms. What function is the circuit now performing?
(h) Replace the 74LS74 D flip-flop with the 74LS75 D latch. Check the pinout of the '75 (Fig. 9.17) and make any necessary alterations to the circuit. Then repeat procedures (b) through to (f). Comment on the different results that are obtained.

Fig. 9.21 *T flip-flop*

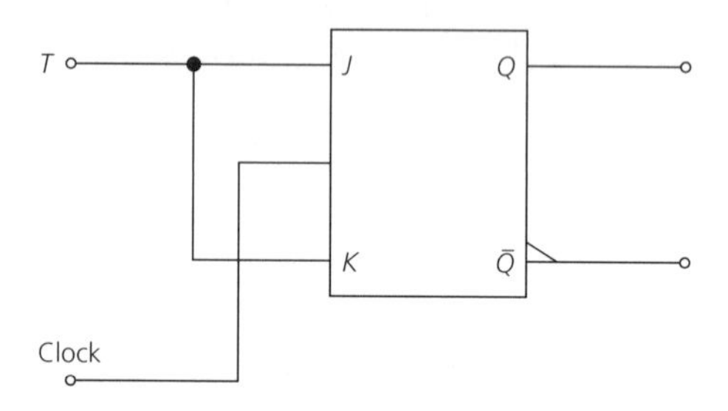

Table 9.7 *T flip-flop*

J	*K*	*Q*	Q^+
0	0	0	0
0	0	1	1
1	1	0	1
1	1	1	0

The T flip-flop

The fourth type of flip-flop is called the trigger or T flip-flop and its symbol is given in Fig. 9.21. The flip-flop is made from a J-K flip-flop merely by connecting its *J* and *K* terminals together. This means that, at all times, $J = K$. Substituting in the truth table of a J-K flip-flop gives Table 9.7. From this table it is apparent that, when the clock is 1, the flip-flop will change state, or toggle, each time there is a trigger (*T*) pulse applied to its input. Thus, when $T = 0$, the *Q* output will not change state when the clock goes to logic 1; when $T = 1$, the *Q* output will change state each time a clock pulse is received (Fig. 9.22). The T flip-flop is not available as an integrated circuit since it is so easily obtained from a J-K flip-flop. It can also be derived from either an S-R latch or a D flip-flop – see Figs 9.23(a) and (b).

Fig. 9.22 *Waveforms in a T flip-flop*

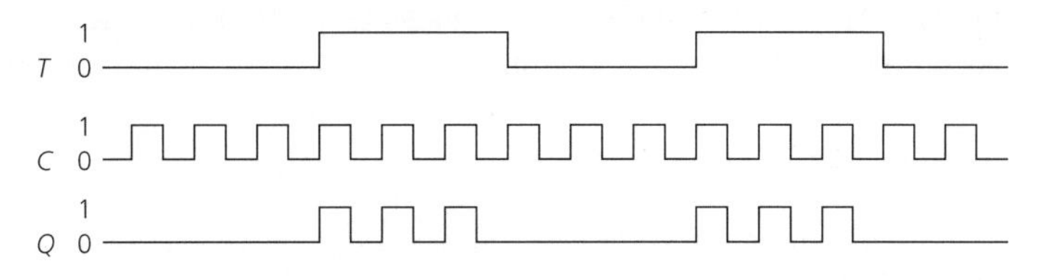

Fig. 9.23 *T flip-flops derived from (a) S-R latch; (b) D flip-flop*

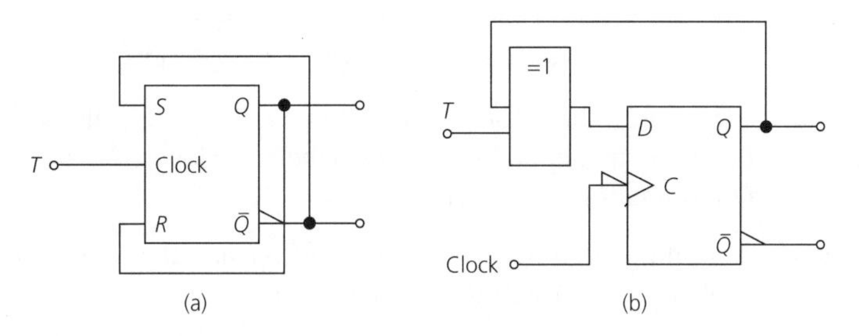

Fig. 9.24 *The operational difference between various types of flip-flop*

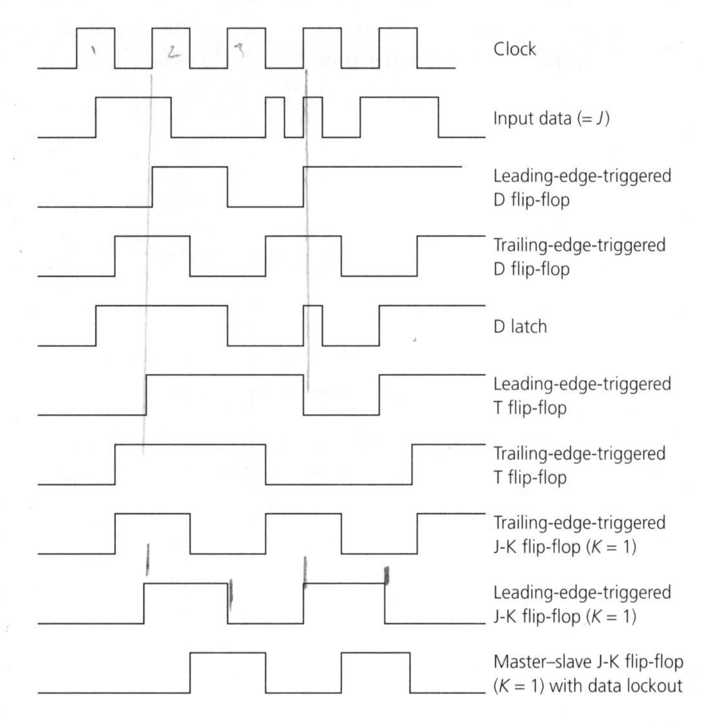

The differences between the various types of flip-flop can be seen with the aid of a timing diagram (Fig. 9.24). The D flip-flops change their output state to be the same as the input state only at the edge of the clock waveform. The D latch has an output that is the same as the input whenever the clock is at the logical 1 voltage level. The T flip-flop changes its state at the leading edge of the clock waveform if the input is at 1. The master–slave J-K flip-flop changes its state at the trailing edges of the clock waveform according to whether the J input is at the 0 or the 1 logical level. Finally, the edge-triggered J-K flip-flops respond to the J input only at either the trailing edge or the leading edge of the clock pulse.

EXERCISES

9.1 An S-R flip-flop has the waveforms shown in Fig. 9.25 applied to its S and R terminals. Sketch the waveform at the Q and $\bar{Q}$ outputs.

9.2 Figure 9.25 shows one latch and two different flip-flops. The waveforms given in Fig. 9.26 are applied to the pins labelled. Sketch the waveforms that appear at the Q terminal of each circuit.

9.3 A leading-edge-triggered J-K flip-flop has the waveforms shown in Fig. 9.27 applied to it. Sketch the Q waveform.

9.4 It is required to display the output of a 4-bit binary adder on four LEDs. The LEDs are to operate simultaneously when a control signal goes HIGH. Draw a possible circuit. [Note: the outputs of the Σ_1, Σ_2, Σ_3 and Σ_4 outputs do not appear at precisely the same time.]

Fig. 9.25

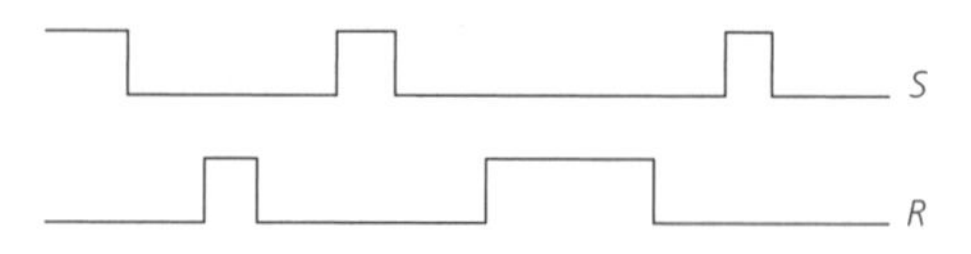

Fig. 9.26

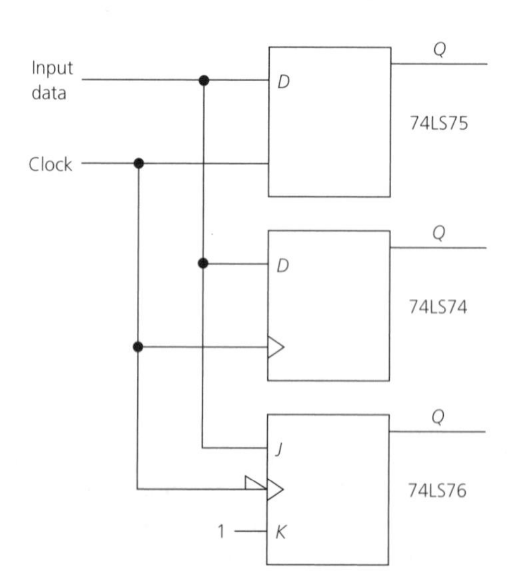

Fig. 9.27

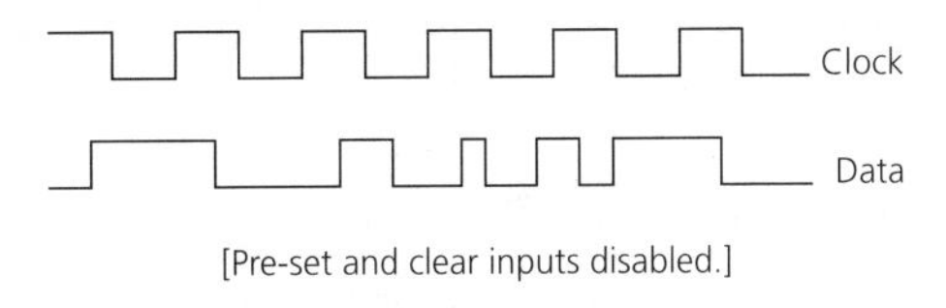

Fig. 9.28

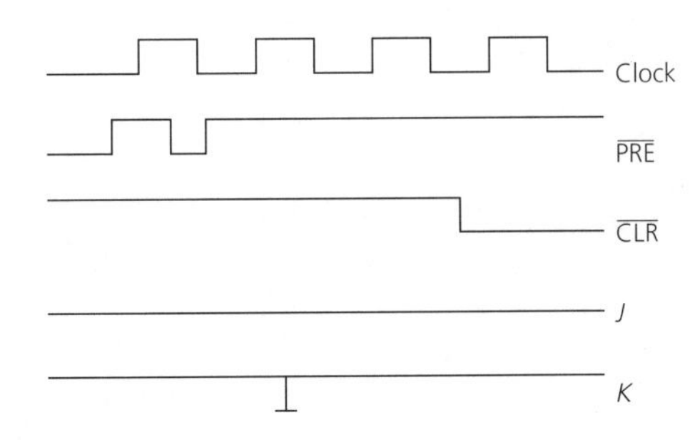

9.5 (a) A J-K flip-flop is in the state $Q = 0$, $\bar{Q} = 1$. What change in state will occur if the flip-flop is
(i) Set.
(ii) Reset
(iii) Cleared?

(b) Write down the truth table of a J-K flip-flop. Let $K = J$ and then determine the corresponding Q values. What kind of circuit does the truth table now represent?

9.6 (a) What are the functions of the (i) clear and (ii) reset terminals of a flip-flop? A J-K flip-flop has active-LOW clear and reset terminals. If $J = 1$ and $K = 0$ and the clear = 0 what is the state of the Q output when reset = 1?

(b) Refer to the function table of the 74LS74 D flip-flop, Table 9.5. (i) If the pre-set terminal is LOW does the flip-flop set whatever the state of the D and clock inputs? (ii) If the pre-set and clear inputs are not used should they be connected to +5 V or to 0 V? (iii) If the D input is HIGH and the pre-set and clear terminals are not used does the flip-flop set or reset when the clock changes from 0 to 1?

9.7 (a) What are the stable states of a flip-flop?
(b) What is the difference between a latch and a flip-flop?
(c) Outline the difference between a master–slave and an edge-triggered J-K flip-flop.

9.8 The 74HC74 D flip-flop is a leading-edge-triggered device with synchronous data and clock inputs and non-synchronous set (pre-set) and reset (clear) terminals.
(a) The inputs are $D = \text{CLK} = \overline{\text{set}} = \overline{\text{reset}} = 1$. What is the state of Q?
(b) The clear terminal goes LOW. What is now the state of Q?
(c) Draw the logic symbol for the device from the above information.

Fig. 9.29

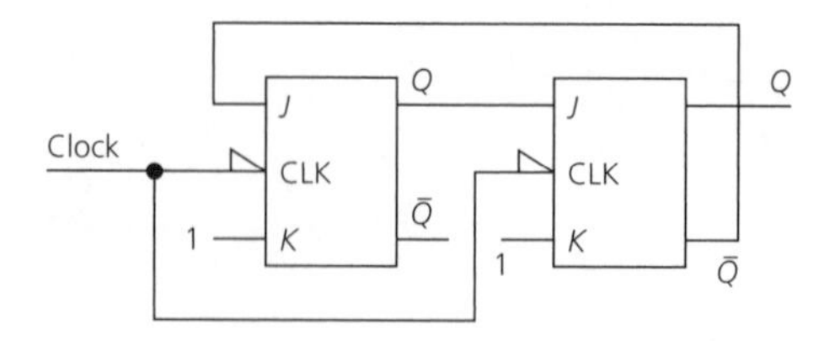

Fig. 9.30

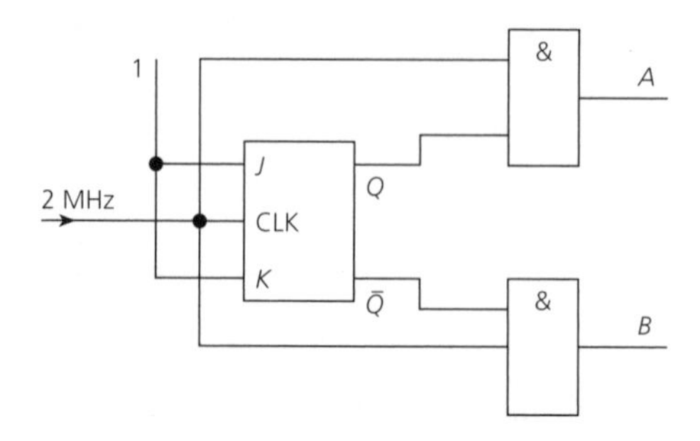

(d) Look in the data sheet for the IC and determine
  (i) The input set-up time.
  (ii) The input hold-up time.

9.9 Write down the truth tables for the D and J-K flip-flops. Hence, determine how a J-K flip-flop can be made to act like a D flip-flop.

9.10 Determine the count of the circuit shown in Fig. 9.29.

9.11 Determine the output frequency and the mark–space ratio of the output waveform for the circuit shown in Fig. 9.30.

10 Counters

After reading this chapter you should be able to:

(a) Use a timing diagram to describe the action of a counter.
(b) Design a ripple counter to give a count of 2^n, where n is the number of stages.
(c) Know how to reduce the count.
(d) Describe the differences between a ripple counter and a synchronous counter.
(e) Use IC counters, both non-synchronous and synchronous.
(f) Reduce the count of an IC counter.
(g) Recognize synchronous counters for divide-by-8, 16, 3, and 5.
(h) Understand the principle of an up-down counter.

Sequential logic circuits follow a pre-determined sequence of digital states and are triggered into operation by a clock pulse. Sequential circuits include counters and shift registers. A *counter* is an electronic circuit that is able to count the number of pulses applied to its input terminals. The count may be outputted using the straightforward binary code, or may be in binary-coded decimal (BCD). Alternatively, the outputs of a counter may be decoded to produce a unique output signal to represent each possible count. A counter that counts from 0 to 15, which is 16 states, is known as a *modulo-16*, or *MOD-16*, counter. A counter may also be used as a frequency divider to reduce the frequency of a pulse waveform.

Essentially, a counter consists of the cascade connection of a number of flip-flops, usually of either the J-K type or the D type which may be operated either synchronously or non-synchronously. With synchronous operation all the flip-flops making up the counter operate at the same instant in time under the control of a clock pulse. In the case of non-synchronous operation each flip-flop operates in turn. The switching of the least-significant flip-flop is initiated by a clock pulse but the remaining flip-flops are each operated by the preceding flip-flop. This means that each stage must change state before the following stage can do so. As a result, synchronous operation of a counter is much faster and the use of a non-synchronous counter is acceptable only when the speed of operation is not of particular importance. A counter can be constructed by suitably interconnecting a number of J-K or D flip-flops and, perhaps, gates, but several different counters are available in the various TTL and CMOS logic families.

The possible applications for counters are many. They are often used for the direct counting of objects in industrial processes and of voltage pulses in digital circuits such as digital voltmeters. Counters can be used as frequency dividers and for the measurement of frequency and time.

Non-synchronous counters

J-K flip-flops

A single J-K flip-flop will act as a divide-by-two circuit. If its J and K terminals are both held at logic 1 and a clock pulse is applied to its clock input terminal, the flip-flop will toggle and so the Q output will switch backwards and forwards between logic 1 and logic 0 (Fig. 10.1(a)). In the drawing of the waveforms shown in Fig. 10.1(b) it has been assumed that the flip-flop toggles at the trailing edge of the clock pulse, i.e. as the clock pulse changes from 1 to 0. This is generally true for non-synchronous counters. Clearly, the number of pulses per second occurring at the Q terminal is only one-half of the number of clock pulses, so the circuit is a divide-by-2 counter.

Fig. 10.1 *(a) Divide-by 2 J-K flip-flop and (b) waveforms in a divide-by-2 circuit*

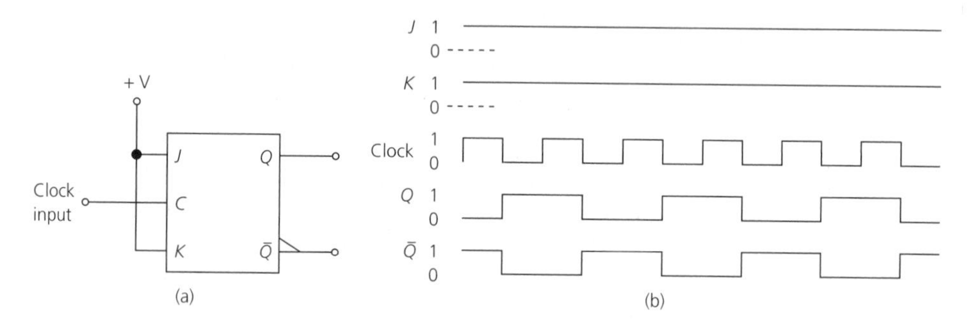

For counts in excess of two, a number n of J-K flip-flops must be connected in cascade to give a count of $2^n - 1$. Since the first state is 0 this means that n flip-flops are able to give 2^n count states. The *count modulus* is 2^n.

Three-bit counter

A 3-bit counter can be built using three 74LS112A or 74HC112 J-K flip-flops connected as shown in Fig. 10.2. Each flip-flop has its J and K terminals connected together and to logic 1 so that it operates in the toggle mode. This means that each flip-flop will change state whenever the voltage at its clock input changes from 1 to 0.

- At the end of the first clock pulse, flip-flop A toggles and Q_A changes from 0 to 1. This change in voltage appears at the clock input of flip-flop B but does not cause it to toggle, hence, $Q_B = Q_C = 0$. The count is now 001 or decimal 1.
- At the end of the second clock pulse, flip-flop A toggles again and Q_A changes from 1 to 0. This trailing-edge clock transition appears at the clock input of flip-flop B and causes that flip-flop to toggle. Now $Q_B = 1$ and $Q_A = Q_C = 0$, which means that the count is now 010 or decimal 2. Obviously, Q_B changes state a little while after Q_A changes its state.

Fig. 10.2 *A 3-bit counter using J-K flip-flops*

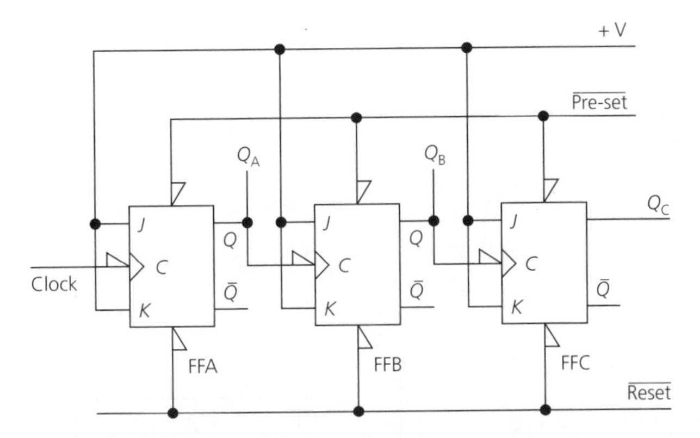

- The trailing edge of the third clock pulse toggles flip-flop A so that Q_A changes from 0 to 1. This transition has no effect on flip-flop B so that Q_B remains at 1 to give $Q_A = Q_B = 1$, $Q_C = 0$, which is a count of 011 or decimal 3.
- When the fourth clock pulse ends, its trailing-edge transition toggles flip-flop A to give $Q_A = 0$. The negative Q_A transition toggles flip-flop B so that $Q_B = 0$ also. The change in Q_B from 1 to 0 toggles flip-flop C to give $Q_C = 1$. Now the count is 100 or decimal 4.
- This action continues for every clock pulse with the result that the circuit counts through the sequence 101, 110, and 111. When the circuit reaches a count of 111 or decimal 7 all three flip-flops are set. The trailing edge of the next clock pulse toggles flip-flop A so that $Q_A = 0$. The negative-going Q_A transition toggles flip-flop B so that Q_B changes to 0 and, finally, the negative transition of Q_B toggles flip-flop C so that Q_C becomes 0 also. Now the count of the circuit is 000 and a new counting sequence can begin.

PRACTICAL EXERCISE 10.1

To investigate the operation of a 3-bit binary counter built using three 74LS112A (or 74HC112) J-K flip-flops. Components and equipment: two dual 74LS76 (or 74HC112) J-K flip-flops. Three 270 Ω resistors. Three LEDs. Pulse generator. Power supply. Breadboard.

Procedure:

(a) Build the circuit shown in Fig. 10.3.

(b) Apply logic 1 to the $\overline{\text{pre-set}}$ line. Apply logic 0 to the $\overline{\text{clear}}$ line to ensure that all three flip-flops are cleared and then connect the $\overline{\text{clear}}$ line to logic 1.

(c) Apply first logic 1 and then logic 1 to the clock input to the counter and then change it for logic 0. Note which LED turns ON. Now alternately connect logic 1 and then logic 0 to the clock input and keep a note each time which LEDs glow visibly. Go through one complete cycle of the count, i.e. 0 through to 7,

Fig. 10.3

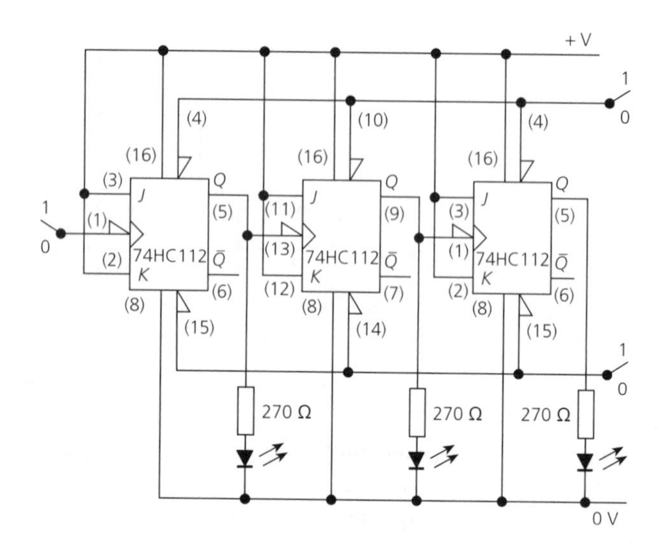

and part of the next count stopping when the count is decimal 3. Write down the truth table of the circuit and compare it with the truth table given in Table 10.2. If the flip-flop will not trigger, use a pulse generator to supply the clock.

(d) Now apply logic 0 to the $\overline{\text{clear}}$ line. What happens to the LEDs?

(e) Now alter the circuit by moving the connections to Q_A, Q_B and Q_C to $\bar{Q}_A$, $\bar{Q}_B$ and $\bar{Q}_C$, respectively. Repeat tests (b) through to (d). Write down the function table of the modified circuit and state what function is now performed.

Four-Bit counter

Figure 10.4 shows how a non-synchronous counter can be constructed from four J-K flip-flops. The *J* and the *K* inputs of each flip-flop are permanently connected to the logic 1 voltage level so that each flip-flop will be toggled by a pulse applied to its clock input.

The *Q* output of the first, second and third stages is connected to the clock input of the following stage. The $\bar{Q}$ terminals are left unconnected. Each J-K flip-flop is switched by the trailing edge of a pulse applied to its clock input.

Fig. 10.4 *Non-synchronous 4-bit counter*

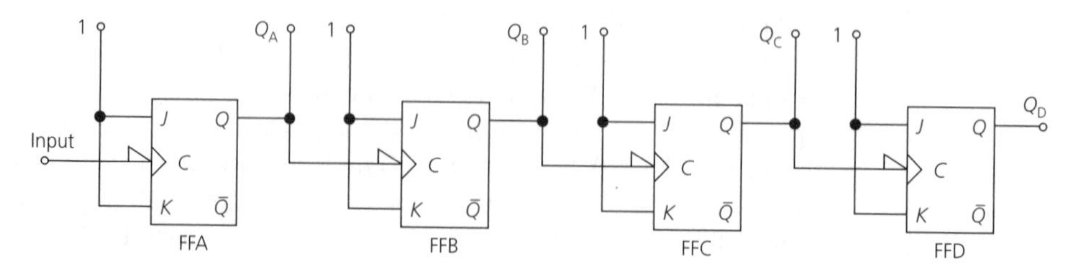

Suppose that initially each stage is reset, i.e. $Q_A = Q_B = Q_C = Q_D = 0$. The trailing edge of the first clock pulse will toggle flip-flop A so that Q_A becomes logic 1. The count will then be 0 0 0 1 (reading from the right) or denary 1. The next clock pulse will toggle flip-flop A and the change of Q_A from 1 to 0 will set flip-flop B. Thus, after two clock pulses have been applied to the counter, only Q_B is at 1. The third clock pulse will set flip-flop A so that Q_A changes from 0 to 1 but such a change will not affect the state of flip-flop B and so this stage remains set. Now both the first two stages are set, $Q_A = Q_B = 1$, and the last two stages remain reset, $Q_C = Q_D = 0$; the count is now 0 0 1 1 or decimal 3. When the fourth clock pulse arrives, the first stage toggles so that Q_A changes from 1 to 0, and this change causes flip-flop B to reset. Thus, Q_B changes from 1 to 0 and in so doing sets flip-flop C; now $Q_A = Q_B = Q_D = 0$ and $Q_C = 1$ and decimal 4 is stored. The operation of the counter as the fifth, sixth, seventh, etc. clock pulses are applied follows the same lines as just described and is summarized by the truth table of the counter (see Table 10.1). Note that the count is 15 and there are 16 different combinations of Q_A, Q_B, Q_C and Q_D.

The operation of the counter can be illustrated by the waveforms given in Fig. 10.5. The Q outputs of the flip-flops seem to *ripple* through the circuit and for this reason this type of circuit is often known as a *ripple counter*. Since 16 input pulses produce one output pulse the circuit is known as a divide-by-16 counter.

The first eight pulses also apply to the 3-bit counter shown in Fig. 10.2.

Table 10.1 ***Four-bit ripple counter***

Clock pulse (count)	0	1	2	3	4	5	6	7	8	9	10	11	12	13	14	15	16
Q_A	0	1	0	1	0	1	0	1	0	1	0	1	0	1	0	1	0
Q_B	0	0	1	1	0	0	1	1	0	0	1	1	0	0	1	1	0
Q_C	0	0	0	0	1	1	1	1	0	0	0	0	1	1	1	1	0
Q_D	0	0	0	0	0	0	0	0	1	1	1	1	1	1	1	1	0

Fig. 10.5 *Waveforms in a 4-bit non-synchronous counter*

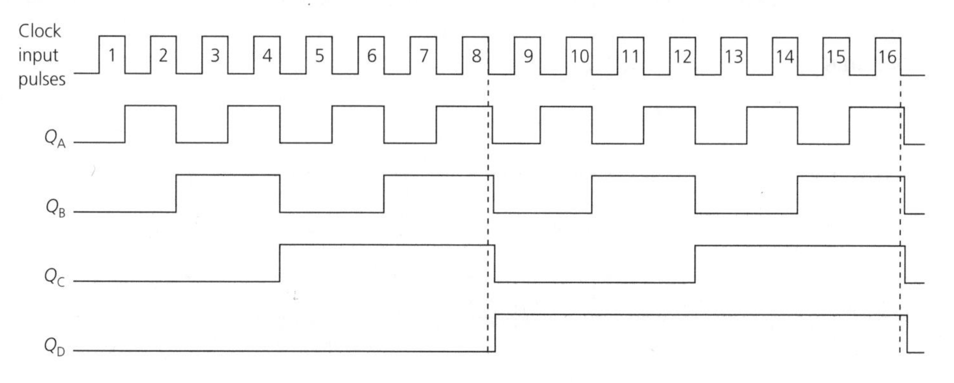

The count is of a binary nature and the counter operates non-synchronously because the flip-flops operate at different times, shown at the end of the eighth and the sixteenth clock pulses by the vertical dotted lines. Flip-flop A operates twice as often as flip-flop B, four times as often as flip-flop C, and eight times as often as flip-flop D. There is a maximum clock frequency that can be used since the periodic time of the clock waveform must be greater than the sum of the propagation times through the counter and the time duration of the output pulse.

EXAMPLE 10.1

The flip-flops in a 4-bit ripple counter each introduce a maximum delay of 40 ns. Calculate the maximum clock frequency.

Solution

Total propagation delay $= 4 \times 40 = 160$ ns
Maximum clock frequency $= 1/(160 \times 10^{-9}) = 6.25 \times 10^{6}$ Hz (*Ans.*)

If the Q output of each flip-flop is connected to the clock input of the next flip-flop, the circuit will count down from 15 to 0.

D Flip-flops

To obtain a count that is a multiple of 2 a number of D flip-flops are each connected to give a count of 2 and are then cascaded. An example of this technique is shown by the divide-by-8 ripple counter shown in Fig. 10.6.

Decoded outputs

Often a binary readout of the count is undesirable; when this is so the outputs of the individual flip-flops of the counter can be decoded so that each count produces a unique output. Decoding can be achieved by connecting the Q and the $\bar{Q}$ outputs of each

Fig. 10.6 *Ripple counter using D flip-flops*

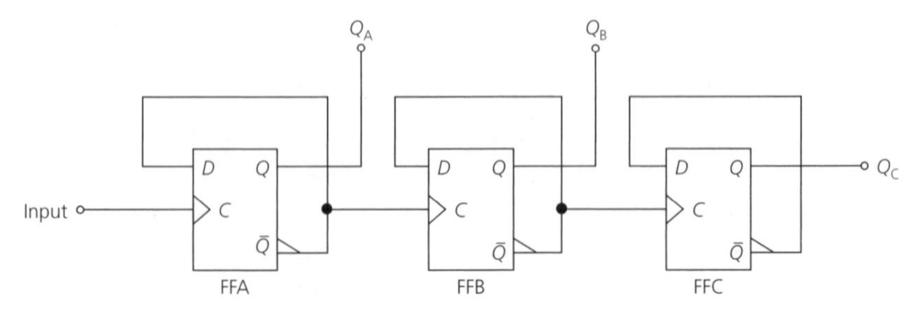

Fig. 10.7 *Two-stage counter with decoded output*

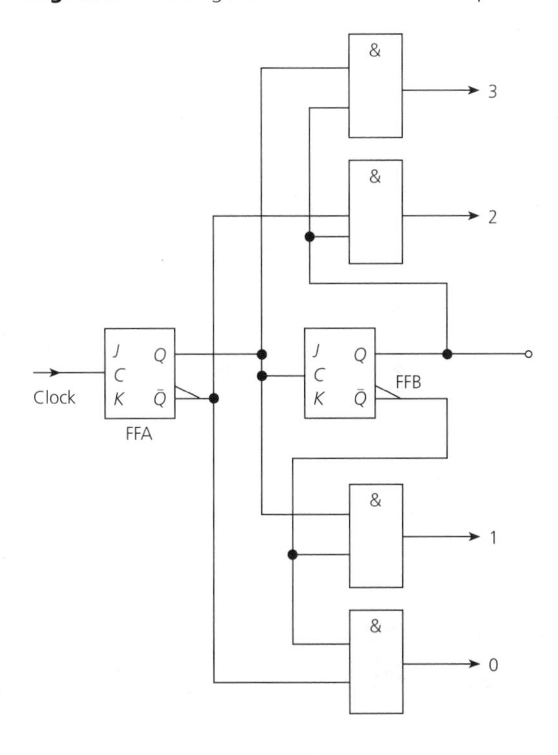

Fig. 10.8 *Clocking of decoding gates in a counter*

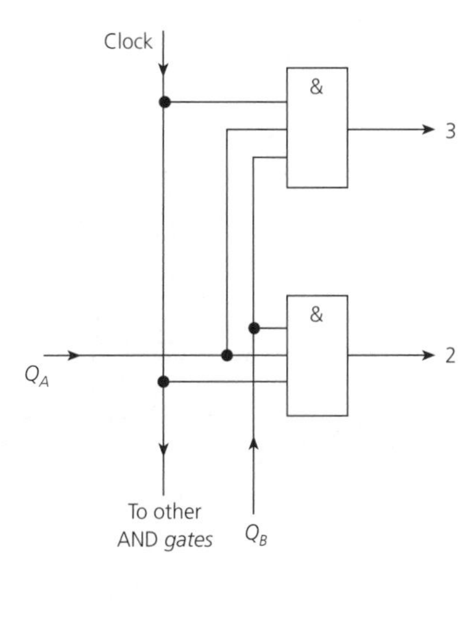

flip-flop to the inputs of a number of gates. This is shown in Fig. 10.7 for a 2-stage counter. The output of a 2-input AND gate is 1 only when both its inputs are at 1. Hence, the top gate, for example, will have an output of logic 1 only when $Q_A = Q_B = 1$ and the count is 3.

It is possible for false 1 signals to appear at the flip-flop outputs which may give false counts when decoding circuitry is used. These *glitches* or *dynamic hazards* arise because not all the flip-flops change state at precisely the same time when the edge of a clock pulse arrives. To prevent this happening the decoding gates can be clocked as shown in Fig. 10.8. The enabling clock pulse is not applied until all the flip-flops have reached their steady (final) values. The glitch-free output is, however, obtained at the expense of a reduction in the speed of operation. Some IC counters include the decoding circuitry within the package.

If the AND gates are replaced by NAND gates the output would be active-LOW.

Reducing the count to less than 2^n

Very often a counter is required to have a count of less than $2n - 1$, where n is the number of flip-flops it contains. The reduced count is obtained by modifying the basic counter circuit so that one or more of the possible states are omitted. Thus, if a count of 7 is required, a 3-stage counter must be used, having 2^3 or 8 (including 0) states; this means that *one* of the counts must be eliminated.

Fig. 10.9 *Non-synchronous decade counter*

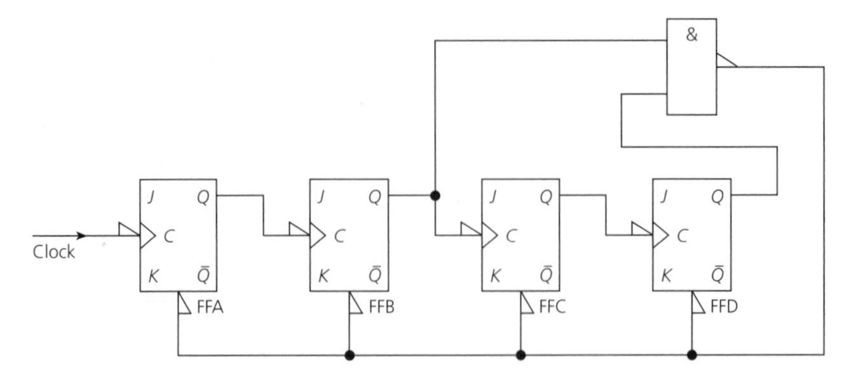

Fig. 10.10 *Timing diagram of a decade counter*

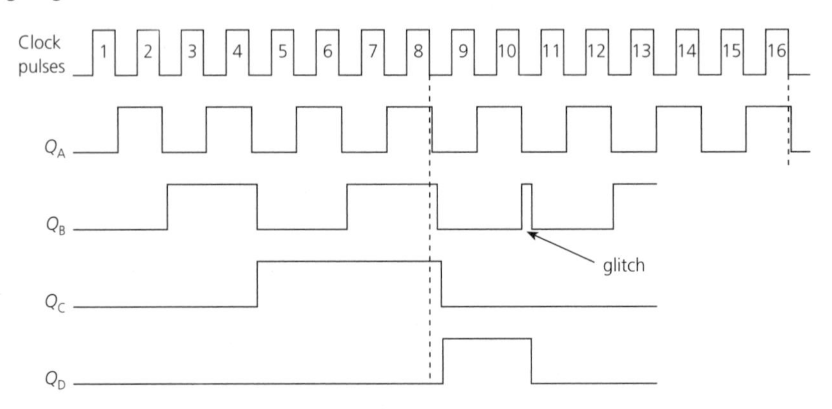

If a count of 6 is required, a 3-stage counter must be used with *two* of its counts eliminated. The reduction of the count is achieved by decoding the maximum count number wanted and using a signal derived from it to clear the counter.

The design of the reset type of counter must start with a decision on the number n of stages that are necessary. A NAND gate is then required whose output is connected to all the reset, or clear, flip-flop inputs in parallel. The inputs to the NAND gate must consist of the Q output of each stage that is at logical 1 when the required count is reached.

Decade counter

Four stages are necessary: for a count of 10_{10} or 1010, then $Q_A = 0$, $Q_B = 1$, $Q_C = 0$, $Q_D = 1$, and, hence, the NAND gate inputs are Q_B and Q_D. The circuit is shown in Fig. 10.9. The timing diagram shown in Fig. 10.10 illustrates the operation of the decade counter.

Divide-by-6 counter

Three stages are needed: for a count of 6 or 110, then $Q_B = 1$, $Q_C = 1$ and these are the necessary NAND gate inputs (Fig. 10.11).

Fig. 10.11 *Divide-by-6 ripple counter*

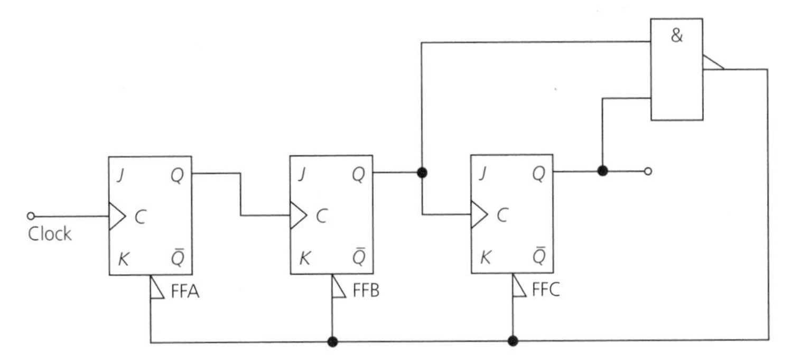

Divide-by-5 counter

Three stages are needed: for a count of 5 or 101, then $Q_A = 1$ and $Q_C = 1$ and so the NAND gate inputs are Q_A and Q_C. This counter is easily converted into a decade counter by preceding it with a D or a J-K flip-flop connected as a divide-by-2 circuit.

IC counters

If a regular counting sequence is required (as is usual), an IC counter can be employed. If an irregular sequence is wanted it will be necessary to build the counter using IC flip-flops. Four-bit non-synchronous counters are available in both the TTL and CMOS logic families. Four examples are the 74LS90, the 74LS92, the 74LS93 and the 74HC393.

74LS93/293 4-bit binary counter

The 74LS93 4-bit binary counter shown in Fig. 10.12(a) has two sections: one, a single divide-by-2 circuit, the other, three J-K flip-flops connected as a divide-by-8 ripple counter. For the circuit to act as a divide-by-16 circuit the Q output of the first flip-flop must be connected to the clock input of the second flip-flop. Referring to Fig. 10.12, terminals 1 and 12 are connected together. Pin 14 is often labelled as C_{p0} or as CKA, and pin 1 is often labelled as C_{p1} or CKB. The terminals labelled R_{01} and R_{02} are the 'reset to 0' inputs. Logic 1 applied to these terminals produces a logic 0 at the output of the associated NAND gate and this is the logical state required to reset, or clear, all four flip-flops. The logic symbol for the IC is shown in Fig. 10.12(b). The 74LS293 and the 74HC393 are electrically and functionally identical to the 74LS93, but their pinouts are different.

The 74LS93 can be modified, using the reset method, to give a count other than 2, 8 or 16. The circuit will reset at the required count if the Q outputs that are then at 1 are fed back to the reset inputs R_{01} and R_{02}. Table 10.2 gives examples of the connections necessary for various division ratios. An external gate may sometimes be necessary.

To obtain a decade counter, pin 9 should be connected to pin 2; pin 11 should be connected to pin 3; and pin 12 should be connected to pin 1. The circuit is then set up as shown in Fig. 10.13. When the count reaches decimal 10, $Q_B = Q_D = 1$ and the output of the NAND gate becomes 0, resetting all four flip-flops. To obtain a divide-by-11 counter, connect Q_D to R_{02}, and Q_A and Q_B to a NAND gate whose output is connected to R_{01}.

Table 10.2

Divide-by-n	R_{01} *to:*	R_{02} *to:*	
$n = 7$	Q_A	Q_BQ_C	(via gating)
$n = 9$	Q_A	Q_D	
$n = 10$	Q_B	Q_D	
$n = 11$	Q_A	Q_BQ_D	(via gating)
$n = 12$	Q_C	Q_D	
$n = 13$	Q_A	Q_CQ_D	(via gating)
$n = 14$	Q_BQ_D	Q_C	(via gating)
$n = 15$	Q_A	$Q_BQ_CQ_D$	(via gating)

Fig. 10.12 *A 74LS93 4-bit binary counter: (a) pinout and (b) logic symbol*

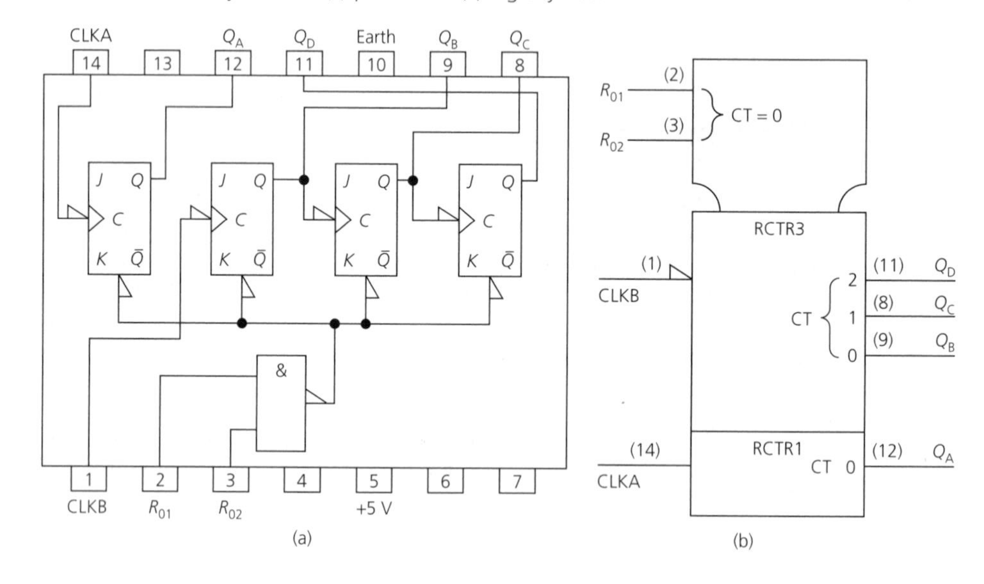

Fig. 10.13 *A 74LS93 connected as a decade counter*

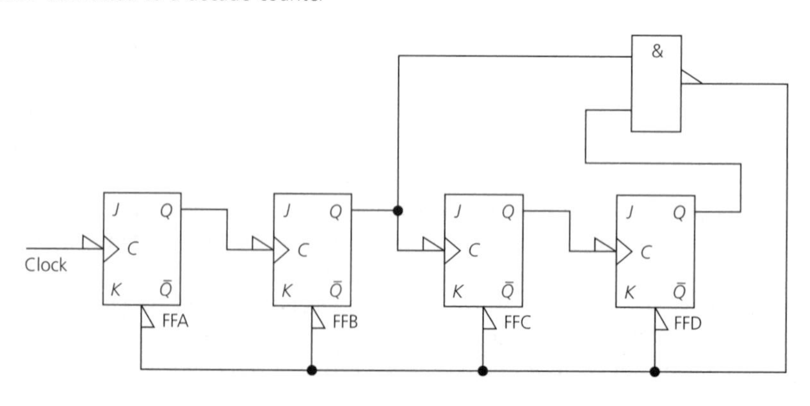

PRACTICAL EXERCISE 10.2

To investigate the performance of the 74LS93 binary counter (or the 74HC293).
Components and equipment: one 74LS93 counter (or 74HC293). Four LEDs. Four 270 Ω resistors. Pulse generator. Power supply. Breadboard. CRO. If the '293 is used, look up the different pin numbers.
Procedure:

(a) Connect pins 1 and 12 of the IC together to make the counter into a 4-bit circuit. Connect the 5 V power supply between pins 5 and 10 and connect the pulse generator to the input terminal of the counter, pin 14. Connect the outputs Q_A, Q_B, Q_C and Q_D via series 270 Ω resistors and an LED to 0 V.
(b) Connect the R_{01} and R_{02} pins (2 and 3) together and connect them to the 5 V line. The counter should now be reset. Check that all four LEDs are OFF. Now move the common R_{01}/R_{02} connection from the 5 V line to the 0 V line to disable the reset. Set the pulse generator to a low frequency and observe the operation of the counter. If the pulse generator frequency is low enough it will be possible to see if the counter goes through its correct sequence of 0000 (0) up to 1111 (15).
(c) Connect one input of the CRO to the clock input, pin 14, and the other CRO input to the Q_D output and observe the input and output waveforms with the pulse generator set, in turn, to three different frequencies. Measure the frequencies of the input and output waveforms and each time calculate the division ratio.
(d) For any one of the pulse generator frequencies used in (c) employ the CRO to measure the frequency of the waveform that appears at the Q_A, Q_B and Q_C outputs.
(e) Now connect the circuit to act as a decade counter (see p. 232) and leave the pulse generator at the frequency used in (d). Use the CRO to check the frequency of the output waveform.
(f) Work out how to make the circuit operate as a divide-by-12 counter with the count sequence 0, 1, 2, 3, 4, 5, 6, 7, 8, 9, 10, 11, 0, 1, etc. Make the required connections and check the operation of the circuit using the CRO.

74LS90/290 decade counter

The two counters are electrically and functionally identical but have different pinouts. Pin 14 is often labelled as C_{p0} or CKA and pin 1 as C_{p1} or CKB.

The 74LS90 is a decade counter that consists of one J-K flip-flop which acts as a divide-by-2 circuit and a 3-stage counter that is internally connected to give a count of 5. The pin connections of the 74LS90 are shown in Fig. 10.14(a). To obtain the maximum count, pin 12 must be connected externally to pin 1. The overall division ratio is thus $2 \times 5 = 10$ but the mark–space ratio is 1:4. If the ÷5 circuit is placed first and followed by the ÷2 circuit an alternative decade counter is obtained. The output is now

Fig. 10.14 *(a) Pin connections of a 74LS90 counter and (b) logic symbol*

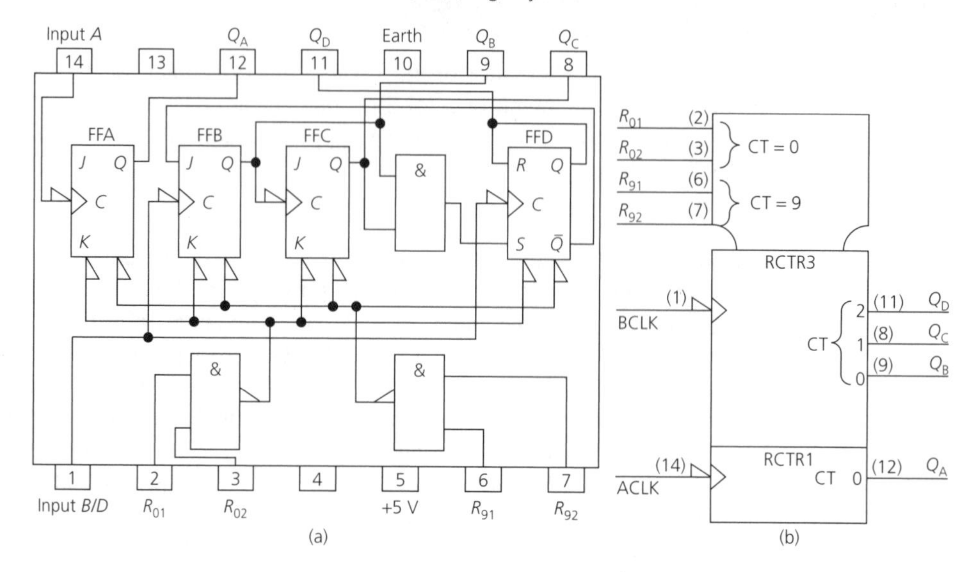

taken from Q_A and the count follows the sequence 0, 1, 2, 3, 4, 8, 9, 10, 11, and 12. This connection has the merit that its output waveform is square. In addition to the clear or reset terminals marked R_{01} and R_{02}, this particular integrated circuit has two other pins, labelled R_{91} and R_{92}, which can be used to set the counter to a count of 9. This feature is useful because it ensures that the circuit will start to count from 0 after the first active clock transition. The logic symbol is shown in Fig. 10.14(b).

When these two pins are not used they should both be connected to earth. They can be used to convert the IC into a divide-by-7 counter. The output pins Q_B and Q_C are connected to R_{91} and R_{92}, respectively. Then, when the circuit reaches a count of 6, it will immediately move to a count of 9.

The 74HC series includes the 74H393 dual 4-bit binary counter, the 74HC4040 12-bit binary counter, and the 74HC4020 14-bit binary counter. All these counters are trailing-edge triggered.

PRACTICAL EXERCISE 10.3

To investigate the operation of the 74LS90 decade counter.

Components and equipment: one 74LS90 decade counter IC. Four LEDs. Four 270 Ω resistors. Pulse generator. Dual-beam CRO. Power supply. Breadboard.

Procedure:

(a) Connect up the circuit shown in Fig. 10.15.

(b) Use the CRO to check the operation of the circuit with the pulse generator frequency at any convenient value. What is the mark–space ratio of the output waveform?

Fig. 10.15

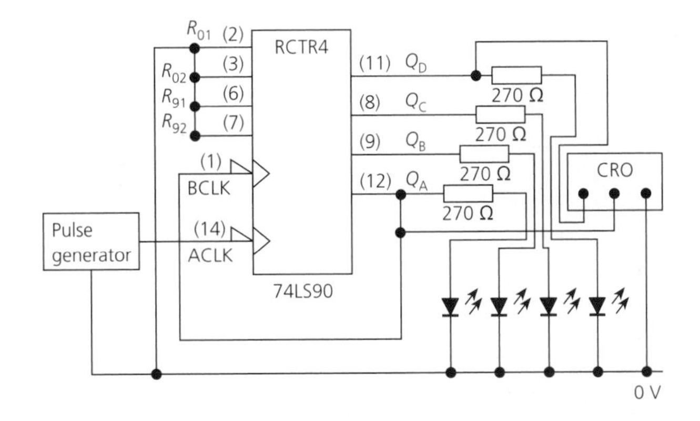

(c) Remove the connection between pins 1 and 12. Connect pin 11 to pin 14 and apply the input signal obtained from the pulse generator to pin 1. Connect the CRO between pin 12 and 0 V. What is now the function of the circuit and what is the frequency and mark–space ratio of the output waveform?

(d) Restore the original circuit. Remove the pulse generator from the input to the circuit and apply alternate 1s and 0s to the input. Observe which LEDs are ON each time and, hence, determine the function of the circuit.

(e) Reduce the frequency of the generator to about 50 Hz and observe the LEDs. Do they go ON and OFF as expected?

Clocks

Most digital circuits use some form of rectangular pulse generator, known as a *clock*, to control the times at which the various stages change state. At lower frequencies the clock may consist of an astable multivibrator, but at higher frequencies the clock is usually some form of crystal oscillator in order to achieve good frequency stability.

Crystal oscillator

When a crystal oscillator is used as the clock, it is necessary to convert its sinusoidal output voltage into the required rectangular waveform; this is easily achieved with the use of a Schmidtt trigger as shown in Fig. 10.16.

Fig. 10.16 *Crystal oscillator clock*

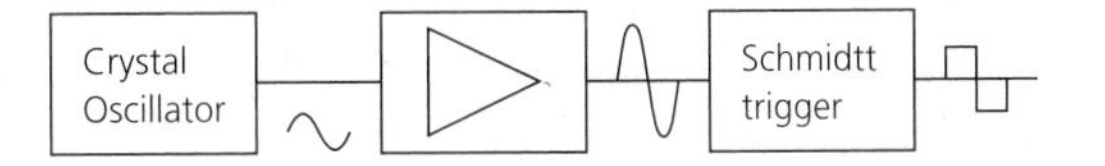

Fig. 10.17 *Logic gate clock oscillators*

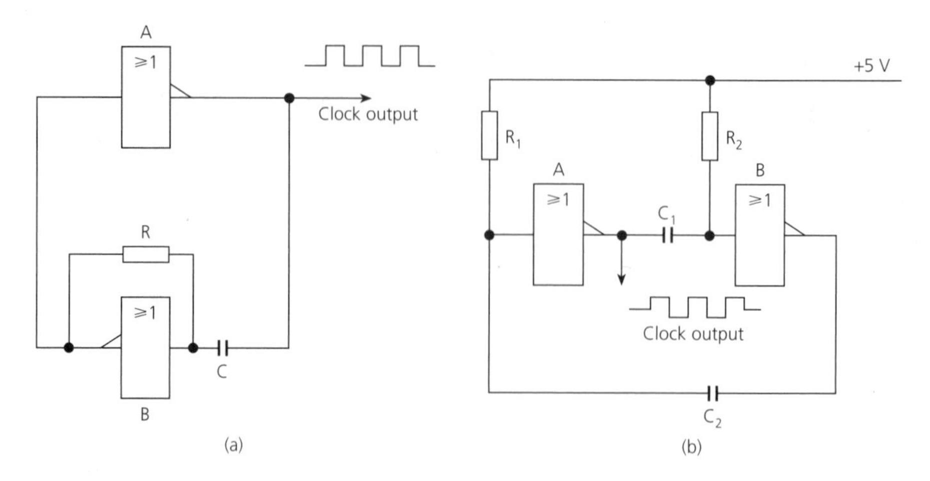

Logic gates

Another way of producing a clock is by the suitable interconnection of logic gates. Since both NAND and NOR gates include amplification, they can be used to form an oscillatory circuit because essentially an oscillator is merely an amplifier that provides its own input signal. One way in which NOR and NAND gates can be connected to form a clock is shown in Fig. 10.17(a). The feedback network is provided by capacitor C and resistor R together with NOR gate *B*. The amplifier is provided by the NOR gate *A*. Suppose that, initially, the capacitor C is discharged so that the clock output is zero. The gate *B* inverts its input voltage and so its output voltage is at the logic 1 level. Capacitor C is therefore charged via R with a time constant of *CR* seconds. As C charges, the voltage across it rises until it becomes equal to the value corresponding to logic 1. Immediately the output of gate *B* goes to logic 0 and so the output of gate *A* – which is the required clock output – goes to logic 1. Now capacitor C commences to discharge via R to the output of gate *B*. Once C has discharged to the level at which the output of gate *B* switches back to logic 1, the output of gate *A* goes to logic 0. The frequency of operation of the clock is determined by the time constant *CR*.

An alternative method of making a clock from NOR or NAND gates is given in Fig. 10.17(b).

Synchronous counters

A ripple counter can be used only for applications in which the speed of operation is not very important. If several stages of counting are employed, the time taken for a clock pulse to ripple through the counter may well be excessive. The operating time can be shortened considerably and glitches avoided by arranging for all the flip-flops to be clocked at the same moment. This is known as *synchronous* operation.

In a synchronous counter all the flip-flops change their state simultaneously, the operation of each stage being initiated by the clock. All flip-flops other than flip-flop A

Fig. 10.18 *Divide-by-8 synchronous counter*

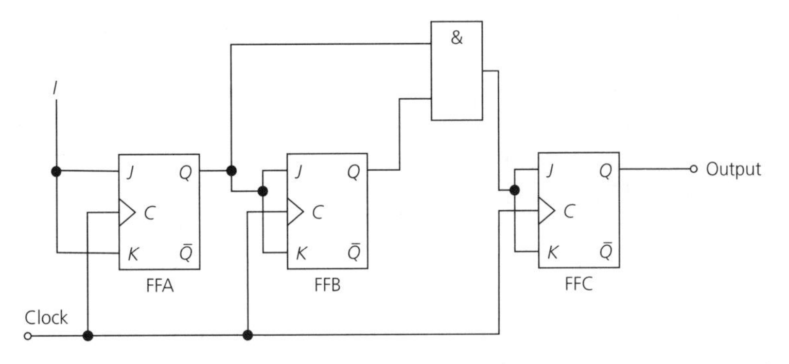

Fig. 10.19 *Synchronous counter*

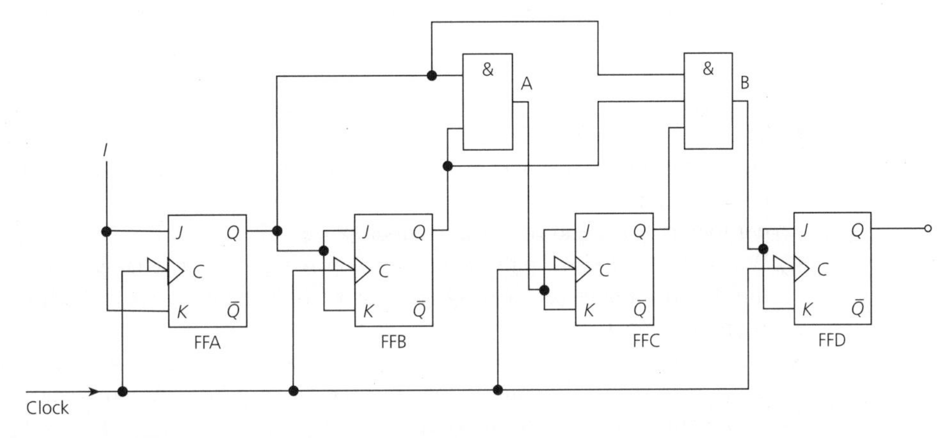

must be prevented from changing state until it is their turn to do so. This is arranged in the following way: the *n*th stage must change its state only for those clock pulses that arrive when all the preceding stages are set, i.e. have $Q = 1$.

Divide-by-8

Three stages are necessary. The first stage must toggle on each clock pulse and so $J_A = K_A = 1$. The second stage must toggle only when $Q_A = 1$ and, to achieve this, Q_A is connected to both J_B and K_B. Stage C must toggle only when *both* Q_A and Q_B are high and so these terminals are connected to the inputs of a 2-input AND gate. The output of the AND gate is then connected to both J_C and K_C. Thus, the required circuit is shown in Fig. 10.18.

Divide-by-16

Figure 10.19 shows the circuit of a 4-bit synchronous counter.

- Suppose that initially all the four stages are reset so that the count is 0, i.e. $Q_A = Q_B = Q_C = Q_D = 0$.
- At the trailing edge of the first clock pulse, flip-flop A toggles so that $Q_A = 1$.

Fig. 10.20 *(a) Divide-by-3 synchronous counter and (b) timing diagram for (a)*

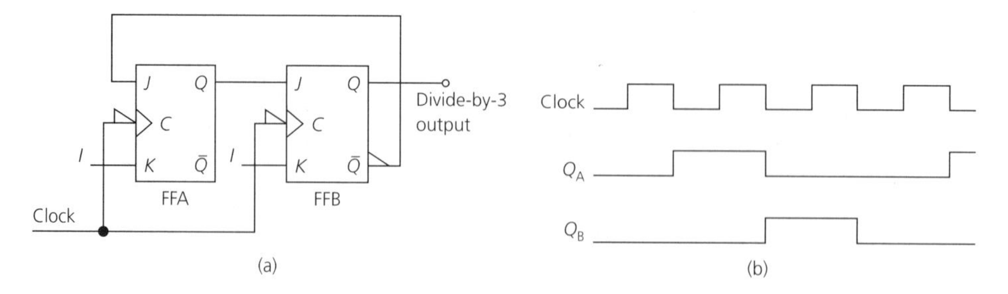

Fig. 10.21 *Methods of obtaining counts of (a) a multiple of 3 and (b) a product of 3 and 2^n*

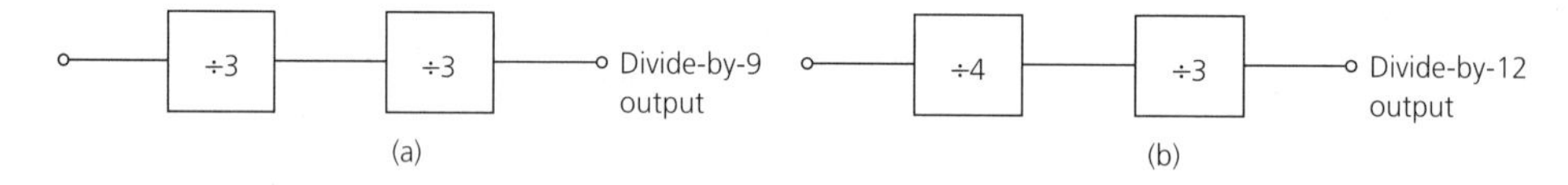

- The second stage now has $J_B = K_B = 1$ and so it will toggle when the second clock pulse ends and Q_A changes from 1 to 0. Now $Q_A = 0$ and $Q_B = 1$. AND gate A now has one input at 1 and the other input at 0 and so flip-flop C has $J_C = K_C = 0$ and will *not* toggle at the end of the next clock pulse.
- When the third clock pulse ends, flip-flop A toggles, Q_A changes from 0 to 1, and flip-flop B remains set. Now $Q_A = Q_B = 1$, so the count is 3, and both inputs to gate A are at logic 1.
- This means that flip-flop C has $J_C = K_C = 1$ and will toggle when the fourth clock pulse ends. Thus, the fourth clock pulse resets flip-flops A and B and sets flip-flop C; now only $Q_C = 1$ and the count is 4.
- The operation of the counter continues in this way as further clock pulses are applied. All three inputs to AND gate B will be at 1 after the *seventh* clock pulse has ended and $Q_A = Q_B = Q_C = 1$.

Divide-by-3

A divide-by-3 counter is easily obtained using two flip-flops with Q_A connected to J_B and $\bar{Q}_B$ connected to J_A, and also with $K_A = K_B = 1$ (see Fig. 10.20(a)). Flip-flop A is unable to toggle when $Q_B = 1$ and so the circuit counts to 3, as indicated by the timing diagram in Fig. 10.20(b).

Counts that are a multiple of 3 *or* the products of 3 and 2^n can also be easily obtained. Two examples are given in Fig. 10.21(a) and (b).

Divide-by-5

The principle can be extended to obtain a divide-by-5 counter as shown in Fig. 10.22. A decade counter can be obtained by preceding the circuit by a divide-by-2 circuit.

The synchronous counter is faster to operate than the non-synchronous counter because the clock frequency is limited only by the delay of *one* flip-flop (since all

Fig. 10.22 *Divide-by-5 synchronous counter*

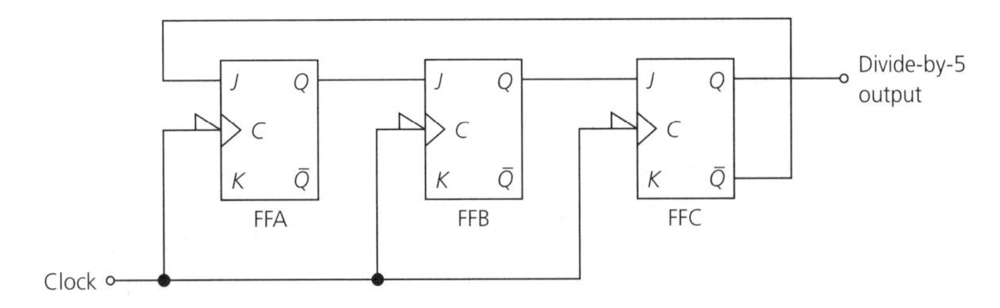

flip-flops operate simultaneously) plus the delays introduced by the AND gates. Other advantages are: (i) a decoded synchronous counter does not usually suffer from glitches, and (ii) since all flip-flops change state simultaneously, a binary (or BCD) output cannot be misread. As for a non-synchronous counter, a synchronous counter will count down if the outputs are taken from the Q outputs.

Reducing the count to less than 2ⁿ

The count of an n-bit synchronous counter can be reduced to less than 2^n.

EXAMPLE 10.2

Design a divide-by-6 synchronous counter using three 74HC112 J-K flip-flops.

Solution

The count will start from 000 and go through 001, 010, 011, 100, and 101. When the next clock pulse is received the circuit is required to reset to 000. Hence, the Q_B and Q_C outputs must be connected to a NAND gate and the output of the NAND gate connected to the active-LOW reset terminals of the three flip-flops. The required circuit is shown in Fig. 10.23 (*Ans.*)

Fig. 10.23

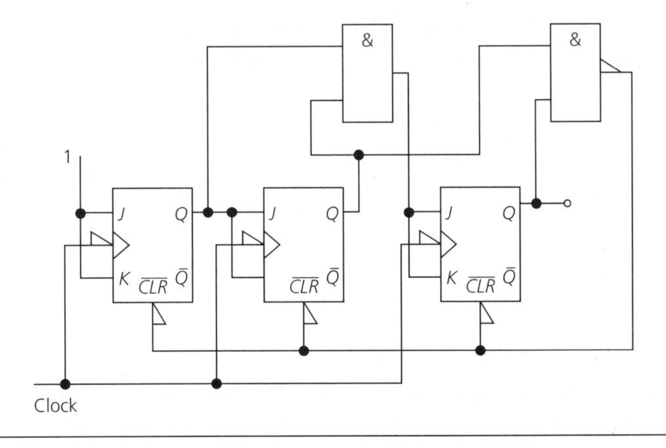

IC synchronous counters

Synchronous counters in the 74LS and 74HC logic families employ a combination of both pre-setting and resetting to modify their count. Four such counters are the 74LS161A/3A and the 74HC161/3 4-bit binary counters. These four counters have the same pin connections and operate in a similar manner to one another. 74LS161A and 74HC161 have non-synchronous clear, a LOW level at the $\overline{\text{clear}}$ input will clear all flip-flops regardless of the levels of the clock, $\overline{\text{load}}$ or enable (ENP and ENT) inputs. The clear function for the 74LS163A and 74HC163 is synchronous. A LOW level at the $\overline{\text{clear}}$ input clears the circuit after the next clock pulse.

74LS161A synchronous counter

The pinout of the 74LS161A synchronous counter is shown in Fig. 10.24(a) and the logic symbol is shown in Fig. 10.24(b).

Fig. 10.24 *A 74LS161A synchronous counter: (a) pinout and (b) logic symbol*

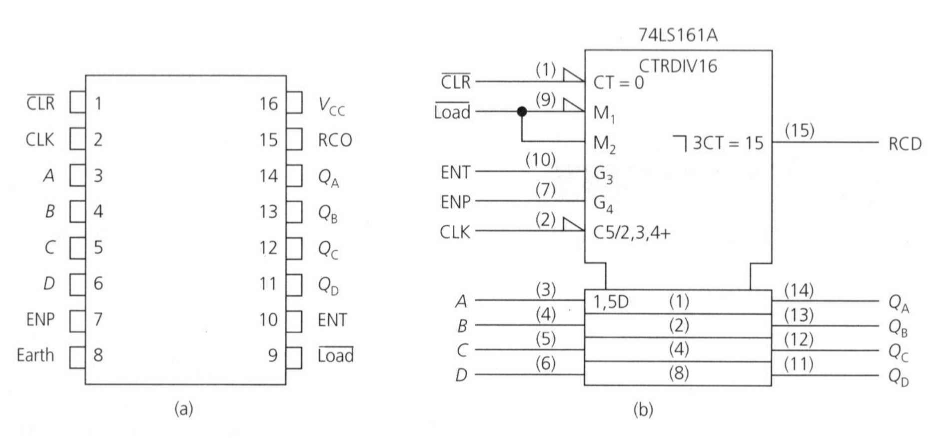

The meanings of the various symbols used are:

CTRDIV16: counter – divide-by-16. CT = 0: non-synchronous clear when reset is LOW. [If the clear were synchronous the CT symbol would be preceded by a number that corresponded to the clock number.]

M_1, M_2: mode dependency; functions (here load) occur for both HIGH and LOW levels.

G_3, G_4: AND dependency; for the count to proceed both the ENT and ENP pins must be HIGH.

1, 5D: input stores data as long as 1 and 5 are active.

3CT: outputs become active during the specified state.

The 74LS161A counter can be programmed to have any initial count in the range decimal 0–15. The application of a LOW voltage to the load terminal disables the counter and causes the four Q outputs to take up the same state as the data on inputs A, B, C and D after the next clock pulse, whatever the levels of the two enable inputs ENP and ENT. A LOW-level voltage applied to the $\overline{\text{CLR}}$ pin will set all four flip-flops to 0 after the next clock pulse, again no matter what the levels of the clock, load or enable inputs.

The operation of the counter is illustrated by the waveforms shown in Fig. 10.25 in which it has been supposed that (a) the flip-flops have first been reset, and (b) the counter has then been pre-set to binary 11. The counter will then count through the decimal sequence 12, 13, 14, 15, 0, 1, 2, 3, 4 → 11, 12, etc.

If the ENP pin is taken LOW at any instant the count will stop at that point. In the Fig. 10.25 ENP goes LOW during the eighth clock pulse and so the count is 12, 13, 14, 15, 0, 1, 2.

The counters have a ripple carry output which is used when the circuit is cascaded with one, or more, other counters. The ripple carry output goes HIGH when the count reaches the maximum, which is either 10 or 15. It will go LOW when the enable T pin goes HIGH even at the highest count to allow multi-stage operation.

Fig. 10.25 *Waveforms in a 74LS161A synchronous counter*

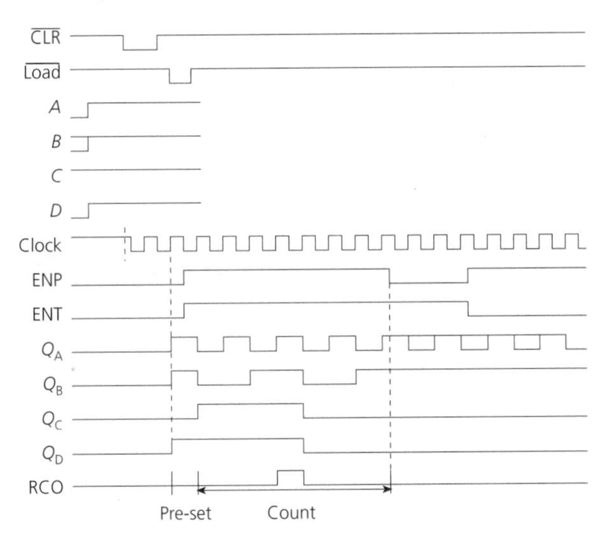

Figure 10.26 shows how the 74161/3 can be cascaded to obtain a higher count using the ripple carry out (RCO) and enable (ENP and ENT) pins. The RCO output goes HIGH when the count of an IC reaches 15 (1111) and this RCO pulse can be used to enable the next more-significant counter IC. There are two ways in which the counters may be cascaded known as *ripple-mode carry* and *look-ahead carry*.

Ripple-mode Carry

With ripple-mode carry, shown by Fig. 10.26(a), the ENP pin is held HIGH and the RCO output of the least-significant stage is connected to the ENT input of the next more-significant stage. The next stage is enabled for the duration of one clock pulse only and hence its count is incremented just the once when its ENT pin goes HIGH.

Look-ahead Carry

The look-ahead carry connection is shown by Fig. 10.26(b). The ENP pin of the least-significant stage is held HIGH and is connected to the ENT input of the more-significant stage. When the RCO output of the least-significant stage goes HIGH it enables the ENP input of the more-significant stage and its count is incremented.

Fig. 10.26

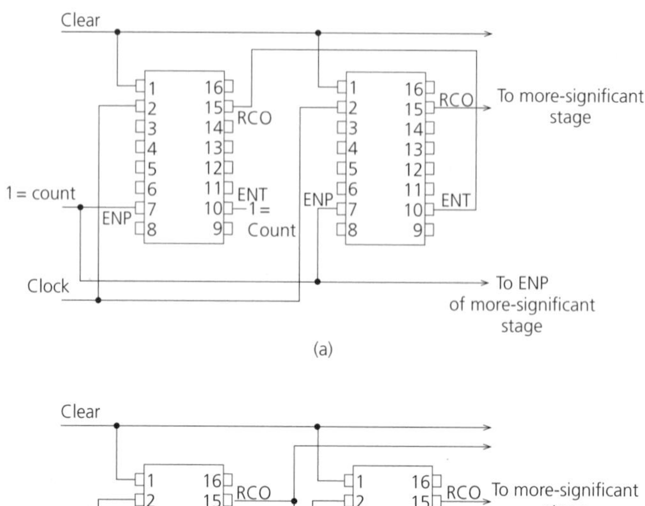

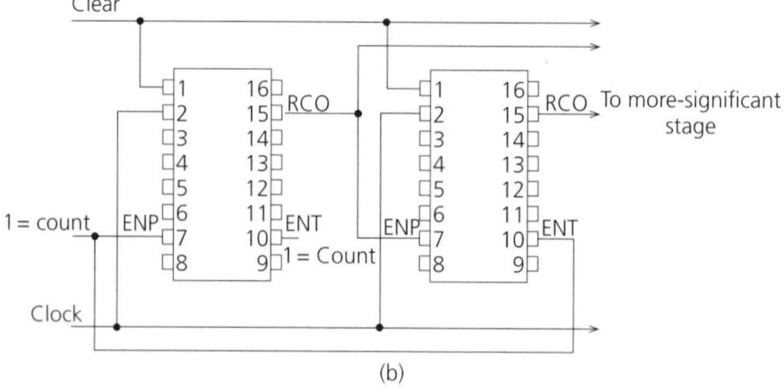

PRACTICAL EXERCISE 10.4

To investigate the 74LS161A (or 74HC161) 4-bit synchronous counter.
Components and equipment: one 741LS161A (or 74HC161) 4-bit synchronous counter IC. One 74LS08 (or 74HC08) quad 2-input AND gate IC. Four LEDs. Four 270Ω resistors. Pulse generator. Power supply. Breadboard.
Procedure:

(a) Build the circuit shown in Fig. 10.27; +5 V on pin 16 and 0 V on pin 8.
(b) Apply logic 0 to the ENP and ENT pins. Apply logic 0 to the $\overline{\text{CLR}}$ pin to reset the counter. Check that all four LEDs are OFF.
(c) Pre-set the counter to decimal 9 by setting the input pins to $A = D = 1$ and $B = C = 0$ and then place logic 0 on the $\overline{\text{load}}$ pin. This loads the data into the counter.
(d) Set the pulse generator to about 10 Hz and apply its output to the CLK input pin. Note what happens to the LEDs.

Fig. 10.27

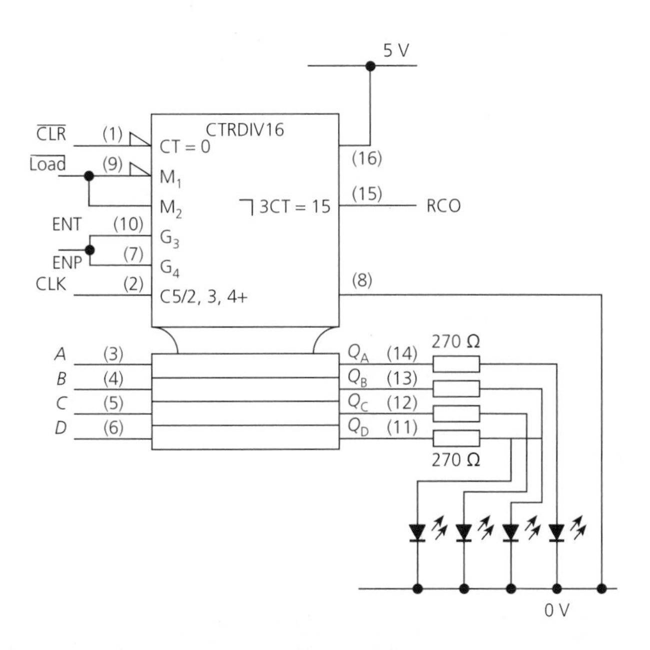

(e) Take the enable pins ENP and ENT to logic 1 and note what happens to the four LEDs. Now connect the ENP pin to logic 0. What happens to the counter? Next, take the ENP pin to logic 1 again and take the ENT pin to logic 0. What happened when this was done?

(f) Modify the circuit so that it counts from decimal 9 through 15 to 3 and then stops automatically. [ENP must go LOW when the count is 3 or $Q_D = Q_C = 0$ and $Q_A = Q_B = 1$.] Devise a gate arrangement that can be connected between one, or more, of the Q outputs and the ENP input to achieve this.

Up-down counters

All of the counters described so far in this chapter have counted from 0 up towards some number and are hence *up-counters*. For some digital applications it is necessary to be able to count downwards, e.g. 9, 8, 7, 6, 5, 4, 3, 2, 1, 0. Many circuits are capable of counting in either direction and the basic arrangement of a non-synchronous *up-down* counter is shown in Fig. 10.28.

If the count-up line is taken to logic 1 level, the AND gates A and D are enabled, connecting the Q outputs of flip-flops A and B to the clock input of the following flip-flops. The circuit then operates as an up-counter having a count of 8.

When the count-down line is at logic 1, and the count-up line is at logic 0, gates B and E are enabled, while gates A and D are inhibited. Now the $\bar{Q}$ outputs of flip-flops A and B are connected to the clock inputs of the following stages. Suppose that initially all three flip-flops are set, i.e. the count is 7. At the end of the first clock pulse, flip-flop A

Fig. 10.28 *Up-down counter*

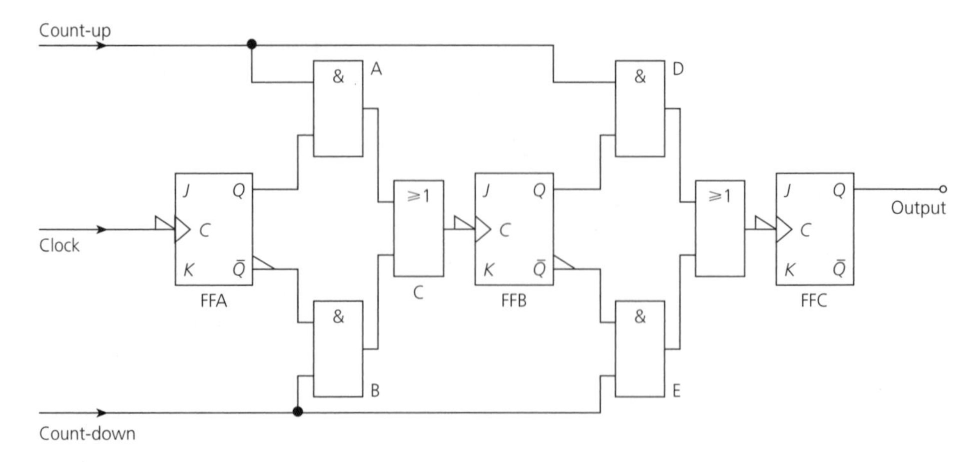

Fig. 10.29 *74LS193/74HC193: (a) pinout and (b) logic symbol*

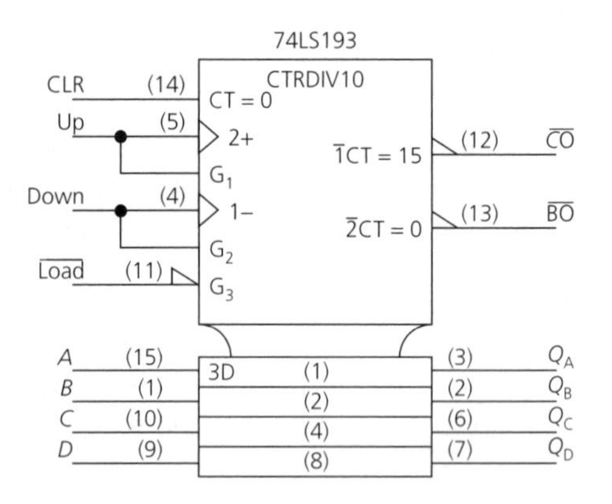

resets so that $\bar{Q}_A = 1$ and the count is 1 1 0 or 6. The next clock pulse causes flip-flop A to toggle and the trailing edge of its $\bar{Q}$ pulse resets flip-flop B; now the count is 1 0 1 or 5. At the end of the third clock pulse, the first stage toggles so that $Q_A = 0$, $\bar{Q}_A = 1$ and the count is 1 0 0 or 4. The state of the counter is now $\bar{Q}_A = \bar{Q}_B = 1$, $\bar{Q}_C = 0$, and so the fourth clock pulse sets flip-flops A and B and resets flip-flop C to give a count of 3, and so on until all three stages are reset. The count is then 0 and the next clock pulse will return the counter to its original count of 7.

IC up-down counters are of the synchronous type, e.g. the 74LS191/3 and the 74HC191/3 which are 4-bit counters.

The 74LS193 and the 74HC193 are 4-bit up-down binary counters. The logic symbol for the '193 is shown in Fig. 10.29. Each IC has two separate clock inputs: up for up-counting and down for down-counting. One clock input must be held HIGH while the other clock input is active. The clear pin is active-HIGH. It is taken HIGH to clear all the

Q outputs to 0. Each counter can be pre-set to any value in its range by placing a binary word on the data inputs A, B, C and D and then taking the load pin LOW. The counter operates in very similar fashion to the 74LS/HC161. The counters are operated in the following way:

- To count up, connect the $\overline{\text{load}}$ input to logic 0 to disable the counter while it is loading data.
- Connect the count down input to logic 1.
- Input the required data word to set the wanted starting point.
- Set the $\overline{\text{load}}$ input to logic 1.
- Apply the input waveform to the count-up terminal.
- To count down instead of up; connect the up terminal to logic 1 and use the down terminal as the input.

EXAMPLE 10.3

Design a 74HC193 up-down counter to give the count sequence 0, 1, 2, 3, 10, 11, 12, 0, etc.

Table 10.3

Q_A	0	1	0	1	0	0	1	0
Q_B	0	0	1	1	0	1	1	0
Q_C	0	0	0	0	1	0	0	1
Q_D	0	0	0	0	0	1	1	1

Solution

When the count reaches 4 the circuit must load 10 and resume counting. When the count reaches 13 the circuit must clear. The function table is shown in Table 10.3. When the count is 0100 it must be decoded and the decoded output used to $\overline{\text{load}}$ 1010. When the count is 1100 it must be decoded and the decoded output used to clear the counter. Hence, a NAND gate with inputs $\bar{Q}_D Q_C \bar{Q}_B \bar{Q}_A$ must be connected to the load terminals. An AND gate with inputs $Q_D Q_C \bar{Q}_B \bar{Q}_A$ must be connected to the clear terminal. The required complements can be obtained using either a hex inverter or a quad 2-input NAND gate. The circuit is shown in Fig. 10.30 (*Ans.*)

The relative merits of TTL and CMOS circuits have been tabulated earlier (p. 171). TTL counters can operate at much higher clock frequencies than 4000 CMOS, e.g. standard TTL 20–30 MHz, low-power Schottky TTL 10–100 MHz, CMOS 10 MHz. But HC and AC circuits are as fast as their TTL alternatives.

Fig. 10.30

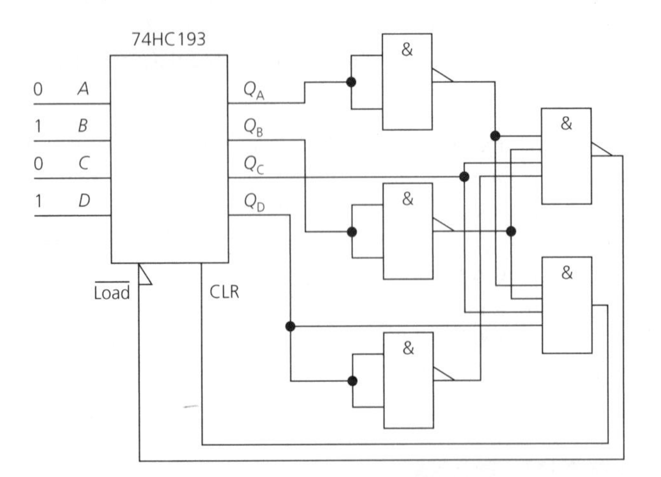

All versions of CMOS counters dissipate much less power than TTL counters. Another advantage of CMOS devices is that a much larger number of stages can be provided within a standard d.i.l. IC package. The 74HC4020 has 14 stages and the 74HC4040 has 12 stages, for example. Because of the limitations on the number of package pins, only a few stages (usually one) can have an output that is externally accessible.

PRACTICAL EXERCISE 10.5

To investigate the operation of the 74LS193 (or 74HC193) 4-bit synchronous up-down counter.

Components and equipment: one 74LS193 (or 74HC193) counter IC. Four LEDs. Four 270 Ω resistors. Pulse generator. Power supply. Breadboard.

Procedure:

(a) Build the circuit shown in Fig. 10.31.
(b) Move the clear connection to logic 0, $\overline{\text{load}}$ to 0, and input the data word 0000. Connect down to 1.
(c) Move the count-up connection between logic 0 and logic 1 a number of times and each time note the state of each LED. Check that the count starts at decimal 0 and proceeds up to 15 before resetting back to 0. Also note the states of the LEDs that are connected to the Borrow $\overline{\text{BO}}$ and Carry $\overline{\text{CO}}$ terminals.
(d) Load the input data word 0111. Do this by first setting $\overline{\text{load}}$ to logic 0 (which disables the circuit) and then applying $D = 0$, $C = B = A = 1$ to the data input terminals. Connect the CLR terminal to logic 1 and the back to logic 0 and note the state of each LED. What count is indicated? Now return the load terminal to logic 1 and repeat procedure (c). Connect the pulse generator to the count-up pin 15 and display the waveforms on the CRO.
(e) Connect the count-up terminal to logic 1 and use the count-down terminal as the counter input. Repeat procedures (b); (c) and (d).

Fig. 10.31

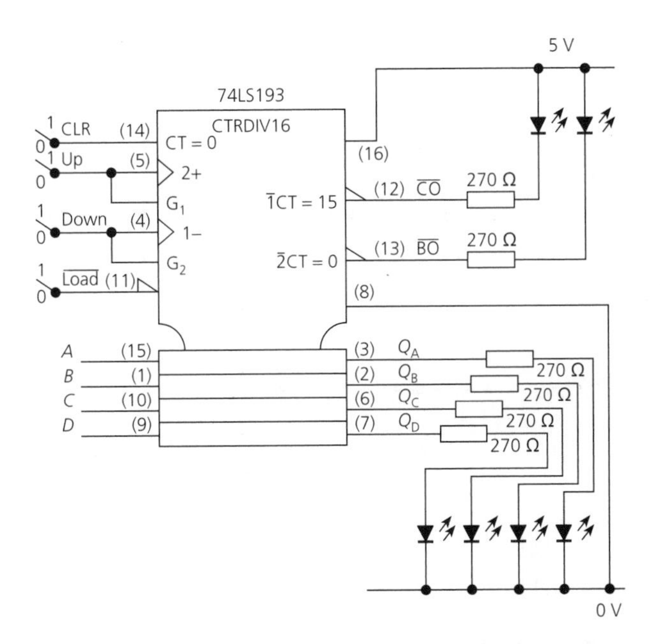

Fig. 10.32

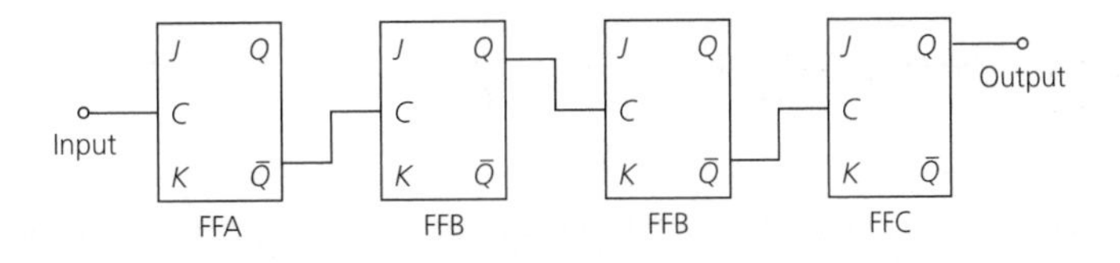

Fig. 10.33

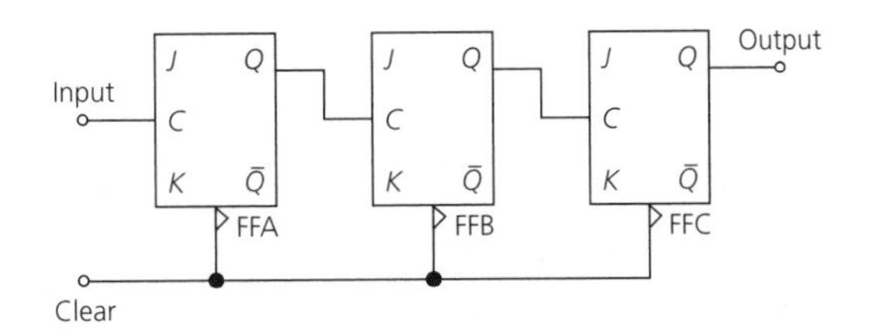

EXERCISES

[In all circuits $J = K = 1$ when not otherwise shown.]

10.1 For Fig. 10.32:

(a) What is the count of the circuit?

(b) Is it a synchronous or a non-synchronous circuit?

(c) To what logic levels must (i) the J, and (ii) the K inputs be connected?

10.2 Modify the circuit shown in Fig. 10.33 to have a count of 6.

Fig. 10.34

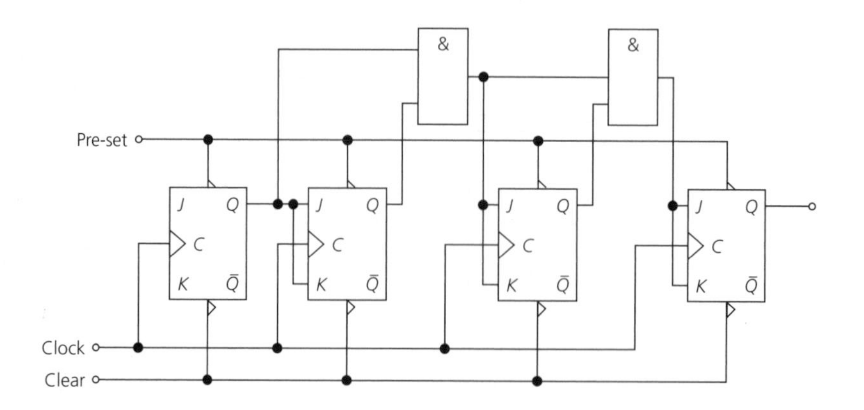

Fig. 10.35

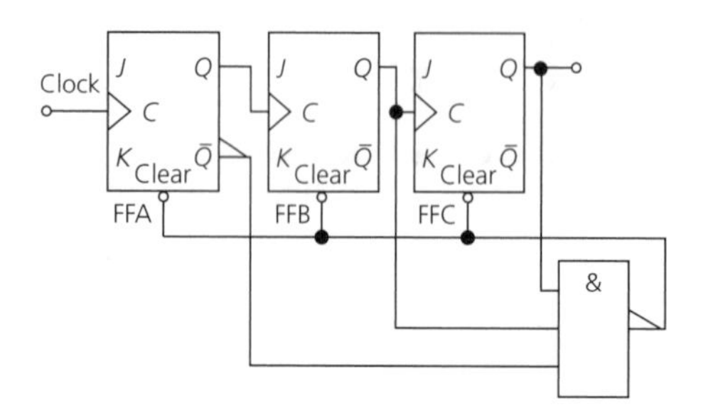

Fig. 10.36

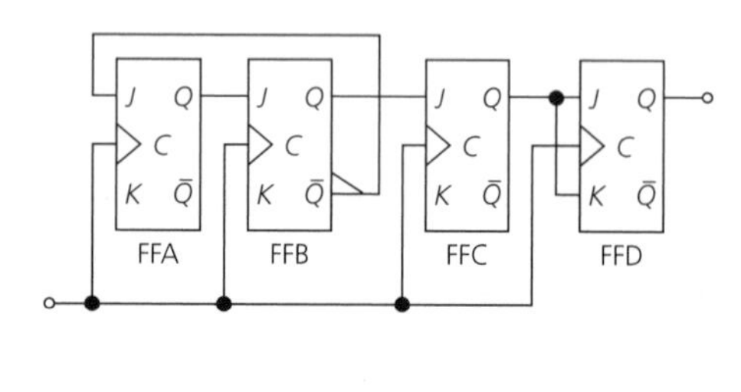

10.3 Give a step-by-step description of the operation of the counter given in Fig. 10.34.

10.4 Show how one 74LS90 and one 74LS93 IC counter can be connected together to give a count of 80. The pin connections of these devices are given in Figs 10.12 and 10.14.

10.5 Determine the count of the circuit given in Fig. 10.35.

10.6 Determine the count of Fig. 10.36.

10.7 (a) Show how a 4-bit synchronous counter can have its count reduced to 9 by using the reset terminal of each flip-flop.
(b) The CMOS 4020 counter has 14 stages. What is its maximum count? Why does it have only 12 Q outputs?

10.8 (a) List the relative merits of TTL and CMOS counters.
(b) Three J-K flip-flops are connected to operate as a ripple counter. Give the counting sequence obtained at (i) the Q outputs and (ii) the $\bar{Q}$ outputs.

10.9 Determine
(a) the maximum count.
(b) the decoded count of the circuit given in Fig. 10.37.

Fig. 10.37

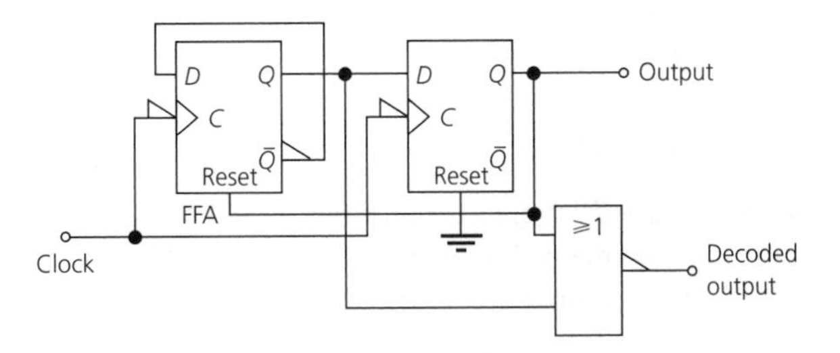

Fig. 10.38

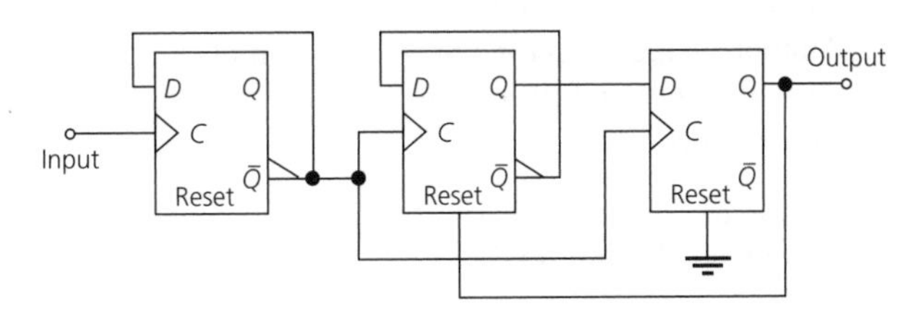

10.10 Determine the count of the circuit shown in Fig. 10.38.

10.11 (a) Two 74LS90 decade counters are to be connected together to give a count of 40. State the necessary connections.

(b) Two 74LS90 counters are connected in cascade. One has pin 12 to pin 1, the other has input as pin 1. Determine the count of the circuit.

10.12 (a) Draw the diagram of a divide-by-8 synchronous counter that uses three trailing-edge-triggered J-K flip-flops.

(b) What is

(i) The modulus.

(ii) The highest number of the counter?

(c) What code

(i) Precedes 000

(ii) Follows 000?

10.13 (a) Describe the meaning of the label C5/2,3,4+ in the logic symbol of the 74LS161A synchronous counter given in Fig. 10.24.

(b) The counter has an input waveform of periodic time 1 ms. Calculate the frequencies of the waveforms appearing at each of the Q outputs.

10.14 A 4-bit binary counter is to be decoded to produce an output when the count is (a) 1, (b) 9 and (c) 13. Determine the Q outputs which must be decoded in each case.

10.15 A 4-bit non-synchronous counter using J-K flip-flops is to have its count reduced to 10. Design the counter if the flip-flops have active-LOW reset terminals.

10.16 Design a divide-by-6 counter using the 74LS90 IC.

10.17 State how the following counters can be obtained using J-K flip-flops.
(a) Divide-by-3.
(b) Divide-by-12.
(c) Divide-by-9.
(d) Divide-by-5.

10.18 A 74HC4040 ripple counter has a count of 4096.
(a) How many stages does it have?
(b) If an output waveform of frequency 66 Hz is required what should be the clock frequency?

10.19 Four T flip-flops are connected to form a 4-bit counter. Each T input is connected to logic 1 and the clock input of each flip-flop is connected to the Q output of the preceding stage. The input is applied to the clock terminal of the first flip-flop and the output is taken from the Q terminal of the last flip-flop. If all the flip-flops are initially reset, determine the count sequence.

10.20 Draw a diagram to show how a 4-bit binary counter can be decoded to have outputs 1 through to 7.

10.21 Show how a 74LS92 can be connected as a divide-by-6 counter.

10.22 Show how a 74LS90 can be connected as a divide-by-6 counter.

11 Shift registers

After reading this chapter you should be able to:

(a) Connect a number of J-K or D flip-flops as a shift register.
(b) Understand the four modes of operation of a shift register.
(c) Explain what is meant by a universal shift register.
(d) Use an IC shift register.
(e) Explain the operation of a ring counter.

A *register* consists of a number of flip-flops connected in such a way that data can be stored. A *shift register* can have its stored data moved, or *shifted*, either to the right or to the left by the application of the clock. A shift register consists of a number of either D or J-K flip-flops connected in cascade. The number of flip-flops used to form the register is equal to the number of bits to be stored. The flip-flops are generally provided with clear or reset terminals so that the register can be cleared.

A shift register can be operated in any of the four ways: (a) serial-in/serial-out, (b) serial-in/parallel-out, (c) parallel-in/serial-out, and (d) parallel-in/parallel-out. These are shown in block diagram form in Fig. 11.1. The main applications for a shift register are (a) the temporary storage of data, (b) serial-to-parallel or parallel-to-serial conversion, (c) digital delay circuit and (d) mathematical operations.

Serial-in/serial-out shift register

In a *serial-in/serial-out shift register* data is entered one bit at a time, moves through the register from left to right, or from right to left, and leaves one bit at a time. The basic circuit of a SISO shift register that uses J-K flip-flops is shown in Fig. 11.2(a). The Q output of each stage is connected to the J input of the next, and each $\bar{Q}$ output is connected to the K input of the following flip-flop. The first stage is connected to act as a D flip-flop. Each new bit is clocked into the first flip-flop and the bit that was stored by that flip-flop moves to the second flip-flop, and so on. Each stage acts as a temporary storage device. Thus, each flip-flop takes up the logical state of the preceding flip-flop at the trailing edge of each clock pulse. The most significant bit of the data word is the first to be entered. It will take four clock pulses to load a 4-bit data word, and another four

Fig. 11.1 *Methods of using a shift register: (a) serial-in/parallel-out; (b) parallel-in/serial-out; (c) serial-in/serial-out and (d) parallel-in/parallel-out*

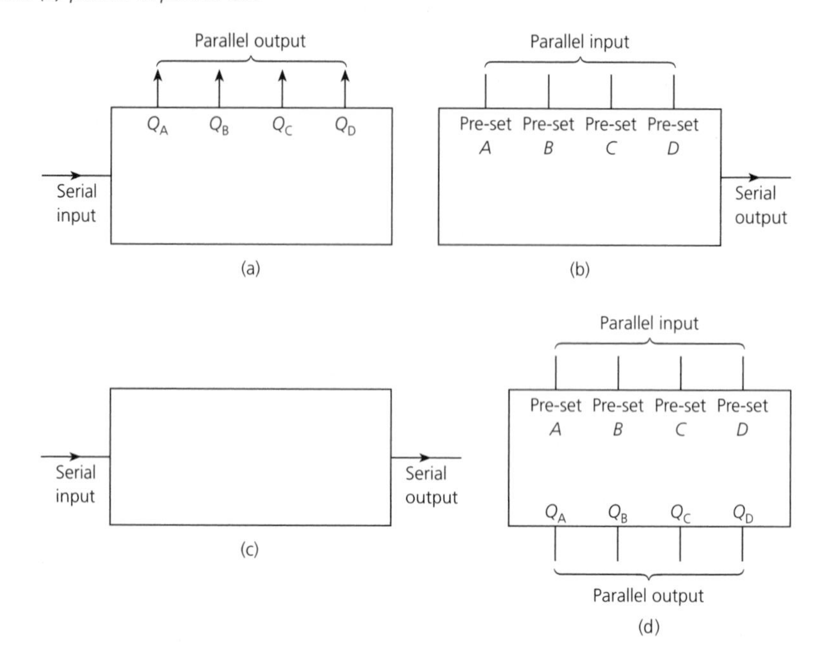

Fig. 11.2 *(a) Shift register using J-K flip-flops and (b) using D flip-flops*

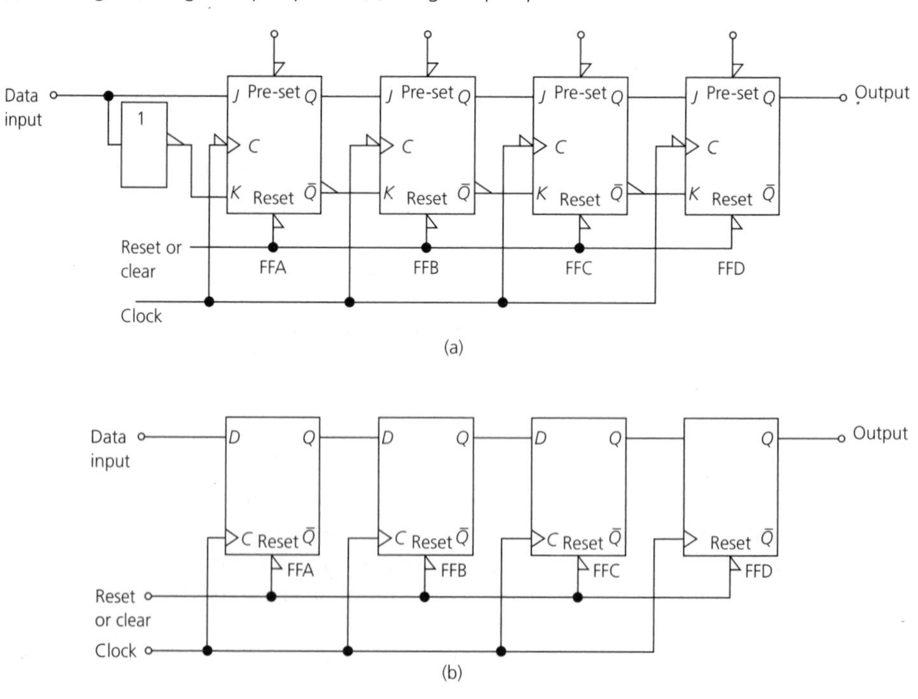

Fig. 11.3 *Timing diagram for a 4-bit shift register*

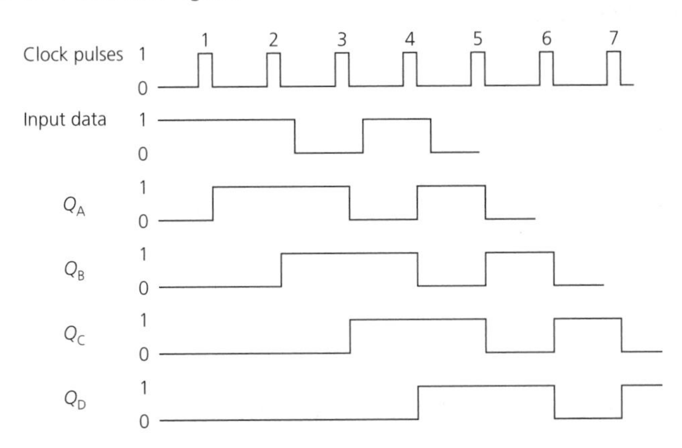

clock pulses to shift the data out of the circuit one bit at a time. A shift register is normally shown as shifting data to the right. If the stored data is a binary number the least significant bit is on the left, i.e. stored in flip-flop A.

If the data to be stored by the register is applied serially to the input terminal, the input data will shift one stage to the right at the trailing edge of each clock pulse. This means that, if the data stored in the register is a binary number, the most significant bit is on the right. If the data applied to the input terminal is 1101, the bits 11 are loaded first and the action of the circuit is as given below.

Suppose that initially all the four flip-flops are cleared, i.e. $Q_A = Q_B = Q_C = Q_D = 0$. At the end of the first clock pulse, $Q_A = 1$. After the second pulse, $Q_A = 1$, $Q_B = 1$. The third clock pulse will cause $Q_B = Q_C = 1$, but the first flip-flop will now reset to have $Q_A = 0$. The fourth bit of data is a 1 and this will be stored by flip-flop A at the end of the fourth clock pulse, when the bits stored by the other flip-flops all move one place to the right. At this stage the data stored is 1101. Now the complete number is stored by the register. At the end of the fifth clock pulse, flip-flop A clears and all the other flip-flops take up the state of the preceding stage. The most significant bit of the data has been shifted out of the register and lost.

After the sixth clock pulse, the two most significant bits have been lost, and so on. After eight pulses all the data has been lost.

The timing diagram illustrating the operation of the register is shown in Fig. 11.3. Note that the Q_A waveform is delayed behind the input data waveform by a time that may not be equal (in the figure it is not) to the periodic time of the clock waveform. However, the Q waveforms of all the other flip-flops are delayed behind the Q waveform of the preceding flip-flop by *exactly* the clock period.

A serial-in/serial-out shift register can be used as a delay circuit or as a temporary store, but the stored data can be accessed only in the order in which it was stored.

A SISO shift register that uses D flip-flops is shown in Fig. 11.2(a). D latches cannot be employed. The Q output of each stage is connected to the D input of the following stage. At the leading edge of each clock pulse the data held in the register is shifted one place to the right and data enters from the left at flip-flop A. The data is outputted serially via Q_D.

Table 11.1

Clock pulse	*Data at D_A*	Q_A	Q_B	Q_C	Q_D
0	1	0	0	0	0
1	0	1	0	0	0
2	0	0	1	0	0
3	0	0	0	1	0
4	0	0	0	0	1
5	0	0	0	0	0

Fig. 11.4 *Timing diagram for a 4-bit D flip-flop connected as a shift register*

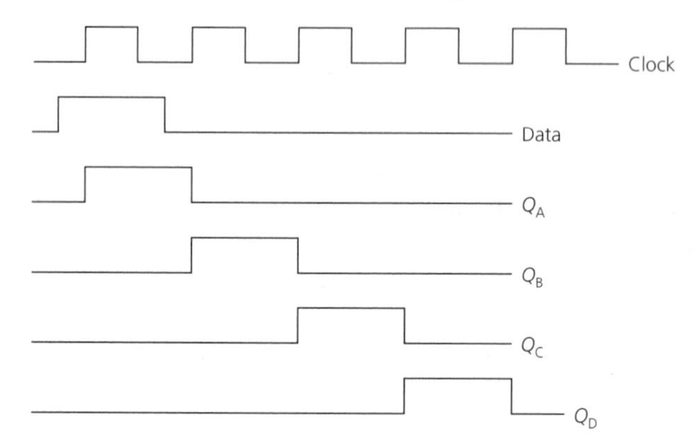

Assume that all the stages have been cleared and that the serial input data is 1 followed by several 0s. The operation of the circuit is tabulated in Table 11.1.

The data bit is transferred, first to Q_A, and then, at the trailing edge of each clock pulse, to Q_B, Q_C and Q_D before dropping out of the register after the fifth clock pulse. The waveforms in the circuit are shown in Fig. 11.4.

Serial-in/parallel-out shift register

The circuit of a serial-in/parallel-out (SIPO) shift register is basically the same as that of a SISO shift register except that all of the Q outputs are externally accessible. Figure 11.5 shows the circuit of a SIPO shift register. The basic SISO circuit is extended by the four AND gates that are used to allow the stored data word to be read out of the register. One clock pulse is needed for each stage to load the input data, i.e. four clock pulses for a 4-bit circuit, but only one clock pulse is needed to output the parallel data. When the complete data word has been stored all the bits can be read out simultaneously by applying logic 1 to the read data line.

Fig. 11.5 *SIPO shift register*

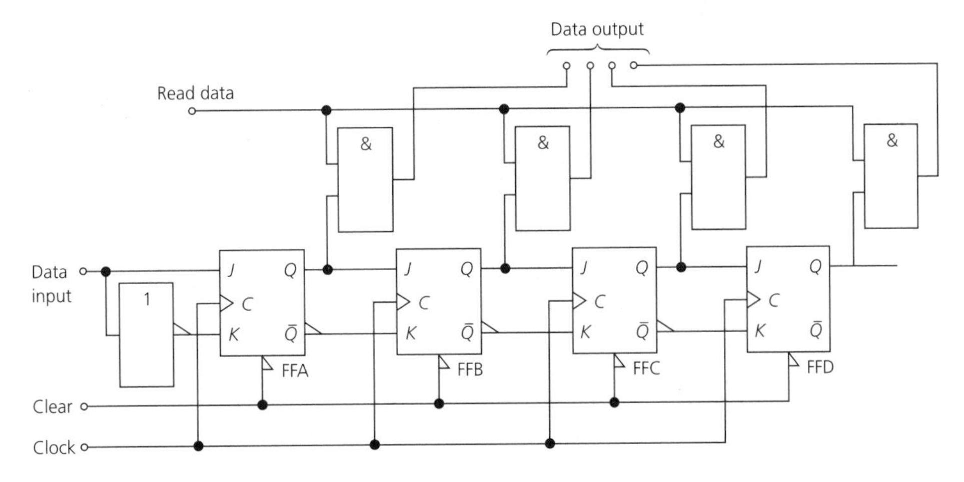

Fig. 11.6 *PISO shift register*

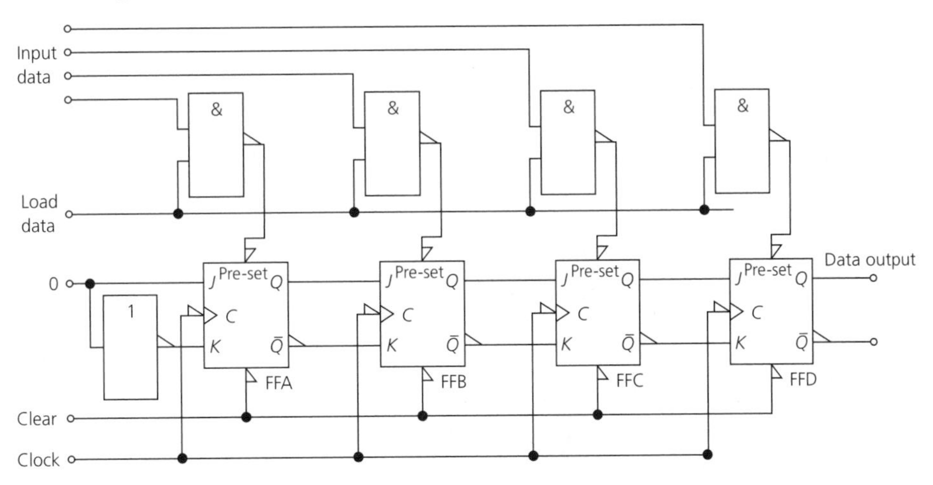

Parallel-in/serial-out shift register

The parallel-in/serial-out (PISO) shift register, shown in Fig. 11.6, operates in exactly the opposite way. The data to be stored is set up by first clearing all the stages and then applying logic 1 to the load data line. If an input data line is at logic 0, logic 1 will be applied to the associated pre-set terminal and the flip-flop remains cleared. If a data input is at logic 1, the associated flip-flop has its pre-set terminal taken LOW and hence that flip-flop is set. Once the 4-bit data word has been loaded it can be shifted to the right by applying successive clock pulses to the register.

Figure 11.7 shows the waveforms in the circuit when the parallel input data word is 0111.

Fig. 11.7 *PISO waveforms for data word 0111*

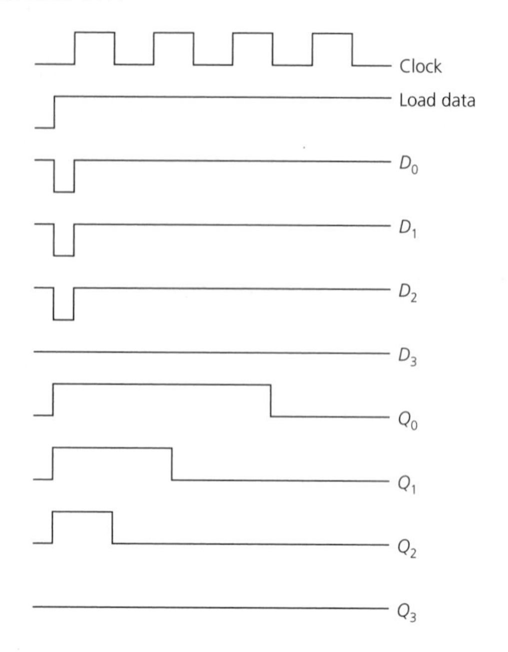

Fig. 11.8 *PIPO shift register*

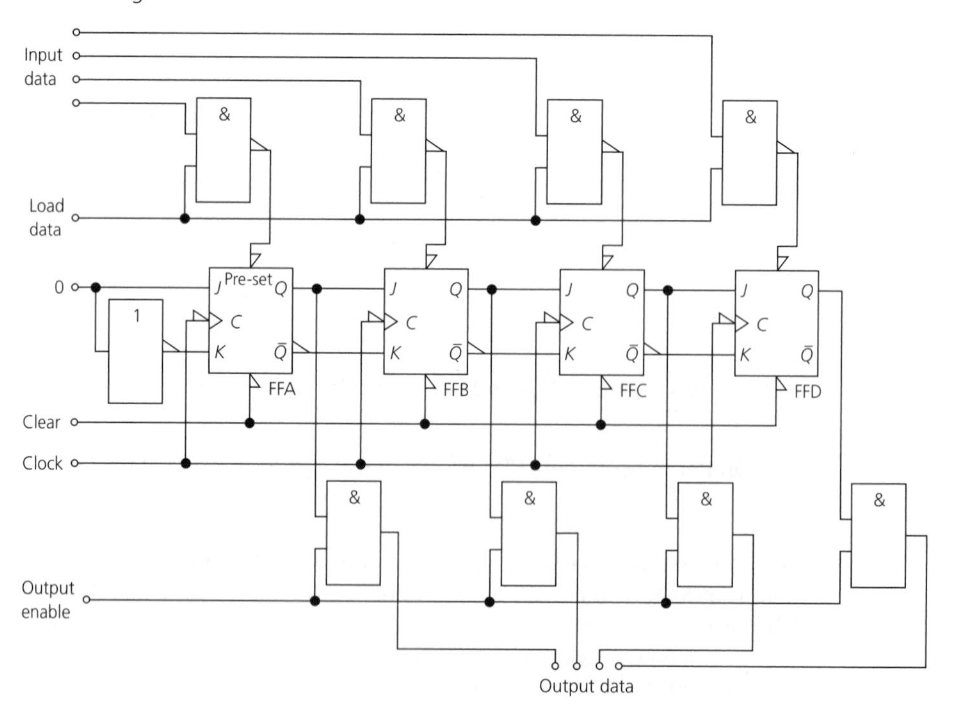

Parallel-in/parallel-out shift register

Figure 11.8 shows a parallel-in/parallel-out shift register and this also acts as a short-term store. All the bits of a data word enter and later leave the circuit simultaneously. Data is loaded into the register in the same way as for a PISO circuit and is outputted when the parallel output enable line is taken HIGH.

PRACTICAL EXERCISE 11.1

To investigate the basic operation of each type of shift register.
Components and equipment: one 74LS00 (or 74HC00) quad 2-input NAND gate IC, two 74LS112A (or 74HC112) dual J-K flip-flop ICs. Four LEDs. Four 270 Ω resistors. Power supply. Breadboard.
Procedure:

Fig. 11.9

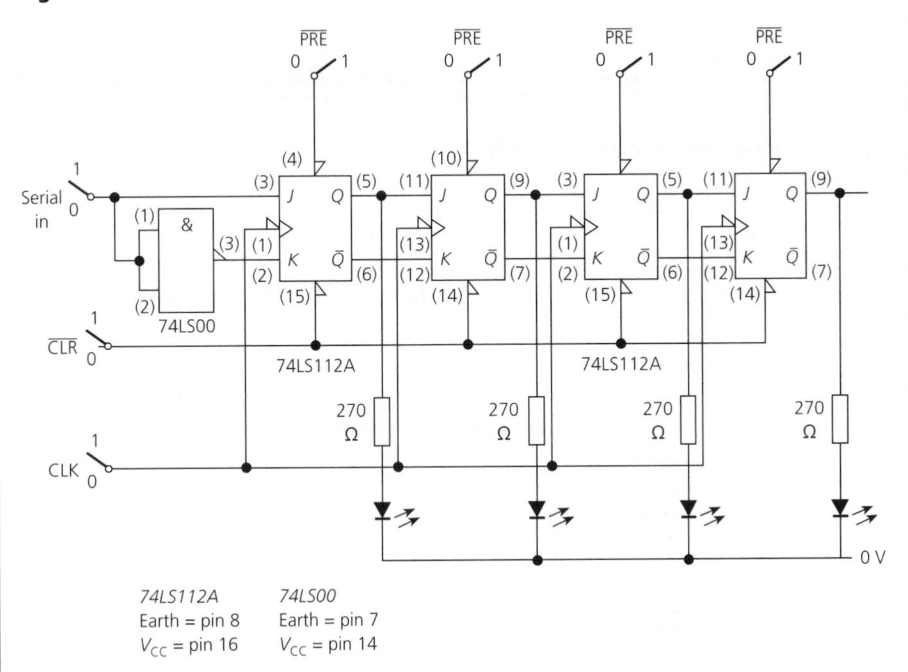

(a) Build the 4-stage SISO shift register shown in Fig. 11.9.
(b) Apply logic 0 to the $\overline{\text{CLR}}$ line and logic 1 to the $\overline{PRE}$ line to clear all the stages. Ensure that the LEDs are dim. Return $\overline{\text{CLR}}$ to 1.
(c) Set the serial input to logic 1 and move the CLK input from 1 to 0. Note what, if anything, happens. Now move the CLK input from 0 to 1 and again note what happens.

(d) Move the serial input to logic 0 and repeat procedures (b) and (c). Next, move the serial input back to logic 1 and again repeat procedures (b) and (c) twice. The data word 1011 has now been entered into the register. Note the logical states of the four LEDs.

(e) Move the CLK input between logical 1 and 0 several times and see whether the stored data is moved to the right and then out of the register. Note the sequence of logical states for the LED connected to Q_D – this indicates the serial output from the register. Say whether or not it was the same as the input data word.

(f) State how the circuit could be used as a SIPO shift register.

(g) Ensure that all the flip-flops are cleared and then move the pre-set input to flip-flops A and C to logic 0. The parallel input data word is then 1010. Note the logical states of the four LEDs.

(h) Connect the serial input to logic 1 and then move the CLK input several times between 1 and 0. Note what happens to the LEDs.

(i) Lastly, connect the serial input to logic 0 and repeat procedure (h). What happens now?

Bi-directional and universal shift registers

Bi-directional shift register

A *bi-directional shift register* is one whose stored data can be shifted either to the left or to the right. The circuit of such a register is shown in Fig. 11.10. There are two data input terminals, one of which is used for serial data that is to be shifted to the right and the other is for left-shifting data. The direction in which the data is shifted is determined by the logic levels on the two lines marked, respectively, as shift right 1/shift left 0 and shift right 0/shift left 1. The circuit is arranged so that the signals applied to these two lines are always the complements of one another. When the top AND gates are enabled, the data-right input is connected to flip-flop A. Q_A is connected to flip-flop B and so on, so that the circuit is similar to that given in Fig. 11.2(b).

Fig. 11.10 *Shift-right/shift-left register*

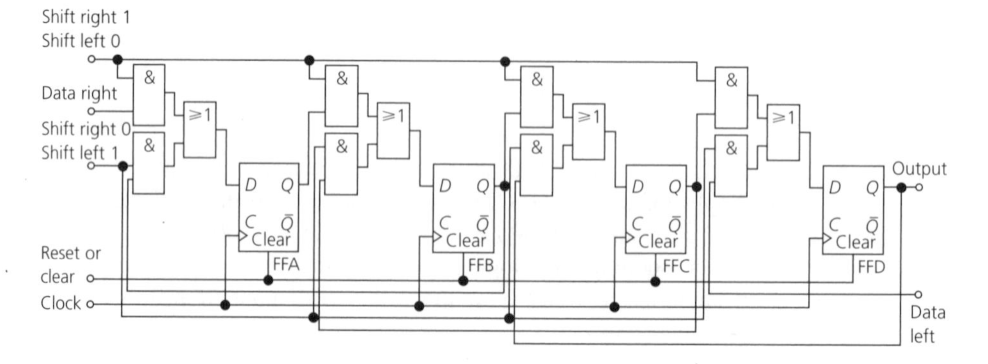

Fig. 11.11 *Universal shift register*

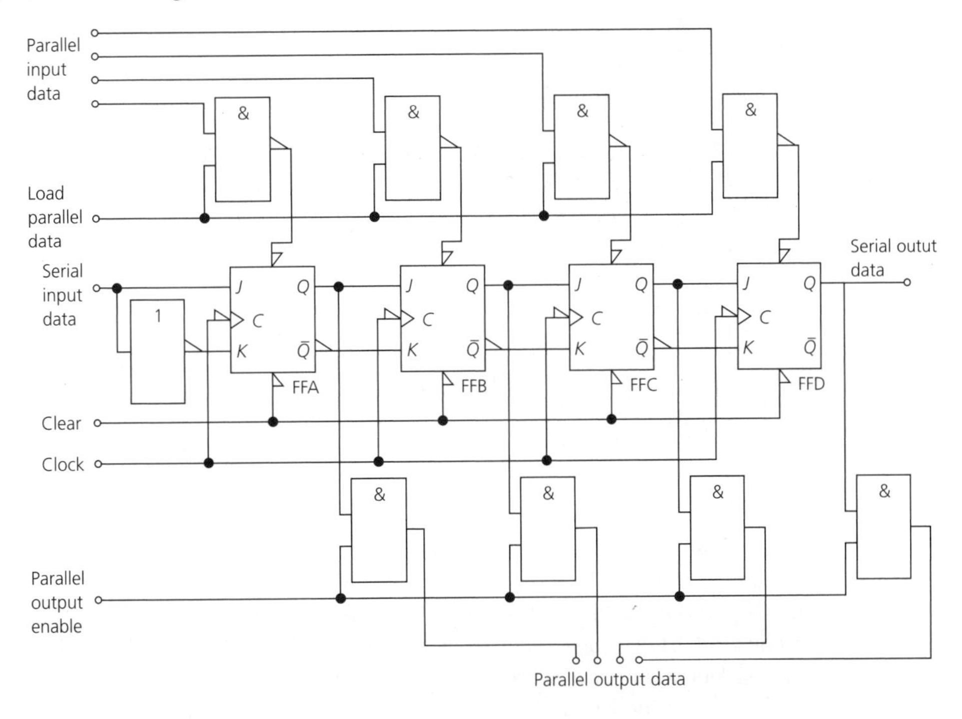

Conversely, when the lower AND gates are enabled, Q_D is connected to the D terminal of flip-flop C, Q_C is connected to flip-flop B and so on. The circuit will then shift data entered serially at the data-left terminal to the left.

Universal shift register

A *universal shift register* is one that can be employed in any of the four modes of operation and is also both left-, and right-shifting. Control inputs are employed to determine the mode of operation. The circuit of a universal shift register which is able to operate in any of the four modes is shown in Fig. 11.11. It can be seen to combine the features of the previous circuits.

IC shift registers

Both the TTL and CMOS logic families include several examples of shift registers; some examples are listed below:

74LS164 and 74HC164 8-bit right-shift SISO/SIPO
74LS165 and 74HC165 SIPO or PISO
74LS194 4-bit universal
74LS195 4-bit PIPO/SIPO

Fig. 11.12 *A 74LS164/74HC164 SISO/SIPO shift register: (a) pinout and (b) logic symbol*

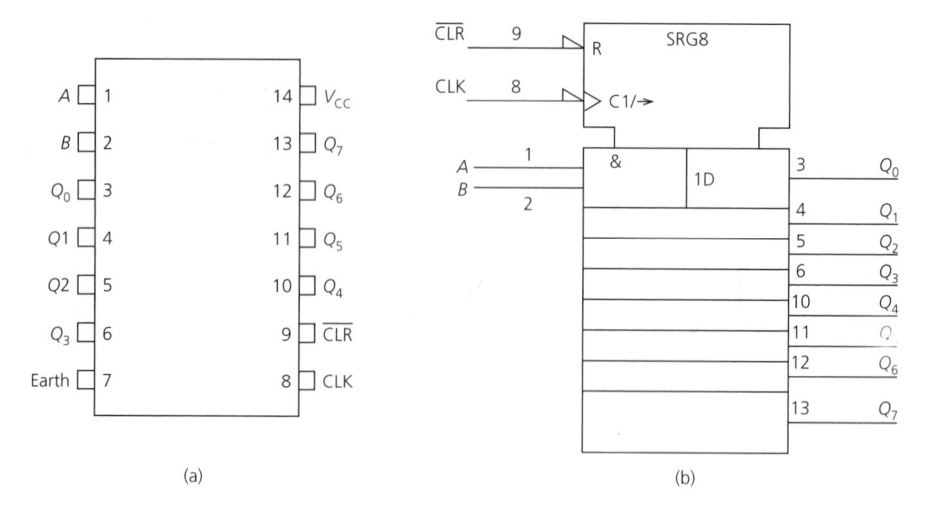

The logic symbol for a shift register uses the identifying symbol SRG*n*, where *n* is the number of bits. The direction of the arrow indicates right-, or left-shifting. The CLR and CLK inputs to the control block show that all states are cleared, or clocked, simultaneously. CLR is non-synchronous and this is shown by R at the CLK input. D represents data input to a storage element. Shift registers can often be operated in any one of several modes, under the control of one or more binary inputs. The letter M is used to show inputs that are affected by the operating mode of a register. If there are four modes it is shown by M0/3.

74LS164 and 74HC164 8-bit SISO/SIPO shift register

The 74LS164 and 74HC164 8-bit shift register has two gated inputs, A and B, and a non-synchronous active-LOW clear. The pinout of the device is given by Fig. 11.12(a) and its logic symbol is given by Fig. 11.12(b). The Q outputs of each of the eight stages are brought out on to pins 3–6 and 10–13 and they can be used to provide a parallel output. A serial output can be taken from pin 13 (Q_7). Serial input data applied to input A is shifted into the register once input B is taken, and held, HIGH. Data at the serial input can be changed while the clock is HIGH or when it is LOW but it must be held steady during the set-up time if it is to be entered into the register. Data is moved to the right at the leading-edge of each clock pulse.

74LS194A 4-bit Universal Shift Register

The pinout of the 74LS194A leading-edge-triggered universal shift register is shown in Fig. 11.13(a) and its logic symbol in Fig. 11.13(b). The function table is given in Table 11.2.

When the reset input is LOW all four outputs Q_A, Q_B, Q_C and Q_D are taken LOW also, whatever the logical states of the other inputs. When the clock input is LOW the four outputs remain in the logical state they were in just before the clock transition.

Fig. 11.13 *A 74LS194 4-bit universal shift register: (a) pinout and (b) logic symbol*

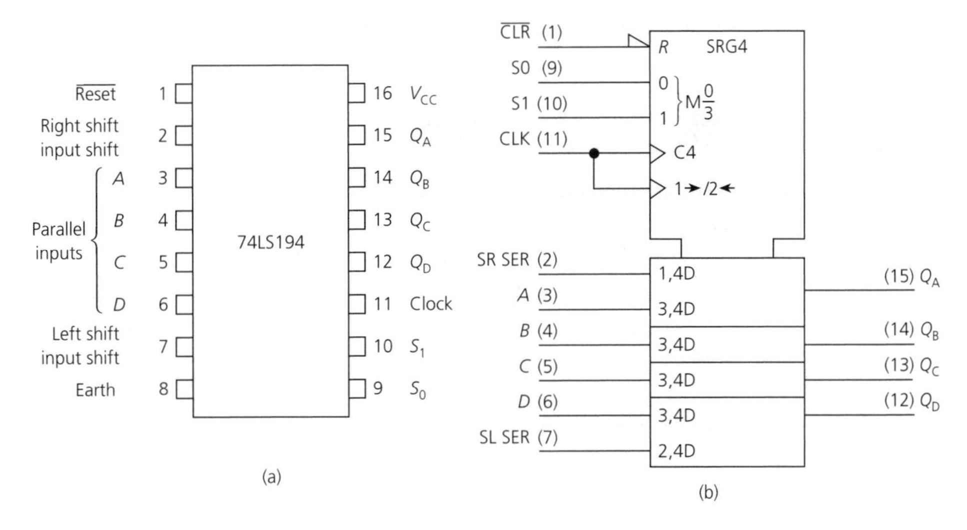

Table 11.2

Inputs										Outputs			
Clear	*Mode*		*Clock*	*Serial*		*Parallel*				Q_A	Q_B	Q_C	Q_D
	S_1	S_0		*L*	*R*	*A*	*B*	*C*	*D*				
0	×	×	×	×	×	×	×	×	×	0	0	0	0
1	×	×	$0 \to 1$	×	×	×	×	×	×	NC	NC	NC	NC
1	1	1	$0 \to 1$	×	×	*a*	*b*	*c*	*d*	*a*	*b*	*c*	*d*
1	0	1	$0 \to 1$	×	1	1	1	1	1	1	1	1	1
1	1	0	$0 \to 1$	1	×	1	1	1	1	1	1	1	1
1	0	0	$0 \to 1$	×	×	×	×	×	×	NC	NC	NC	NC

- $S_0 = S_1 = 1$. The register acts in the PIPO mode. Whatever the data at the serial input, the data entered at the parallel inputs *A*, *B*, *C* and *D* will appear at the parallel outputs Q_A etc. at the leading edge of the next clock pulse.
- $S_0 = 1$, $S_1 = 0$. Data is loaded into the shift-right serial input and previously loaded data is shifted one stage to the right and the data that was at Q_D is moved out of the register. The data at the parallel inputs *A*, *B*, *C* and *D* has no effect upon the circuit. The movement of data within the register occurs at the leading edge of each clock pulse.
- $S_0 = 0$, $S_1 = 1$. Data is loaded into the shift-left serial input and previously loaded data is shifted one stage to the left. The data at Q_A is moved out of the register. The parallel inputs have no effect.

- $S_0 = S_1 = 0$. The Q_A, Q_B, Q_C and Q_D outputs remain unchanged regardless of the logical states of the other inputs. This is known as the *hold mode*.

PRACTICAL EXERCISE 11.2

To investigate the action of the 74LS194 universal shift register.
Components and equipment: one 74LS194 universal shift register IC. Four LEDs. Four 270 Ω resistors. Power supply. Breadboard.
Procedure:

Fig. 11.14

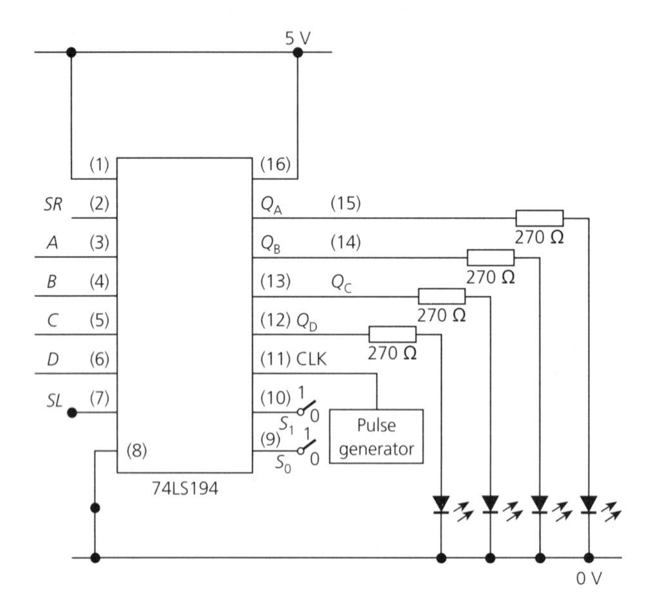

(a) Build the circuit shown in Fig. 11.14. Clear the register by taking pin 1 LOW.
(b) Connect a 4-bit data word to the parallel inputs *A*, *B*, *C* and *D* and connect the mode control inputs S_0 and S_1 to logical 1. The circuit is now in its PIPO mode of operation. Note the logical state of each LED. Does the digital word move to the left or the right? Set SR and SL to 1 or 0 as appropriate.
(c) Set $S_0 = 1$ and $S_1 = 0$ and connect input SR to either 1 or 0 via a switch. Note the logical state of each LED. Does the digital word stored move left or right? Repeatedly move the connection to pin 2 (SR) from logical 0 to logical 1 and note the logical states of the LEDs. Hence, describe what action the circuit is performing.
(d) Set $S_0 = S_1 = 0$ and repeat procedure (c). Did anything happen?
(e) Connect pin 7 (SL) to the switch and pin 2 (SR) to 0 V and then repeat procedures (c) and (d). Now what happens?

Fig. 11.15 *Ring counter*

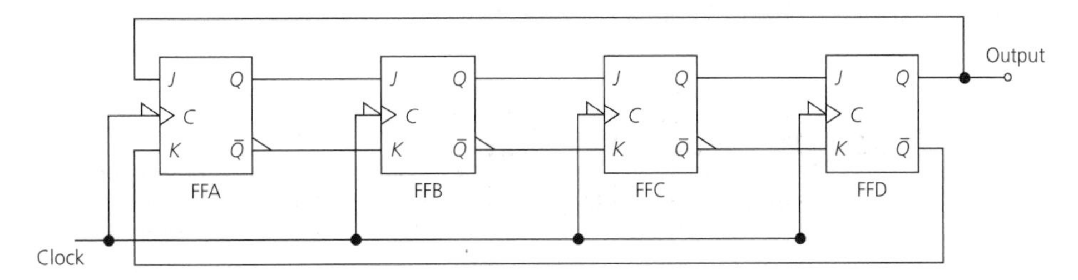

Table 11.3 *Ring counter*

Q_D	Q_C	Q_B	Q_A	*Decoded output*
1	0	0	0	$Q_D = C_0$
0	1	0	0	$Q_C = C_1$
0	0	1	1	$Q_B = C_2$
0	0	0	1	$Q_A = C_3$

Ring counter

A 4-bit *ring counter* consists of a shift register which has its J_A input terminal connected to its Q_D output terminal, and its K_A input terminal connected to its Q_D terminal. Figure 11.15 shows the circuit of the basic ring counter. If the shift register employs D flip-flops then Q_D is connected to D_A.

A ring counter produces a bit pattern that moves through all the stages of the register and that has a modus that is less than 2^n, where n is the number of stages. A ring counter is not self-starting so some gating is required to load just one stage with logical 1 and all the other stages with logical 0. When the clock is applied to the circuit the 1 bit will circulate continuously around the loop. Thus the ring counter goes through the sequence 1000, 0100, 0010, 0001, 1000, etc. The basic count sequence is given in Table 11.3.

Johnson counter

A *Johnson counter* is similar to a ring counter except that the connections between the first and last stages of the shift register are *twisted*. Figure 11.16 shows a Johnson counter.

Fig. 11.16 *Johnson or twisted-ring counter*

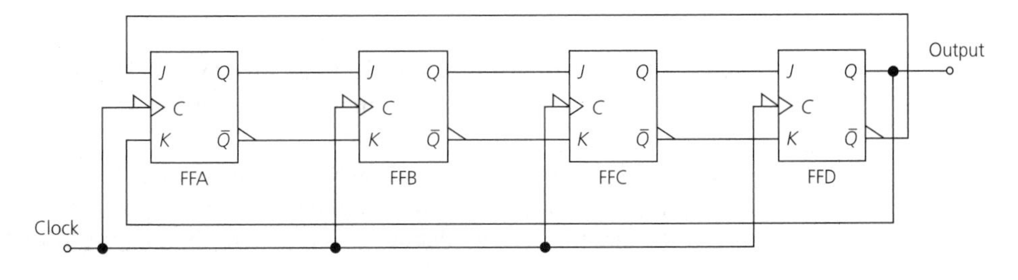

PRACTICAL EXERCISE 11.3

(a) To build a self-starting ring counter and to investigate its operation. (b) To build a Johnson counter and investigate its operation.

Components and equipment: one 74LS164 shift register IC. One 74LS04 (or 74HC04) hex inverter IC. Four LEDs. Four 270 Ω resistors. Power supply. Breadboard. CRO pulse generator.

Procedure:

(a) Build the circuit shown in Fig. 11.17. A ring counter is not self-starting so the switch connected to inputs *A* and *B* is employed to input a single bit into the shift register. The switch must be operated twice (quickly) to input a HIGH voltage and then removed. If it is operated too slowly, two or more bits will be inputted.

Fig. 11.17

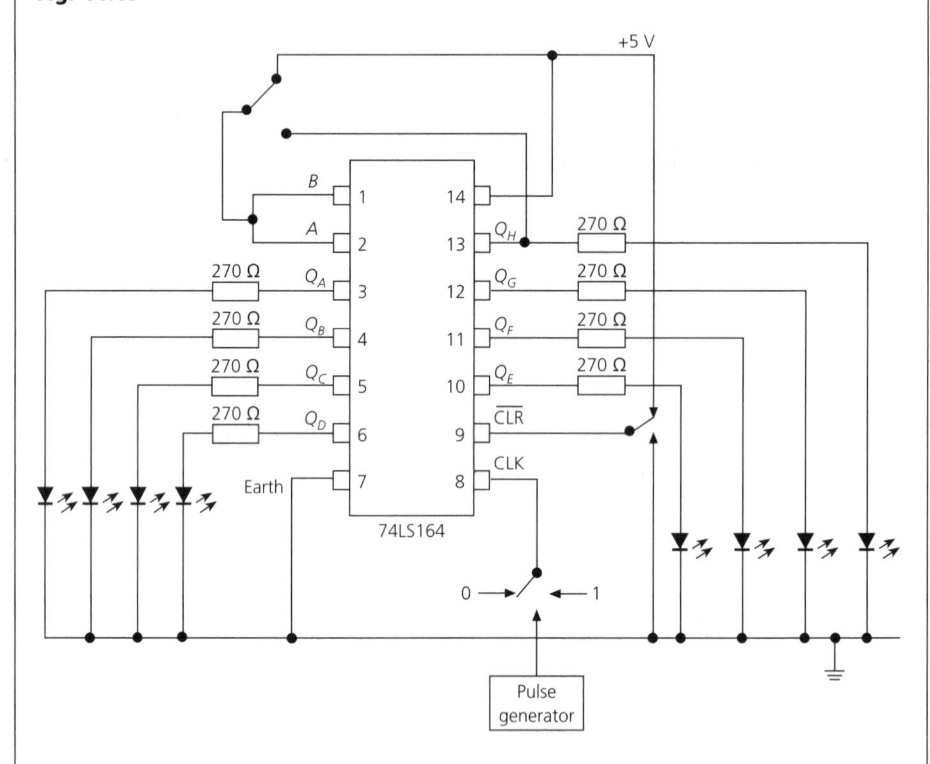

(b) Connect the CLK terminal to first logic 1 and then logic 0, and note the logical states of the LEDs. Note whether the inputted data is shifted through the circuit. Repeat this a number of times.

(c) Switch the CLK terminal to the pulse generator and set the generator's frequency to a convenient low value. Use the CRO to observe the input and output waveforms of the circuit. Try circulating a 2-bit pattern through the circuit.

(d) Change the circuit into a twisted-ring or Johnson counter by inverting the signal applied to the inputs A and B. The modified circuit is shown in Fig. 11.18.

Fig. 11.18

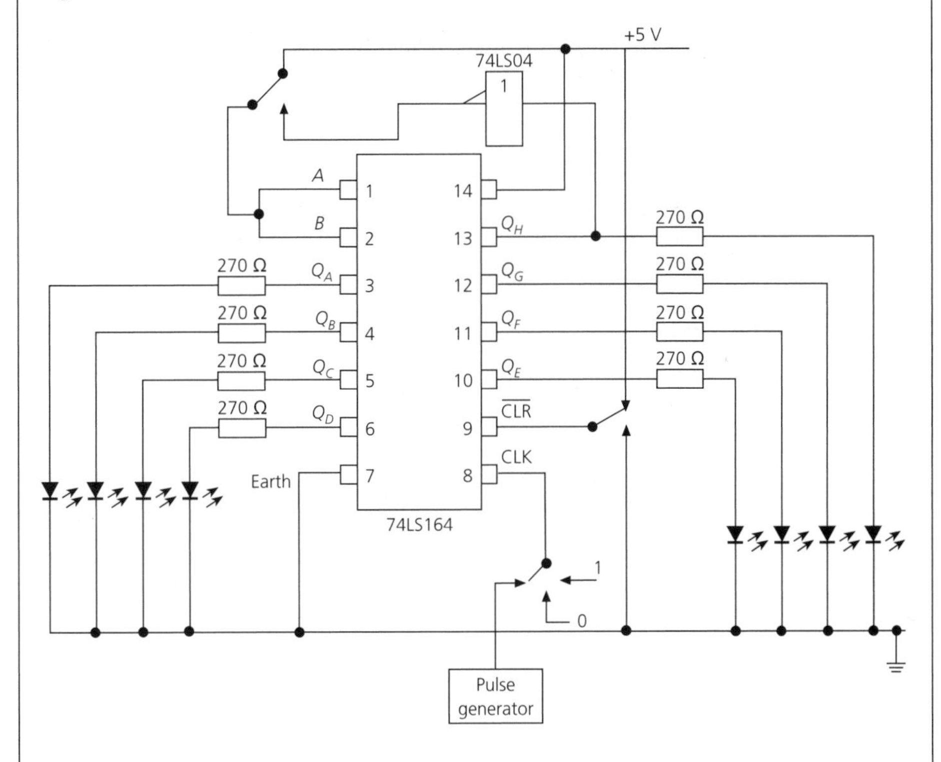

(e) Repeat procedures (b) and (c).

(f) Comment on the output bit pattern in each circuit. Determine the modulus of each counter.

EXERCISES

11.1 (a) A shift register has eight stages. If the clock frequency is 2 MHz calculate the time needed to load the register (i) serially and (ii) in parallel.

(b) A right-shift 4-bit shift register stores the binary word 0101. Write down its contents after (i) one clock pulse and (ii) two clock pulses.

11.2 (a) Draw up a table to show how a SISO shift register using D flip-flops handles the data word 1011. Assume right-shifting.

(b) The data word 0110 is entered into a shift register. The data is (i) shifted one place to the right, and (ii) shifted one place to the left. What is now the new word in each case and what mathematical operations have been performed?

11.3 Digital numbers are often processed in parallel for the utmost speed but then are converted into serial form for transmission from one point to another. Show how a shift register can be employed for this purpose.

11.4 Figure 11.12(b) shows the logic diagram of the 74LS164 or 74HC164 4-bit PIPO/SIPO shift registers. Explain the meaning of each symbol.

11.5 Describe how a shift register may be employed to receive data with the most-significant bit first and later transmit the data with the least-significant bit first.

12 Memories

After reading this chapter you should be able to:

(a) Explain how memory addressing and data storage are carried out.
(b) Discuss the operation and application of the various types of RAM.
(c) Discuss the operation and applications of the various types of ROM.
(d) Explain why a DRAM needs to be periodically refreshed.
(e) Implement a Boolean equation using a ROM.

Many digital systems include memory for the temporary or long-term storage of information. In a digital computer, this information will include numerical data, the intermediate results of computations, and the programs which control the operation of the system. In a telephone exchange, a memory may be used to store code translations and information about each line, such as its number and the nature of the equipment connected to it. A memory should have sufficient capacity to be able to satisfy all the demands made for data and program storage and should be so fast to operate that undue delay is not caused to the main processor. It is also necessary for a memory to be of the minimum possible cost and to be reliable.

Hard and/or floppy discs are used for permanent data storage, and these possess the advantage of being *non-volatile*; this term means that they are able to retain stored data after the power supply has been turned off. A *volatile* memory will lose its stored data when the power supply is removed.

Semiconductor memories are either *random access memories* (RAM) or *read-only memories* (ROM). A ROM is used to store non-volatile, permanent, or semi-permanent data such as the basic input/output system (BIOS) of a personal computer (PC). A BIOS is a set of program instructions that enables a microprocessor to retrieve or *read* data from, and store in or write to, peripherals and memory. A RAM is employed to store data temporarily and it may be either *static* or *dynamic*. Dynamic RAM is much cheaper per bit stored than static RAM. Very short-term memory is provided by a register.

A semiconductor memory consists of a matrix of memory cells and a number of digital circuits that provide such functions as address selection and control. Cells may be single or may be in groups of 4 or 8 bits. Each location in the matrix has a unique address and is able to store 1, 4 or 8 bit(s) of information. The basic requirements for both a ROM and a RAM are that (a) any location in the memory can be addressed, and

(b) data can be read out of an addressed location. Further, for a RAM only, data must be able to be written into any location.

A RAM can have data read out of it, or written into it, with the same access time for all locations. The *access time* is the time that elapses between the start of a memory request and the data appearing on the data bus.

The capacity of a RAM or ROM is the number of bits of information that it is able to store and this is usually quoted in either kilobits or megabits. The prefix kilo does not stand for 1000, which is usual in most other contexts, but for $2^{10} = 1024$. The prefix mega stands for 2^{20} or 1 048 576.

Memory ICs are manufactured in various sizes that are always some power of 2 multiple of kilobits or megabits, such as 32k, 64k, 256k, 512k, 1M, 4M and so on. Thus, a 64k memory has a capacity of $64 \times 1024 = 65\,536$ bits and a 4M memory has a capacity of $4 \times 1\,048\,576 = 4\,194\,304$ bits. Data can be stored (written) or retrieved (read) to/from memory one bit at a time, or several bits at a time, depending upon the *organization* of the memory. If only one bit is written/read at a time the organization is '×1': if four bits are written/read at a time the organization is '×4', and so on. The organization of a memory is always quoted by the manufacturer as m locations × n bits per location. For example, a 64k memory could be organized as 64k × 1, as 16k × 4, or as 8k × 8. The organization of a 4M memory could be 4M × 1, 1M × 4, or 512k × 8.

When a memory is organized as a number of 1-bit words, any one of the locations in the memory can be individually addressed at any time. The number of address lines which are necessary is equal to the power of 2 that gives the storage capability of the memory.

If, for example, there are 1024 1-bit word locations, then 10 address lines will be needed, since $2^{10} = 1024$. If the memory is organized as 4 bits per word, there will be 256 different locations that can be addressed and this will necessitate 8 address lines ($2^8 = 256$). At each location a 4-bit word will be accessed, but individual bits cannot be separately addressed. A 128k × 8 memory has 131 072 addressable locations and so it needs 17 address lines.

Many memories store only 1-bit words because there is then only a need for *one* input/output data terminal. This is particularly true for the larger memories since the number of input and output terminals that can be provided is limited by the number of IC package pins available.

EXAMPLE 12.1

A memory is organized as (i) 64k × 8 and (ii) 256k × 4. Calculate (a) the number of bits stored at each location, (b) the number of locations, and (c) the total number of bits stored.

Solution

(a) (i) 8, (ii) 4 (*Ans.*)
(b) (i) $64 \times 1024 = 65\,536$, (ii) $256 \times 1024 = 262\,144$ (*Ans.*)
(c) (i) $65\,536 \times 8 = 524\,288$, (ii) $262\,144 \times 4 = 1\,048\,576$ (*Ans.*)

EXAMPLE 12.2

Calculate the number of locations that may be addressed in a RAM IC having (a) 10 address pins, (b) 16 address pins, and (c) 20 address pins.

Solution

(a) $2^{10} = 1024 = 1\text{k}$ (*Ans.*)
(b) $2^{16} = 65\,536 = 64\text{k}$ (*Ans.*)
(c) $2^{20} = 1\,048\,576 = 1\text{M}$ (*Ans.*)

All memory ICs used to operate from a 5 V power supply but, for the same reasons given in chapter 7, modern devices usually operate from a 3.3 V power supply.

Random access memory

The memory cells within a RAM IC are arranged in the form of a matrix having m rows and n columns. The cells are located at the intersections of the rows and the columns as shown in Fig. 12.1 in which $m = n = 5$. Each location in the matrix has a unique address so that any one memory cell can be selected at a time. Suppose, for example, that the memory cell located at the intersection of row 2 and column 13 is to be selected. Then the input to the row decoder must be 00010 and the input to the column decoder must be 01101.

Figure 12.2 shows the block diagram of a 1k × 1 memory. Since, for a square memory matrix, 1k = 1024 = 32 × 32, a 5-to-32 line decoder is employed for both the row select and the column select decoders.

The figure also shows a block marked as input/output circuits and control. This circuit performs the functions either of writing new data into an addressed location, or

Fig. 12.1

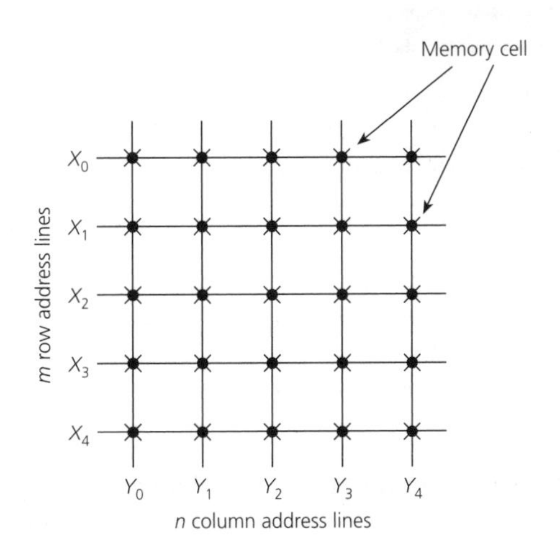

Fig. 12.2 *A 1k × 1 random access memory*

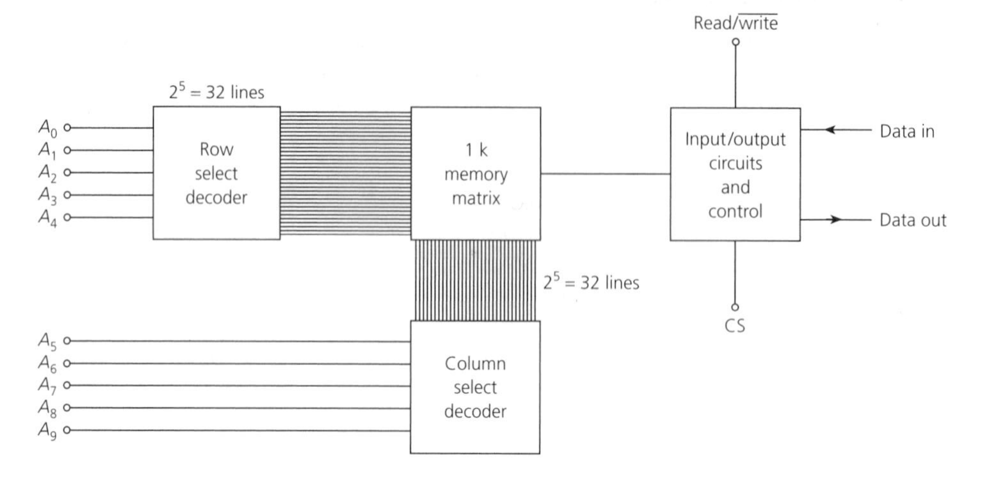

of reading out existing data. The read/$\overline{\text{write}}$ input determines which of the two functions is performed. The chip select $\overline{CS}$ input must (usually) be LOW to enable the memory; this facility is used when two or more chips are combined to produce a larger capacity memory. When the $\overline{\text{CS}}$ pin is HIGH the memory has a high output impedance. Some RAMs employ an active-HIGH *CS* input.

When the data stored at a location is to be read, the read/$\overline{\text{write}}$ line is set to the logic 1 voltage level, and the address of the required location is fed into the address decoders. The data held at that location then appears at the data-out terminal. The read-out process is *non-destructive*.

When new data is to be written into the memory, the read/$\overline{\text{write}}$ line is set to the logic 0 voltage level. The data present at the data-in terminal will then be written into the addressed memory cell. Often, the same terminals are employed for both input and output data.

EXAMPLE 12.3

Data held in the 1k × 1 memory shown in Fig. 12.2 is to be accessed. If the data is at location (a) row 2, column 13, and (b) row 26, column 2, what are the required inputs to the row and column decoders?

Solution

(a) Row decoder: $A_5 = A_4 = A_3 = A_2 = A_0 = 0$, $A_1 = 1$
Column decoder: $A_5 = A_4 = A_1 = 0$, $A_3 = A_2 = A_0 = 1$ (*Ans.*)

(b) Row decoder: $A_3 = A_4 = A_1 = 1$, $A_5 = A_2 = A_0 = 0$
Column decoder: $A_5 = A_4 = A_3 = A_2 = A_0 = 0$, $A_1 = 1$ (*Ans.*)

Two examples of RAM are given in Figs 12.3(a) and (b). These show how the same pins can be used for both input and output data under the control of the read/$\overline{\text{write}}$ input

Fig. 12.3 *(a) 16k × 4 memory matrix and (b) 8k × 8 memory matrix*

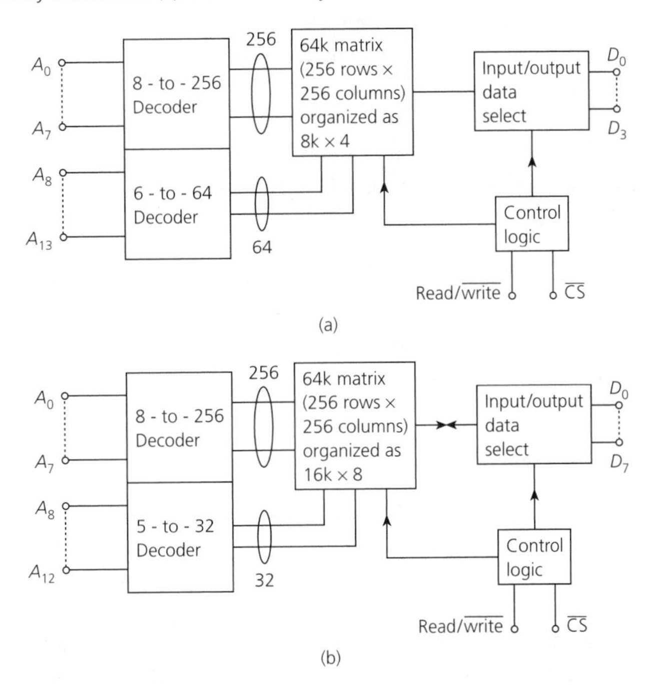

Fig. 12.4 *Time-division multiplexing of address lines*

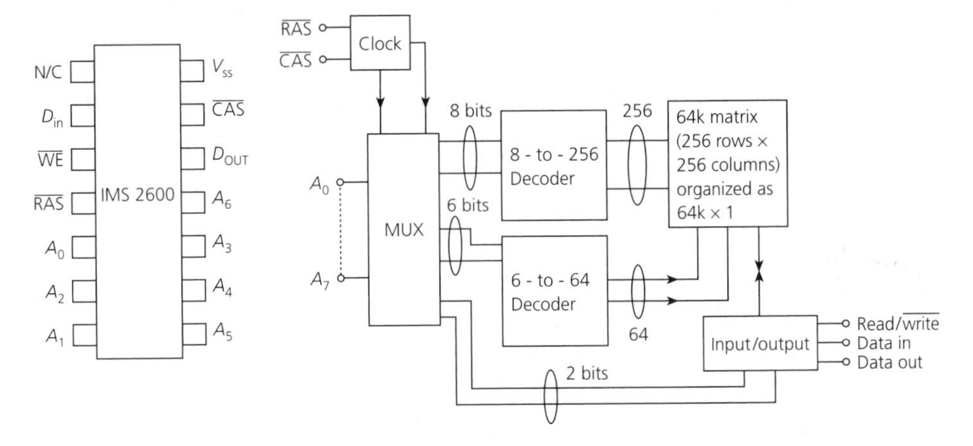

signal. The number of pins needed for an IC package accommodating the RAM of Fig. 12.3(a) would be 14 address pins, 4 data input/output pins, 1 read/$\overline{\text{write}}$ pin, 1 $\overline{\text{CS}}$ pin, and two pins for $+V_{CC}$ and earth. This is a total of 22 pins, so that a 24-pin package would probably be employed. Clearly, the number of pins required for a 64k memory organized as 64k × 1-bit words would be even larger because of the increase in the number of address lines necessary.

For this reason some of the larger memories employ time-division multiplexing of the address lines. An example of this technique is shown in Fig. 12.4 for a 64k × 1 dynamic RAM. The matrix is arranged as 256 rows × 256 columns and the address inputs are time-division multiplexed under the control of the $\overline{RAS}$ and $\overline{CAS}$ clocks. Normally, $\overline{CAS}$ is HIGH when $\overline{RAS}$ goes LOW. Then the 8 address inputs are latched and decoded to select one of the 256 rows in the matrix. This row address must be held for a minimum period and then it can be switched to the wanted column address. Once the column address is stable $\overline{CAS}$ can be taken LOW and then the column address is latched and used to select the wanted column. The address cycle ends when $\overline{RAS}$ goes HIGH.

A 1Mb memory IC needs 20 address pins to be able to address uniquely all its memory locations. Usually, a form of multiplexing is used in which the 20 address bits are split into two groups. Ten bits are used for the row address and 10 bits for the column address. This means that the memory will have $2^{10} = 1024$ rows and 1024 columns.

Static RAM

In a static RAM (SRAM) the memory cells are actually flip-flops. Modern devices mainly use CMOS technology because CMOS RAMS dissipate less power than bipolar versions but their access time is longer. SRAMs are volatile but, since a SRAM cell is a flip-flop, it will retain its logic state as long as the power supply is switched on.

Figure 12.5 shows the circuit of a CMOS SRAM. The flip-flop transistors T_2, T_3, T_4 and T_5 are held in their logical state by the voltages applied to their gate terminals. In one pair of transistors the upper transistor is OFF while the lower transistor is ON; in the other pair it is the other way around, the upper transistor is ON and the lower transistor is OFF. The cell switches to the particular state that is determined by the voltages on the data lines when the word select line is active. During a read cycle the two data lines are floating and when the word line goes HIGH the state of the cell is transmitted via the data lines to the sense amplifier. The power dissipated is very small but still larger than the power dissipated by DRAM. Some versions have the upper transistors T_2 and T_4 replaced by high-value (>1 GΩ) resistors. The access time for a CMOS static RAM is typically between 50 and 150 ns. The logic symbol for a 1k × 4 static RAM is shown in

Fig. 12.5 *CMOS static memory cell*

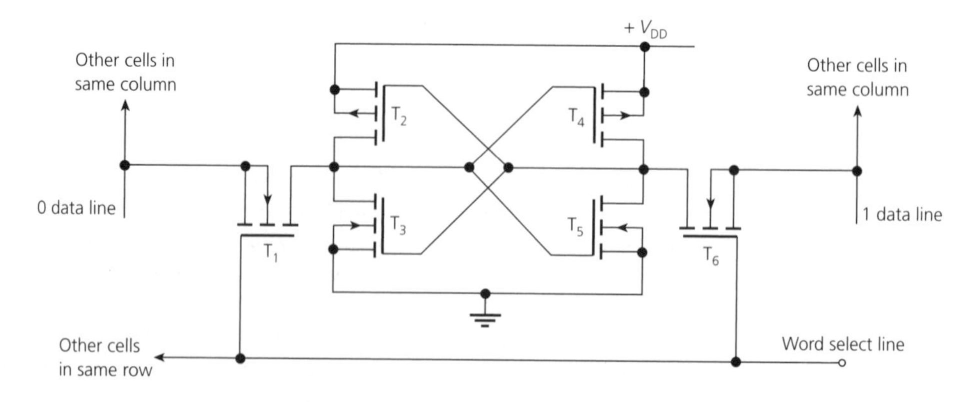

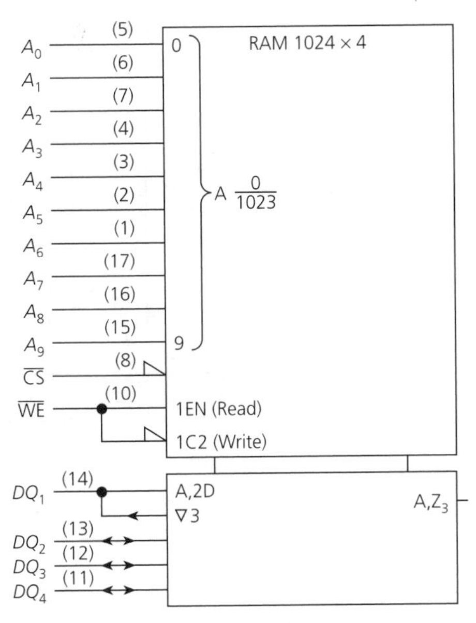

Fig. 12.6 *Logic symbol for a 1k × 4 static RAM*

Fig. 12.6. All the input lines are on the left and the output lines are mainly on the right. When an output line is on the left it is indicated by an arrowhead. Input/output lines D_0 through to D_3 show this. Address inputs are given in binary order, A_0 through to A_9, and the data inputs are D_1 through to D_3. Outputs are indicated by Q_1 etc. But when, as with this IC, common input/output terminals are used they are labelled DQ_1, DQ_2, DQ_3 and DQ_4. Common input/output terminals are indicated by the Z in the A,Z3 label which shows also that the output depends upon the input address A. Thus, Z_3 transfers to 3 at the left to give an input/output port. The lowest and highest addresses in the device are indicated by the A(m/n) symbol, where m is the decimal number of the lowest address and n is the decimal number of the highest address, here A(0/1023). The G_1 label shows AND dependency and means that data is written in when $\overline{CS}$ and $\overline{WE}$ are both LOW (note the 1C2 label); for a read to occur $\overline{CS}$ must be LOW and 1EN HIGH. The ∇3 symbol indicates three-state outputs. A,2D means that the data input D is dependent upon both the address A and the state of C2 (the write input). The numbers given in brackets are the pinout numbers.

The basic block diagram of a 1M × 4 static RAM is shown in Fig. 12.7. The RAM has 20 address pins and, hence, 2^{20} = 1 048 576 4-bit word locations. $\overline{CS}$ (sometimes just $\overline{S}$ or $\overline{E}$) is an active-LOW chip enable pin. $\overline{CS}$ must be LOW for the memory to be able to input or output data; when it is high the data outputs are in their high-impedance state so they have no effect upon the data bus. Address pins A_4 through to A_{13} specify the row in the memory matrix and the remaining address pins specify the required column. The row select block is a 10-to-1024 line decoder for selecting the wanted row out of 1024. The column select circuit is also a 10-to-1024 decoder for identifying the wanted 1-out-of-1024 column. Once the location has been selected the input/output circuitry either allows the input data to enter that location or the data at the location to

Fig. 12.7 *A 1M × 4 static RAM*

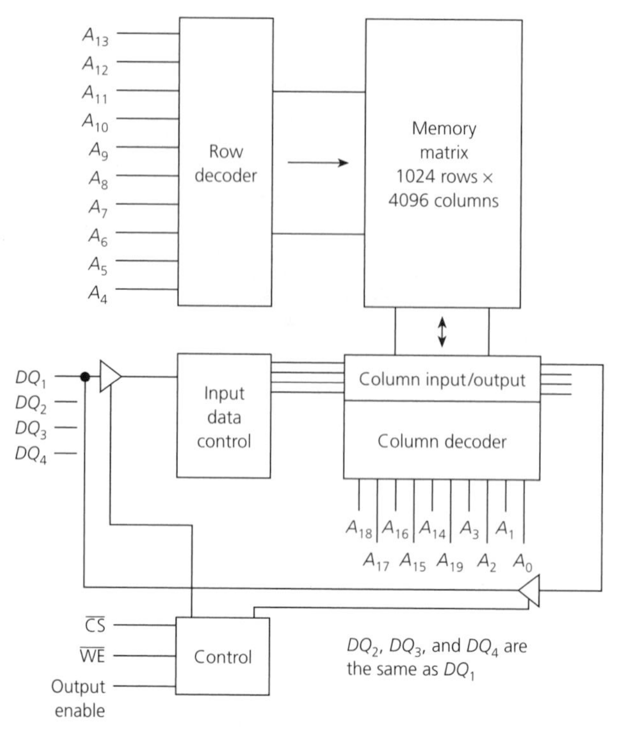

Fig. 12.8 *Dynamic memory cell*

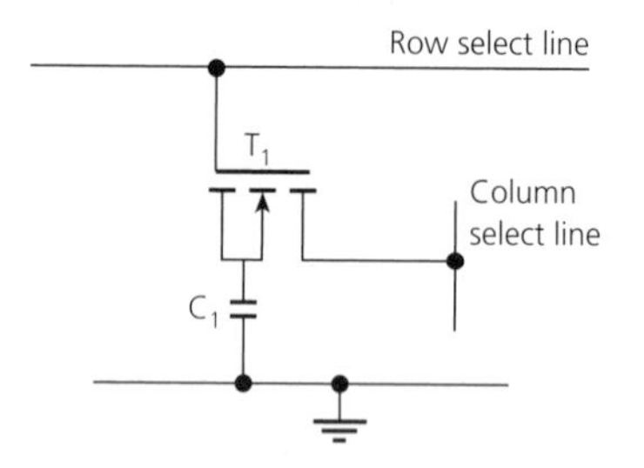

leave the memory. The circuit is told what is required by the logical state of the write enable ($\overline{WE}$ or sometimes read/$\overline{write}$) pin. If $\overline{WE}$ is LOW, data is written into the selected location and, if pin $\overline{WE}$ is HIGH, data is read out.

A number of different organizations are available for static RAMs including: 1M × 32, 1M × 4, 1M × 1, 512k × 32, 512k × 8, 128k × 16, 64k × 32 and 64k × 8.

Dynamic RAM

Dynamic RAMs (DRAMs) employ the one-transistor, one-capacitor memory cell shown in Fig. 12.8. The term 'dynamic' means that the electrical state of the memory cells is continuously changing because the charge stored in a capacitor gradually leaks away. This means that a DRAM has to be continuously *refreshed* to ensure that the electrical

Fig. 12.9 *Organization of DRAM cells*

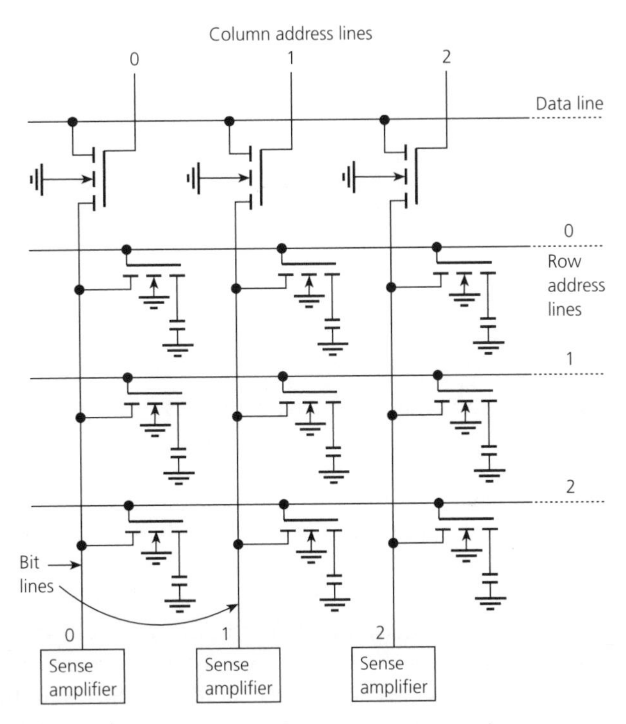

charges that represent bits of information are not lost. The refresh speed is expressed in nanoseconds and is typically either 60 or 70 ns. To write data into the cell the address line is taken HIGH to turn the transistor ON; the voltage on the data line then either charges (logic 1) or discharges (logic 0) the capacitor through the ON transistor. To read a cell the address line is taken HIGH, again turning on the transistor, and this allows the capacitor voltage (1 or 0) to appear on the data (column) line. The readout is destructive so that every readout must be followed by a write operation to put the lost data back into the cell.

The way in which the memory cells in a DRAM are arranged in an array is shown in Fig. 12.9. The MOSFETs connected to the upper data line are turned ON and OFF to connect the selected bit line to the data line. When a row and a column line are both taken HIGH, only the data stored at the cell located at the intersection of the selected row and column is transferred to the data line (read), or data on the line is transferred into the cell (write).

The block diagram of a DRAM is similar to that of a SRAM (p. 271) with the addition of RAS and CAS inputs (see Fig. 12.4).

As the size of a DRAM increases the number of address pins that are required increases also. To reduce the number of address pins required, time-division multiplexing is employed. The usual method employed uses one-half of the number of address pins that would be required without multiplexing and allocates one-half of them for row selection and the other half for column selection. In the first half of the addressing cycle the row address is inputted into the memory and latched; then the column address is

Fig. 12.10 *Use of a dummy cell to detect 0 or 1*

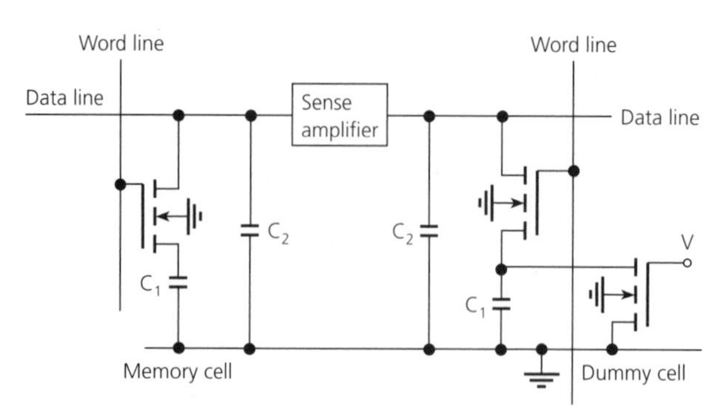

inputted and also latched. The total address is then applied simultaneously to select the wanted cell. Two selection pins are normally employed: (a) *row address select* ($\overline{\text{RAS}}$) which tells the memory that the row address is being inputted, and (b) *column address select* ($\overline{\text{CAS}}$) which indicates the column address.

In a DRAM a pair of conductors, often called the *bit line pair* links all the switching transistors in a column. Before a memory access can take place these conductors are pre-charged to one-half of the power supply voltage. The *memory controller* generates a *row access select* ($\overline{\text{RAS}}$) signal. The $\overline{\text{RAS}}$ signal tells the row decoder to decode the input address to select the wanted row. Then a *column access select* ($\overline{\text{CAS}}$) signal is generated that makes the column decoder decode the rest of the address bits to select the wanted column. A voltage is placed on the selected column but it only turns ON the MOSFET that is located at the intersection of the column and the selected row.

The logical state of the selected cell will make the voltage of one of the bit line pair conductors change by about 0.1 V and this voltage change is applied to the sense amplifier. The sense amplifier amplifies the change in voltage to approximately 5 V and the amplified voltage is transferred to a bus driver. The bus driver is a high impedance circuit that is able to supply the necessary current to the data bus.

Figure 12.10 shows how the sense amplifier is connected between the selected memory cell and a dummy cell. The word line is selected from the row address and data from all memory cells located on this word line are connected to the sense amplifier. During a read cycle the transistor turns ON and the charge stored in C_1 is transferred to the data line. The charge stored in C_2 is combined with the charge from C_1 and this may cause the data line voltage to change. If the cell voltage is zero (0) the data line voltage falls, but if the cell voltage is HIGH (1) the data line voltage is unchanged. The capacitance of the dummy cell is one-half the capacitance of the memory cell and its voltage is cleared to 0 V when the memory is pre-charged. Both data line capacitances have the same value and so they are charged to the same voltage ($\approx V_{CC}$). The voltage of the data line is, therefore, fixed at the average value of the 1 and 0 voltages. The sense amplifier detects the change, or the absence of a change, in the voltage on the data line and uses it to produce an output. The data is decoded as binary 1 when the memory cell has the higher voltage and as binary 0 when the dummy cell voltage is the higher. When the $\overline{\text{CAS}}$ line goes HIGH the data buffer is moved into its HIGH output impedance state effectively disconnecting the memory from the data bus.

Refreshing

A DRAM suffers from the disadvantage that *periodic* refreshing is necessary. If a write is not carried out within a certain time the charge stored in the cell capacitance will leak away. This leakage of charge leads to the difference between a stored 1 and a stored 0 becoming vanishingly small. Eventually, the sense amplifier would be unable to detect a change in voltage on the data line and then all bits would be read as 0. Because of this it is essential that all the memory cell capacitances that are storing logic 1 are periodically re-charged, or *refreshed*. Refreshing is carried out automatically by accessing each row in the memory matrix, one at a time. When each row has been accessed, a voltage is applied to it to charge each capacitor to its original voltage. Refreshing is carried out at regular intervals of time, typically 8 ms.

The time that elapses between the memory controller setting the memory address and the wanted data appearing on the data bus is known as the *memory cycle time*. Typically this time is about 70 ns. The bit line pairs must be pre-charged between each memory access and the time this takes is known as the *RAS pre-charge time*. It is approximately 80% of the cycle time. This means that the minimum permissible time between two successive accesses to a memory is 1.8 times the cycle time. For a 70 ns device this is 125 ns, which allows a clock speed of just 8 MHz.

The logic symbol for a DRAM is similar to that of a SRAM with the addition of several extra labels and blocks which make it rather difficult to follow.

EXAMPLE 12.4

A DRAM has (a) 11 multiplexed address pins and one data input/output pin, (b) 14 non-multiplexed address pins and 4 data input/output pins. Determine the organization of the DRAM.

Solution

(a) Total number of unique addresses = 2^{22} = 41 943 304. Hence, the organization is 4M × 1 (*Ans.*)

(b) 2^{14} = 16 384 = 16k. Hence, organization = 16k × 4 (*Ans.*)

Single in-line memory modules

The RAMs fitted in a PC are usually inserted on to a single in-line memory module (SIMM). A SIMM is a small printed circuit board (PCB) that has several memory ICs mounted on it. Along the lower edge of the board are a number of small metal pads that connect with connectors on the computer motherboard. There are two types of SIMM in general use; the older 32-pin SIMM that has an 8-bit wide array plus a ninth parity bit, and the more modern 72-pin SIMM that provides for 32-bit words. SIMM sockets are commonly employed because they allow a large number of memory ICs to be packed into a small physical area. A 72-pin SIMM may contain 80, or even more, DRAM ICs.

Fig. 12.11 *DRAM ICs mounted on SIMM sockets*

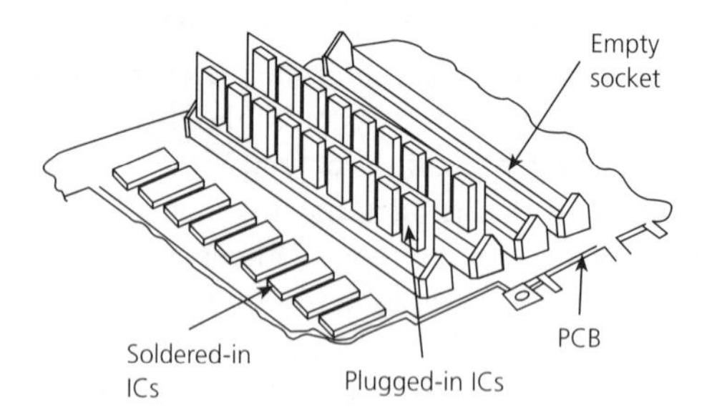

Figure 12.11 shows how a number of DRAM ICs may be mounted in SIMM sockets, alongside some soldered-in DRAM ICs. One of the SIMM sockets is shown empty; this is frequently the case and it allows for future memory expansion to be carried out by inserting more DRAM ICs.

Dual in-line memory modules

64-bit computers often employ dual-in-line memory modules (DIMMs) instead of SIMMs. Most DIMMs are installed vertically into expansion sockets but, like SIMMs, contact pins on opposite sides of the board are not in electrical contact with one another.

Faster access DRAMs

Most computer systems require the use of a higher clock frequency than 8 MHz and, hence, various methods of giving faster access to a DRAM have been developed.

Interleaving

Two banks of memory are provided and successive memory accesses use alternate banks. This system of memory management demands the use of pairs of identical DRAM memory modules. Data can be read from one memory bank while the other bank is being pre-charged. Interleaving gives an almost two-fold increase in access speed because the RAS pre-charge time is no longer a factor. A problem exists, however, if it is required to access two memory locations in the same bank in quick succession, since then the expected increase in speed does not materialize.

Page mode access

Each successive 1024 rows in a memory matrix can be taken together to form a *page*. A 1Mb 30-pin SIMM can have eight ICs arranged to give a page of 1024 bytes. A 72-pin 32-bit SIMM would have a page size of 4kb. When memory accesses occur within the same page – having the same row address – the row address need not be decoded each time. This eliminates one-half of the memory cycle. Also, there is then no need to pre-charge the bit line pairs because the pairs that have not been involved in a memory access are already charged. This means that page mode access more than doubles the access speed.

Page mode DRAM is used with SIMM memory modules to give fast access within the same block of memory.

Fast page mode

Fast page mode is used with 72-pin SIMMs. The memory controller holds the CAS signal active during memory accesses within a page and only the address data is changed. This gives a further increase in speed allowing a 25 MHz clock frequency.

Extended data out DRAM

Extended data out DRAM (EDO DRAM) gives a further increase of about 75% in the speed of memory access compared with ordinary DRAM. EDO DRAM provides a wider effective bandwidth by using separate circuits to pre-charge the memory cells. The memory cycle time is about 20 ns, which allows for a 50 MHz clock frequency. EDO DRAM must be operated by a TRITON system controller. When $\overline{CAS}$ goes HIGH the EDO DRAM data buffers are left active and they remain so until the end of the next CAS pulse goes LOW. Hence, the trailing edge of the pulse can be used to latch the output data. This means that the output data is held valid until the trailing edge of the next CAS pulse, unlike standard DRAM, in which the outputs become high impedance at the end of the CAS pulse ready for the next pre-charge. With an EDO DRAM, the CAS pre-charge overlaps the data valid time which allows the CAS pulse to fall earlier. A further increase in clock frequency to about 60 MHz can be obtained with page mode EDO DRAM.

Page/interleaved mode

The interleaved and page modes of operation can be combined to give the page/interleaved mode. With this method one memory bank contains all the even pages of the memory and the other bank contains all the odd pages. This method of operation eliminates the need to wait for a RAS pre-charge when a memory access crosses a page boundary and gives a further increase in access speed.

Burst DRAM

In a burst DRAM the start address is put onto the address bus by the microprocessor and it is then clocked into an address register situated in the DRAM. The address can then be removed from the address bus by the microprocessor after which it can deal with other work. A burst DRAM uses the start address to output the first data word and then it internally increments the address three times and each time outputs the data word which is stored at the new address.

The time taken for the first data to be outputted is three clock cycles but, at the following three addresses, the data is outputted after each clock cycle. Hence, the time taken for four addresses to output data is $3 + 1 + 1 + 1 = 6$ clock cycles, as opposed to $2 + 2 + 2 + 2 = 8$ clock cycles for ordinary DRAM.

Burst EDO DRAM

A burst EDO DRAM (BEDO) is an EDO DRAM with a fast page mode cycle time. Four-cycle burst access is controlled by an on-chip counter. Only the start address is supplied to the DRAM; the CAS signal is then toggled to latch in the start address and the counter is incremented through the next three addresses. This method of operation allows a page cycle time of 15 ns giving a clock frequency of 66 MHz.

Fig. 12.12 *Synchronous DRAM*

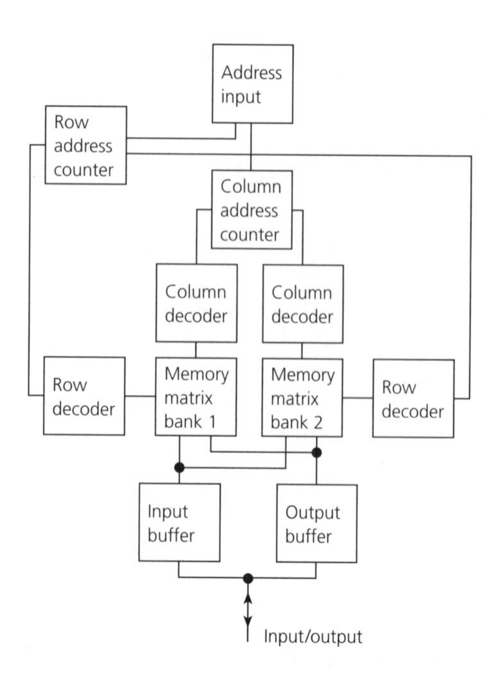

Synchronous DRAM

A synchronous DRAM (SDRAM) is one in which all the operations carried out are synchronized to a clock. Data can be transmitted on every clock cycle and this gives an improvement in performance over EDO DRAM of about 10%. The basic block diagram of an SDRAM is shown in Fig. 12.12. The address, data and control signals are all read into the on-chip registers on the leading edge of a clock pulse. Since these signals only need to be present at the leading edge of the clock a wider bandwidth is obtained than with non-synchronous DRAM. The operation of the SDRAM is controlled by input words such as read and write instead of RAS and CAS signals. The memory in a SDRAM is divided into banks. This allows the use of burst interleaving giving a pre-charge time that is effectively zero, since one bank can be pre-charged while the other bank is being accessed. Like BEDO DRAM, an on-chip counter increments the column addresses after an input start address to allow very fast access. Clock frequencies of up to 100 MHz are possible.

VRAM

Essentially, a VRAM is a DRAM that has extra input/output ports that enable it to have a faster data transfer rate. *Data pipelining* is employed to increase the rate at which data is presented to the memory. At the start of a data transaction a three-stage data pipeline is loaded so that new data becomes available with each successive clock pulse, once the pipeline is full. The access time for the first data byte is the same as for a DRAM. Data becomes valid at the output after the data buffer has been clocked and it remains valid throughout the clock period. The increase in speed given by pipelining is illustrated by Fig. 12.13.

Fig. 12.13 *(a) Pipelining and (b) conventional DRAM*

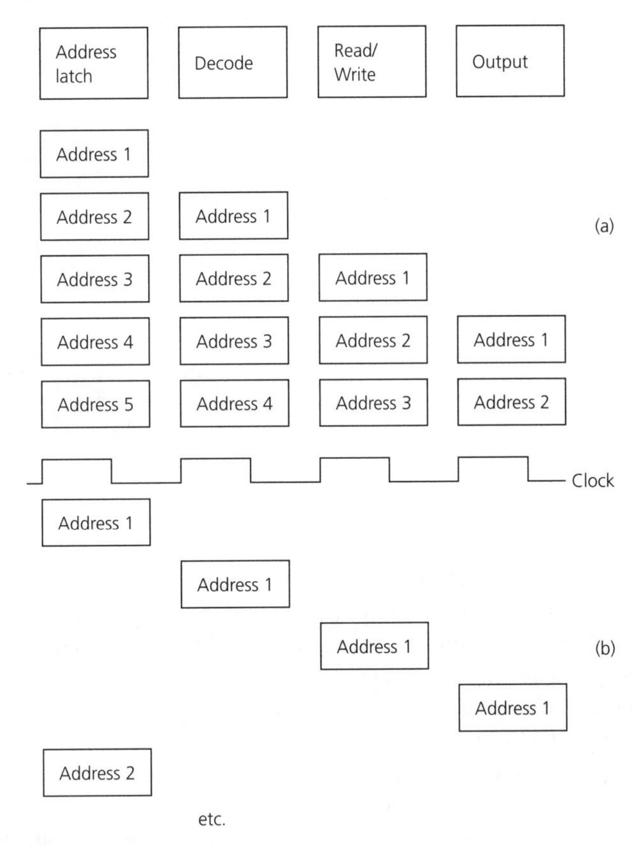

Read-only memory

A read-only memory (ROM) has data written into it, in *permanent form*, by either the manufacturer or the user. When in use, data can *only* be read out of the memory; new data *cannot* be written in. A ROM is non-volatile. ROMs are employed for the permanent storage of computer programs, for dedicated applications like microprocessor-controlled washing machines and microwave ovens, and in look-up tables.

The organization of a ROM is very similar to that of a RAM. Data is stored at different locations within the memory matrix and each location has a unique address. When a particular location is addressed, the data stored at that address is read out of the memory. The read-out is non-destructive. Address decoders are employed to reduce the number of address pins needed, but a $R/\overline{W}$ pin is not necessary. The data held in the ROM is permanent and is used for a wide variety of purposes, such as code conversion, computer programs and logic functions. The truth table of a required circuit operation is implemented by a matrix, with the required connections at matrix intersections being achieved by means of suitably connected diodes or transistors.

Fig. 12.14 *Diode ROM*

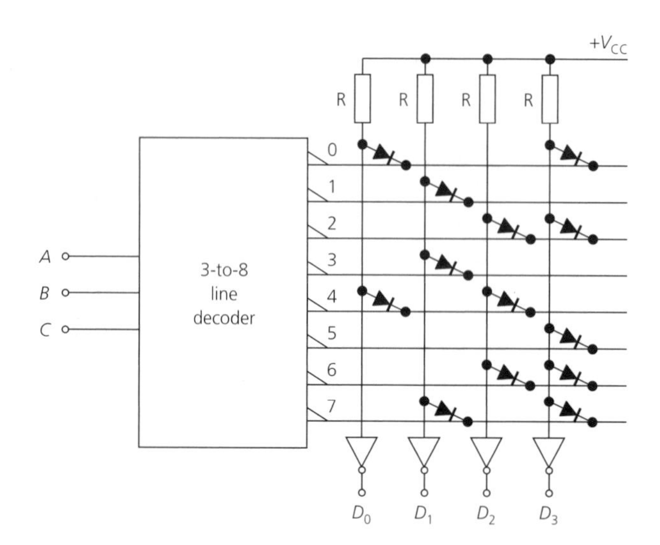

At each cell where a 1 is stored, a voltage-controlled switch is connected between the row line and the column line at that location. At a cell location where logical 0 is stored there is no connection between the row and the column lines. To read the logical state of a memory location a voltage is applied to the row line which closes all the switches connected to that row line. This causes a logic 1 voltage to appear at the output of the column line.

The voltage-controlled switch can be a diode, a bipolar transistor, or a MOSFET. In Fig. 12.14, diodes are shown.

Diodes are connected between some of the row and some of the column lines. When a decoder line, 0 through to 7, is selected and goes LOW, the associated diodes turn ON. Each output line D_0, D_1, D_2 or D_3 connected to an ON diode is then taken LOW also. The voltage on each column line is inverted to give the outputs D_0 etc. Hence, each output connected to an ON diode goes HIGH. When, for example, the input address is $A = 0$, $B = C = 1$, row 6 goes LOW and the output of the ROM is $D_0 = D_1 = 0$, $D_2 = D_3 = 1$. Hence,

$$D_0 = \bar{A}\bar{B}\bar{C} + \bar{A}\bar{B}C,$$
$$D_1 = A\bar{B}\bar{C} + AB\bar{C} + ABC,$$
$$D_2 = \bar{A}B\bar{C} + \bar{A}\bar{B}C + \bar{A}BC,$$
$$D_3 = \bar{A}\bar{B}\bar{C} + \bar{A}B\bar{C} + A\bar{B}C + \bar{A}BC + ABC$$

A ROM is programmed during manufacture and this data cannot be subsequently altered. This means that the intending user must inform the ROM manufacturer of the particular data that each location is to contain. This is acceptable to the large-scale user but it is much less convenient for the user of much smaller quantities. Mask programmable ROMs are not economic unless a large number of them are required. To provide some flexibility in the possible applications of ROMs, user-programmable devices are often used.

Fig. 12.15 *ROM implements $F = \bar{A}\bar{B}C + A\bar{B}C + AB\bar{C}$*

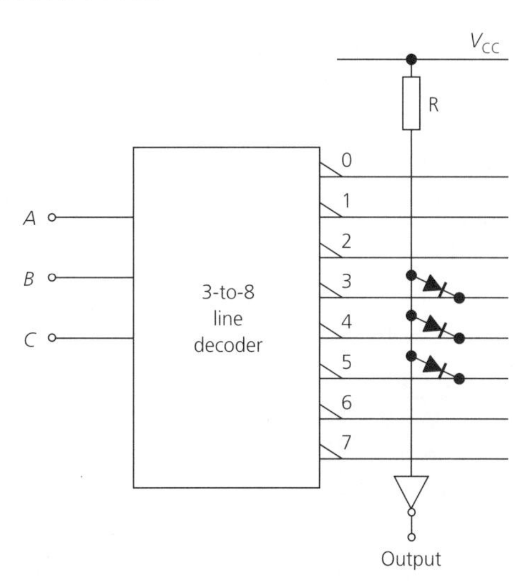

Implementing a combinational logic function using a ROM

Whenever a combinational logic function that is to be implemented has a large number of input variables and/or outputs, it will often prove to be more economic to employ either a ROM or a *programmable logic array* (PLA) (see chapter 13).

A ROM can be made to implement a logic function by using it to store the truth table of that function. The function should be written in the sum-of-product form in which each term includes all of the input variables. Each input variable will be applied to an address line of the ROM. Then, each combination of the input variables will address a particular memory location and thereby produce the required output signal.

Suppose that the function $F = \bar{A}\bar{B}C + A\bar{B}C + AB\bar{C}$ is to be implemented. There are three input variables, so a ROM with three address pins is required, together with a 3-to-8 line decoder. The required ROM is shown in Fig. 12.15

When any of the lines 3, 4 or 5 is selected by an input address, that line is taken LOW and the associated diode conducts. The output line then goes LOW also. Since the outputs are all inverted an active output is indicated by the logical 1 state.

EXAMPLE 12.5

The equations

$$F_0 = A\bar{B}\bar{C}\bar{D} + AB\bar{C}\bar{D} + A\bar{B}C\bar{D} + ABC\bar{D} + A\bar{B}\bar{C}D + \bar{A}B\bar{C}D + \bar{A}\bar{B}CD + \bar{A}BCD$$

$$F_1 = \bar{A}B\bar{C}\bar{D} + AB\bar{C}\bar{D} + \bar{A}BC\bar{D} + ABC\bar{D} + AB\bar{C}D + \bar{A}\bar{B}CD + ABCD$$

and

$F_2 = \bar{A}\bar{B}C\bar{D} + A\bar{B}C\bar{D} + \bar{A}BC\bar{D} + ABC\bar{D} + A\bar{B}CD + \bar{A}BCD + ABCD$

are to be implemented using a ROM. Obtain the desired circuit.

Solution

The required ROM is shown in Fig. 12.16 (*Ans.*)

Fig. 12.16

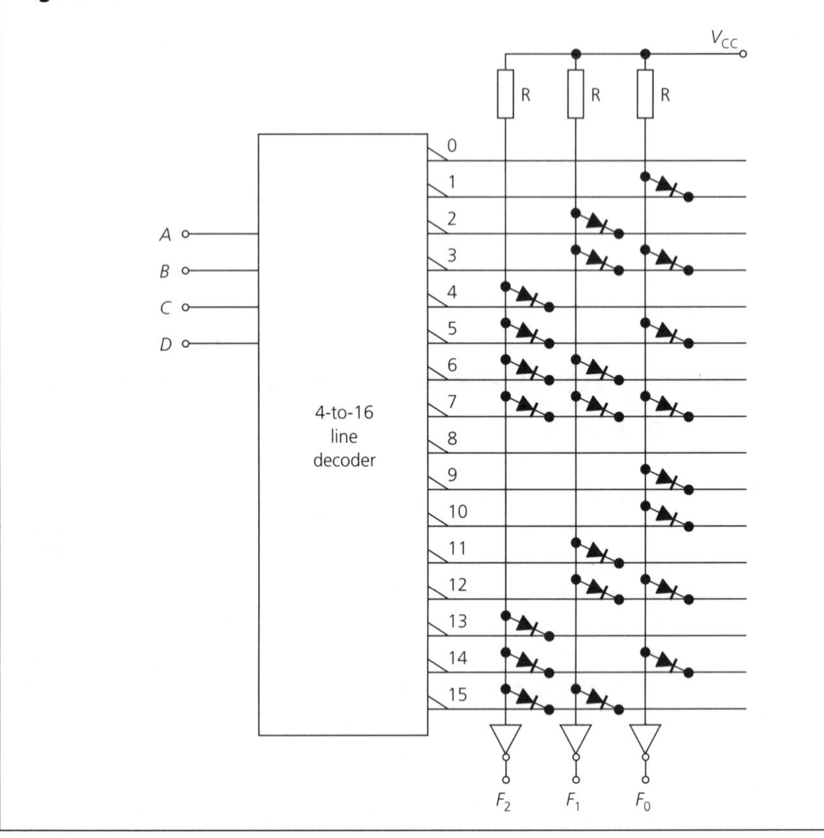

Programmable ROMs

A programmable ROM (PROM) is designed to be programmed by the user to meet the user's specification. When first manufactured all the intersections in the memory matrix are linked together by either a fusible diode or a fusible transistor. Figure 12.17 shows the basic concept. Some PROMs are supplied with all outputs at logic 1 and fuses must be blown in selected cells to produce an output logic 0. Other PROMs initially have all outputs at logic 0 and then fuses must be blown to produce logic 1 at selected cells.

Programming of a PROM is accomplished by addressing a particular location in the memory that is to store a 1, and then passing a sufficiently large current through the

Fig. 12.17 *PROM cell*

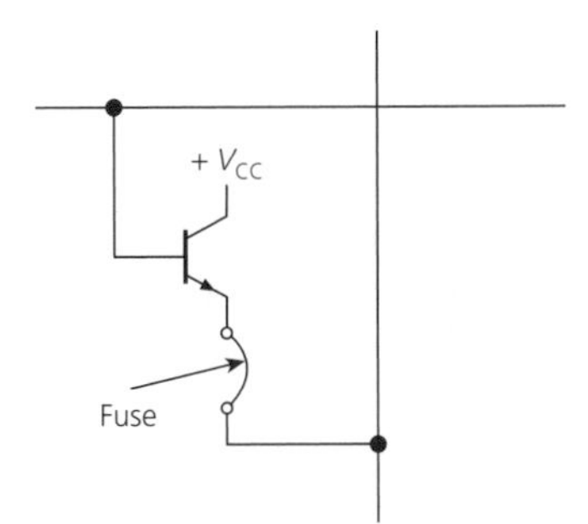

transistor to blow the fuse. The transistor then no longer links the row and column lines at that location. Once programmed a PROM cannot be reprogrammed. PROMS are used widely in the control of electrical equipment such as washing machines and ovens.

Erasable and electrically alterable PROMs

Some PROMs can have their programs altered and a new program written into the memory. In an *erasable PROM*, or *EPROM*, the logic 1 state is stored at a location by the storage of an electrical charge and not by the blowing of a fuse. Programming takes about 10 μs/byte. A read takes about 100 ns. The charge may be held for many years. When a program is to be erased, the chip is exposed to ultra-violet radiation that is directed through a transparent quartz window in the chip package. This radiation removes the stored charge at *every* location in the memory, so that all locations store binary 0. This takes approximately 20 minutes. Reprogramming is carried out by addressing each cell that is to store a logic 1 bit and then causing that cell to store a charge.

EPROM has some advantages over PROM in that it is more reliable, it is user-programmable, and it can be erased and re-programmed many times. In each memory cell a small conductive area, or *floating gate*, is formed inside silicon dioxide insulation between the gate and the body of the FET. The control gate is connected to the word line but the floating gate is left disconnected. To store logical 1 at a location a high voltage (≈12.5 V) is applied to the gate and this produces an electric field that is intense enough to allow electrons to pass through the insulation and on to the floating gate. When the high voltage is removed these electrons are trapped on the floating gate. If a logic 1 voltage of about 5 V is applied to the gate, the trapped electrons will prevent it inducing a drain-source channel in the MOSFET. Since zero drain current flows, the drain voltage will not fall to 0 V and, hence, logical 0 is stored in that cell. A cell that was not programmed with the 12.5 V will have an induced channel when 5 V is applied to its gate and the 5 V will make the drain voltage fall to 0 V; such cells, therefore, have logical 1 stored in them.

Electrically erasable PROM

An alternative to the EPROM is known as the *electrically erasable PROM* or *E^2PROM*. Again, programming a memory cell to store logic 1 is accomplished by charging that cell. With the E^2PROM, however, the erasure procedure is carried out by applying a reverse-polarity voltage to a cell that removes any stored charge. The E^2PROM offers an advantage over the EPROM in that the erasure process can be applied to an individual cell in the matrix and without removal of the IC from the circuit. Writing takes less than 10 ms/byte, and reading is faster at about 100 ns/byte. E^2PROMs are used for the storage of data tables such as the set-up parameters for printers.

The EPROM has two main disadvantages. (a) It is not possible to just reprogram a selected few locations; if any changes are wanted all data must be erased and all locations reprogrammed. (b) An EPROM must be removed from its circuit before reprogramming can be carried out. The E^2PROM does not suffer from these disadvantages.

However, each time an E^2PROM is erased some damage is done to its insulating region and eventually it may become unable to store data. E^2PROM is more expensive than EPROM.

Flash ROM

A *flash ROM* is similar to an EPROM in its operation except that it can be wiped clean electrically instead of by using ultra-violet light. The erasure method is very similar to that used by the E^2PROM but all the data stored in a flash ROM must be wiped simultaneously, while an E^2PROM can be selectively wiped. Flash memory is much quicker to erase than E^2PROM because all cells are wiped together, but it is about 50% more expensive. Flash ROM is employed in computer systems where software and reference tables often need to be updated.

Ferro-electric random access memory

Ferro-electric RAM (FRAM) is a non-volatile form of DRAM which provides fast read and write speeds, low power consumption, and access speeds of the same order as static RAM. The memory cell consists of a capacitor that uses a ferro-magnetic material as its dielectric. When an electric field is applied to a FRAM it switches it into one of its two stable states and, when the field is removed, the FRAM remains in that state.

A ferromagnetic capacitor has the hysterisis loop shown in Fig. 12.18. When data is written into the cell, a voltage V_{CC} is applied that moves the cell to point A on the hysterisis loop. When the voltage is removed the cell settles down at its stable position B, which is the logical 0 state. To write logical 1 into the cell, the voltage V_{CC} must be applied with the reverse polarity to move the cell first to point C and then, after the voltage has been removed, to stable point D, which is the logical 1 state.

To read the state of a cell, voltage V_{CC} is applied with the polarity to bring the cell to position A. Whether a 1 or a 0 is stored in the cell is indicated by the magnitude of the *change* in the stored charge. Clearly, the change is larger if the cell has moved from point D to point A than if it has moved from B to A. After the state of a cell has been read, the cell is left in its logical 0 state. This means that if the stored data is a 1 then a re-write is necessary to restore the data, but no re-write is required if the stored data is 0.

Fig. 12.18 *FRAM principle*

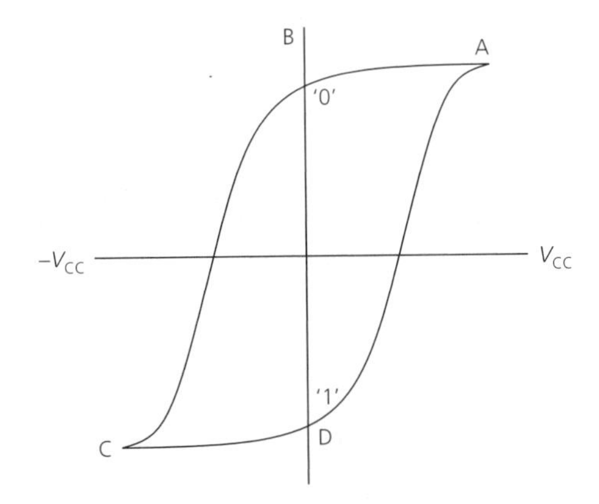

NOVRAM

A NOVRAM consists of a static RAM that has been overlaid bit-for-bit by an E^2PROM. It includes an autostore feature that automatically saves the contents of the static RAM to the E^2PROM when the power supply is turned off. Data can easily be transferred from the RAM to the E^2PROM – known as a store – on shut-down, or from E^2PROM to RAM – known as recall – on power-up.

Memory addressing

Most memory systems include both RAM and ROM. The memory is mapped so that some sets of addresses are allocated to ROM and other sets are allocated to RAM. The mapping is carried out electronically by the use of the $\overline{CS}$ input pin of each IC. Large-capacity memory is obtained by interconnecting several smaller-sized memory ICs. There are two main ways in which this can be done: (a) linear decoding, and (b) fully decoding.

Linear decoding

Linear decoding employs the unused address lines of a microprocessor as chip select lines for the memory. Figure 12.19 shows the basic concept. Address lines A_0 through to A_{12} are used to access locations in the selected memory IC, and address line A_{13} is used as a CS signal to select which IC is to be active. Figure 12.20 shows a memory system that consists of three RAMs and one ROM each of which has a capacity of $2^{11} = 2$k addressable locations. A 16-bit address bus is used and address lines A_0 through to A_{12} are used to address locations in the RAMs and ROM. Address lines A_{12} through to A_{15} are used to select one of the four memory ICs. The microprocessor is programmed to ensure that only one of the four ICs is selected at any one time.

This method has the advantage that it does not require the use of any decoding ICs but it also has some disadvantages.

Fig. 12.19 *Basic linear decoding of memory*

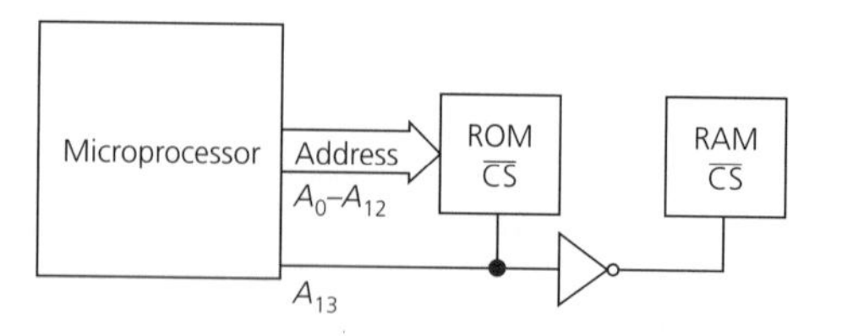

Fig. 12.20 *Linear decoding of a four IC memory system*

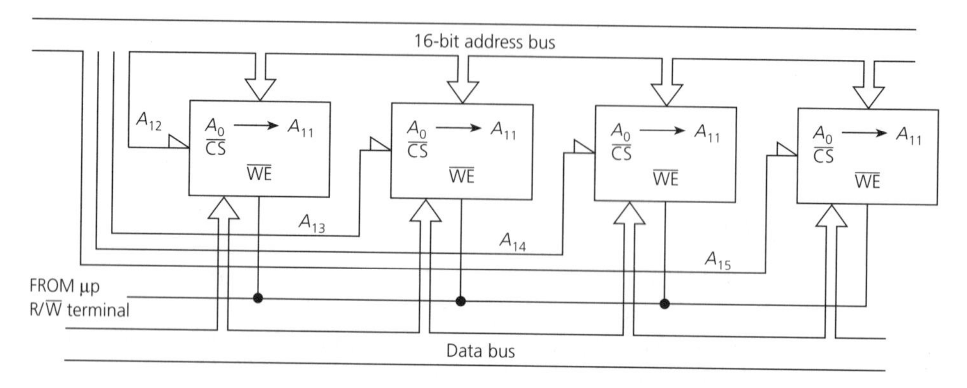

Fig. 12.21 *Full decoding of a four IC memory system*

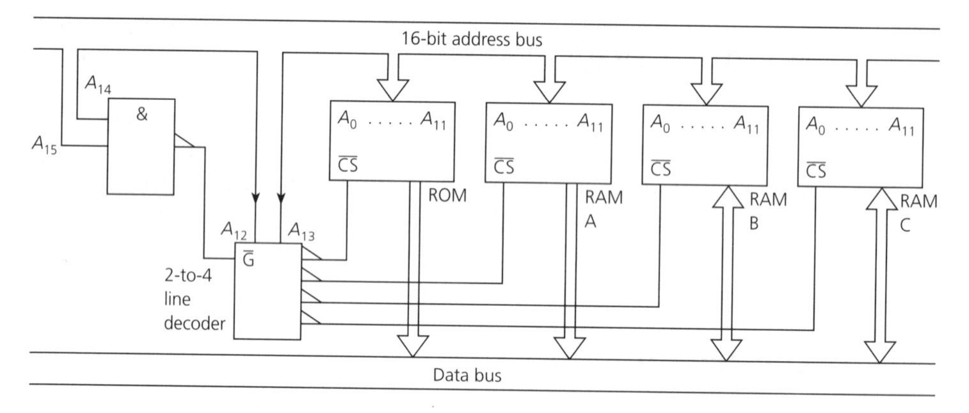

Fully decoding

The fully decoded method of addressing memory ICs makes use of one, or more, decoding ICs. The basic arrangement is shown in Fig. 12.21.

Bits A_{14} and A_{15} are applied to a 2-to-4 line decoder whose four outputs are each applied to the $\overline{CS}$ pin of a memory. Only one of the $\overline{CS}$ pins is taken LOW by the decoder at any time and only that memory is then connected to the data bus and able to be read, or for a RAM be written to. The outputs of the non-selected memories are in a high impedance state, so they have no effect upon signals travelling along the bus. The decoder is disabled if either one, or both, of the address lines A_{14} and A_{15} go HIGH.

Fig. 12.22

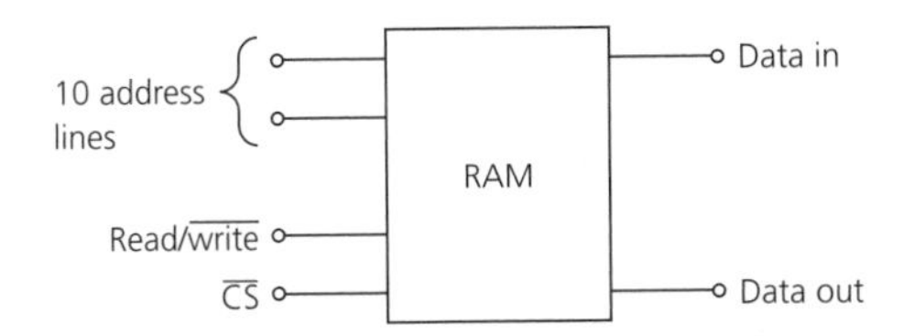

EXERCISES

12.1 A ROM has 11 address lines and 8 data lines. Calculate
(a) The number of bits stored.
(b) The organization of the memory.

12.2 Figure 12.22 shows the basic block diagram of a 1024 × 1 RAM. Draw a diagram to show how four such RAMs could be connected to give a 1024 × 4 memory.

12.3 (a) A ROM has 12 address lines. Calculate the number of memory locations.
(b) A ROM is organized as 8k × 8. List the function of the necessary IC pins. What is the minimum number of pins required?
(c) A 64-bit square memory matrix is addressed by the binary number 110100. In which row and in which column is the wanted location?

12.4 (a) A RAM has 4096 addressable locations. How many address pins does it have? If there are four data input/output pins what is the organization of the RAM? What other pins are also required?
(b) Explain the functions of the $\overline{\text{CS}}$ and R/$\overline{\text{W}}$ pins on a RAM chip. Why does a ROM not have an R/$\overline{\text{W}}$ pin? Why is a decoder employed in the addressing of a location in both a RAM and a ROM?

12.5 Draw a ROM to implement the Boolean functions

$$F_0 = ABCD + A\bar{B}C\bar{D} + \bar{A}B\bar{C}D + AB\bar{C}\bar{D}$$
$$F_1 = A\bar{B} + \bar{A}B$$

12.6 Determine the Boolean equation that is held by the ROM shown in Fig. 12.23.

12.7 A DRAM has inputs $\overline{\text{CS}}$ (chip select), $\overline{\text{OE}}$ (output enable), and $\overline{\text{WE}}$ (write enable). Write down a truth table showing the functions of the device.

12.8 Briefly state the differences between
(a) SRAM, DRAM, EDO DRAM, BEDODRAM, VRAM, NOVRAM and FRAM.
(b) ROM, PROM, EPROM, E^2PROM and flash ROM.

12.9 Draw figures to show how
(a) Four 1k × 1 DRAMs can be connected together to form a 1k × 4 memory.
(b) Four 1k × 8 DRAMs can be connected to form a 4k × 8 memory.

Fig. 12.23

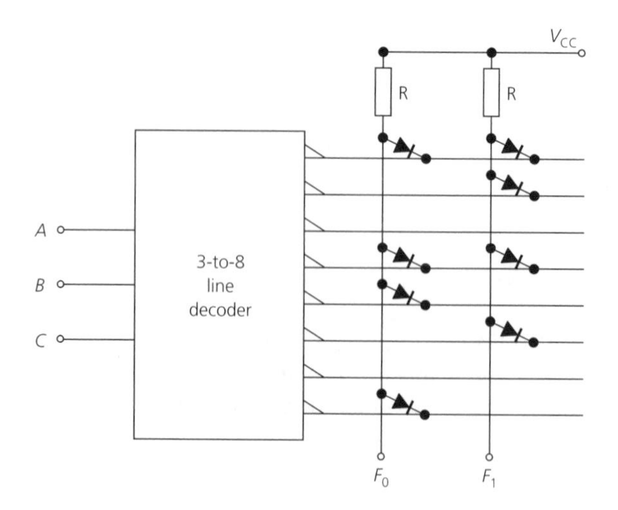

Fig. 12.24

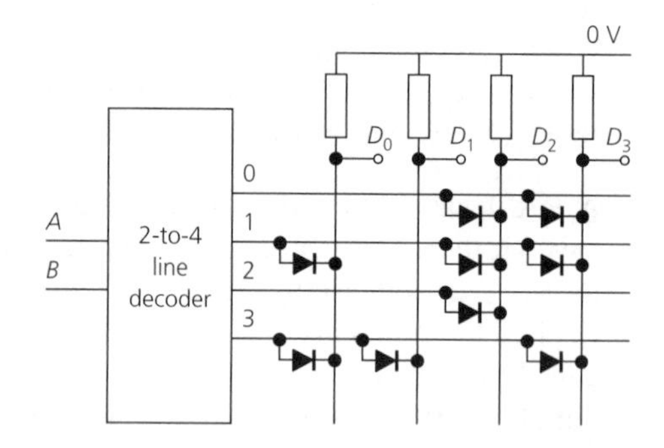

12.10 A ROM has 15 address pins.
(a) How many words can it store?
(b) A ROM can store 128k words. How many addresses must it have?

12.11 (a) Determine the organization of the ROM shown in Fig. 12.24.
(b) Write down a truth table showing the content of each address.
(c) What logical functions are provided?

12.12 A number of 32k × 4 DRAMs are available. How many must be interconnected to form
(a) A 512k × 16 memory.
(b) A 1M × 8 memory?

12.13 (a) A memory is organized as 1M × 4. Calculate
(i) The number of locations in the memory.
(ii) The total number of bits stored.
(b) Repeat for a 64k × 16 memory.

Fig. 12.25

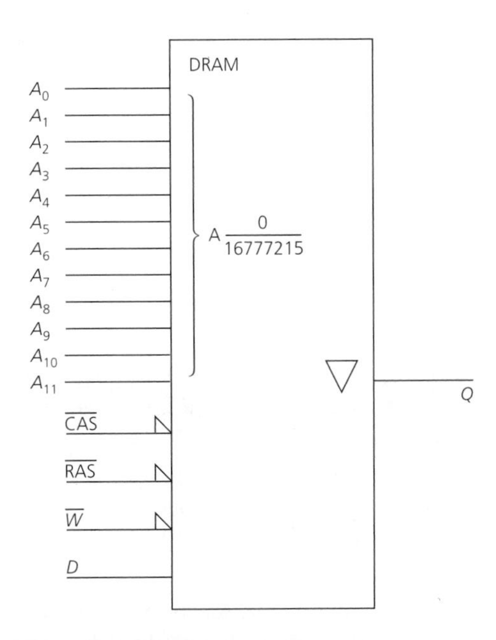

12.14 (a) Express in hexadecimal the lowest and the highest addresses of a (i) 32×4 E^2PROM, and (ii) a $1M \times 4$ DRAM.
(b) Calculate the size of each address.

12.15 Design a 64×16 memory using $64k \times 4$ ICs. The start address is 0H.

12.16 Figure 12.25 shows the simplified logic symbol for a DRAM.
(a) What is its organization?
(b) What kind of output has it?
(c) How many rows and columns has the memory matrix?
(d) How many data inputs are there?

12.17 Implement the Boolean functions $F_0 = AB\bar{C}\bar{D} + \bar{A}\bar{B}C\bar{D} + A\bar{B}\bar{C}D$, $F_1 = A\bar{B}\bar{C}\bar{D} + \bar{A}B\bar{C}\bar{D}$, and $F_2 = ABCD + \bar{A}\bar{B}\bar{C}D + \bar{A}B\bar{C}D + A\bar{B}C\bar{D}$, using a ROM.

13 Programmable logic devices

After reading this chapter you should be able to:

(a) State the difference between a PLD and a PROM.
(b) Understand the advantages of using PLDs over the use of SSI/MSI devices.
(c) Explain the operation of a PAL device.
(d) Explain the operation of a PLA device.

Modern electronic circuitry makes considerable use of *programmable logic devices* (PLDs) and this chapter presents an introduction to their use. A PLD is an IC that can be programmed to perform a wide variety of specific combinational and/or sequential logic functions, although this chapter will consider only combinational logic circuits. Essentially, a PLD consists of two arrays of gates, firstly of AND gates, and secondly of OR gates, with perhaps some flip-flops and/or registers as well. Most random logic designs, except for the simplest, can be implemented using a PLD more cheaply than would be possible using SSI/MSI devices. One PLD can perform the functions of four to twelve SSI/MSI devices.

The design of a circuit that is to perform a required complex logical operation can be carried out using traditional methods such as Boolean algebra and the Karnaugh map. Once a design has been finalized, its implementation can be achieved in one of two ways: either the required design is sent, using custom-designer forms, to the chip manufacturer for implementation; or the design can be implemented by the designer. PLDs of the first type are often known as *mask-programmable* devices. A gate array is manufactured in a standard unprogrammed form and is later (on receipt of a particular circuit design) 'wired' to give the wanted circuit by the application of a metallization conductive pattern. All gate arrays are standard devices until they have been programmed to perform some particular circuit function.

Other PLD chips, often known as *field-programmable*, can be programmed by the user. Mask- and field-programmable arrays are, in this sense, analogous to the ROM and to the PROM that were discussed in chapter 12. The programming of a PLD may be carried out by the selective blowing of fuses in the programmable array(s). Some PLDs (EPLD) are erasable; with these the programming is erased by exposing the device to ultra-violet light after which the IC can be re-programmed. Other PLDs (E^2PLD) are electrically erasable and they can be erased and re-programmed many times. The relative merits of bipolar fused-link and erasable PLDs are as follows:

Fig. 13.1 *Block diagram of a PLD*

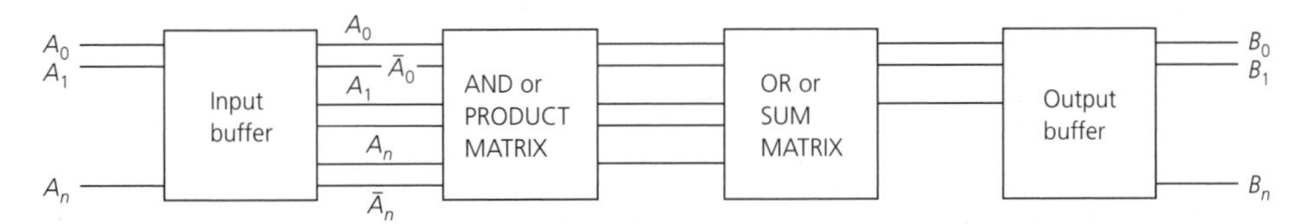

- Bipolar devices have a much higher power consumption and cannot operate from voltages as low as 3 V, as can EPLDs and E^2PLDs.
- Bipolar devices are of larger physical size, mainly because of the size of the internal fuse.
- Bipolar devices are cheaper and easier to use but cannot be re-programmed. This does not give easy prototyping and means that any mistakes which occur will result in the IC having to be discarded.

Field-programmable devices are well suited to small-scale production. For large-volume production of a circuit it is generally more economic to employ mask-programmable devices.

The basic block diagram of a PLD is shown in Fig. 13.1. The input digital codeword is applied to the input buffer and the complement of each variable is generated. The AND matrix is able to handle any one of 2^{n+1} input combinations and each of these can be programmed to produce a required result. In some types of PLD the OR matrix is programmed to produce particular combinations of the words held in the AND matrix and these are passed, via the output buffers, to the output terminals. Other types of PLD have a programmable AND matrix but the OR matrix is fixed and not programmable. The dimensions of a PLD are $m \times n \times p$, where m is the number of inputs, n is the number of product terms, and p is the number of outputs.

In the AND matrix each column pair is allocated to the true and complemented states of a different input. Each row constitutes an AND or product term. Connections are made between rows and columns to determine which combination(s) of inputs will take the product term(s) high. More than one product term is usually applied to the input of an OR gate.

One example of a PLD is the PLS 100; this is a $16 \times 48 \times 8$ programmable logic array and it has 16 inputs, 48 product terms, and 8 outputs. The number of inputs is sometimes known as the *width* of the AND array.

The use of SSI/MSI devices to implement a logic design requires different types of IC to be inter-connected. This often leads to some ICs being under-utilized as well as posing printed circuit board (PCB) layout problems. The use of PLDs enables complex logic functions – that would be uneconomic using SSI/MSI devices or a PROM – to be designed. This greatly reduces the number of ICs used in a design, and this in turn simplifies the PCB layout. Other advantages which derive from the use of PLDs are:

- Smaller system size. The reduced IC count and PCB space result in smaller physical dimensions.
- Higher performance. The reduced number of ICs allows a system to operate faster as well as reducing the overall power consumption.

- Higher reliability. The probability of failure of a digital system is directly related to the number of ICs in that system. Since there are fewer ICs when PLDs are employed, with consequent fewer connections, socket or soldered, the reliability is increased.
- Design security. A system designed with SSI/MSI devices can be replicated fairly easily but this is not so for a PLD design.
- Increased flexibility. Customized components allow for the tailoring of a system to meet the specific requirements of the end-user. Up-grading is also easier.

A PLD differs from a PROM in that, whereas the PROM employs an address decoder, the PLD employs a *programmable address matrix* (the AND matrix). The size of the address decoder in a PROM *doubles* for each additional address bit: a 256×8 PROM will require 8 address bits and a 512×8 PROM will require 9 address bits. In each case the address decoder must be able to access all of the locations in the PROM even though they may not all be used. This means that the use of an address decoder is inherently inefficient.

Since a PLD employs a programmable AND matrix instead of an address decoder, it is able to select any one of a number of input states. Because of this, a PLD does not need to store, and be able to access, unused sum-of-product terms.

The AND matrix does not have a fixed size in relation to the number of its inputs. For example, a 16-input PROM would require an address decoder able to decode 2^{16} words; a PLD with the same number of inputs need only be large enough to store (say) 48 words or product terms. This means that *any* 48 out of 2^{16} or 65 536 words can be used. A PLD permits the programming of simplified Boolean equations, whereas the canonical form of an expression is necessary for a PROM. A consequence of this is that a PLD can have more inputs than a PROM.

Programmable logic devices are available from various manufacturers in the TTL, CMOS and ECL logic families. One manufacturer, for example, offers: TTL arrays with 250, 550, 1000 or 2000 gates; CMOS arrays with 400, 800, 1300, 2100, 4100, 6500 or 11 000 gates; and ECL arrays with 300, 1200 or 2000 gates. The most common form of CMOS PLD is the *generic array logic* (GAL) device.

Generic Array Logic (GAL) devices are electrically erasable CMOS devices that combine re-configurable logic, CMOS low-power dissipation and bipolar high-speed performance. Each GAL device incorporates an *output logic macrocell* (OLM) that can be programmed to allow the device to emulate a bipolar PAL device.

Programmable array logic

The basic *programmable array logic* (PAL) device consists of an array of AND gates whose inputs can be programmed and whose outputs are connected to a fixed array of OR gates. All inputs are buffered. The following explanation refers to bipolar transistor devices. The user is able to program the AND array to perform a wanted sum-of-products logic function by blowing certain fuses in the array. The fuses that are not blown make the required connections in the array for the particular application being implemented. The OR array cannot be programmed. Figure 13.2 shows the basic architecture

Fig. 13.2 *Basic PAL architecture*

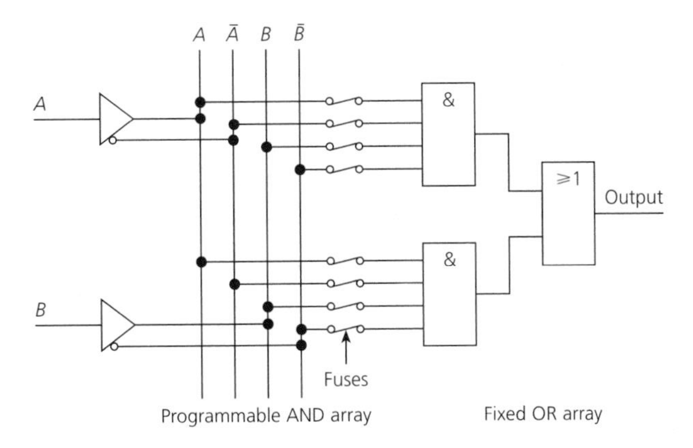

Fig. 13.3 *Alternative method of drawing PAL architecture*

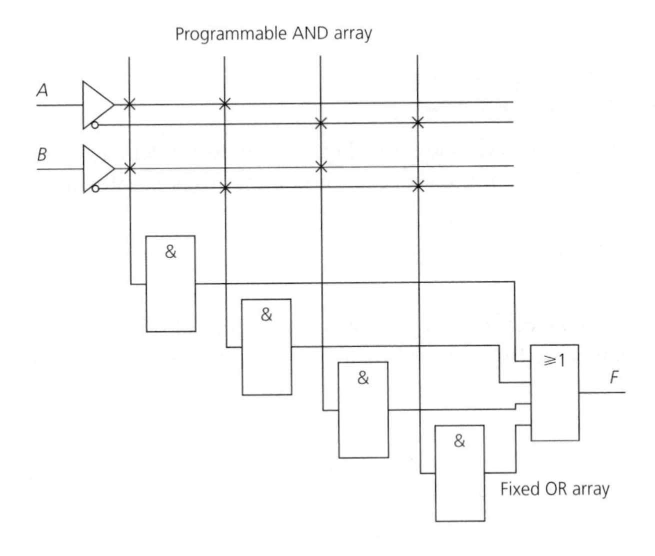

of a PAL device. Every input and its complement can be connected to, or disconnected from, every AND gate.

Instead of drawing fuses, intact or blown, a × is used to indicate the points in the matrix at which the fuse is intact, and the absence of a × at a row/column intersection indicates that the fuse at that point has been blown. This is shown in Fig. 13.3. Any unused AND gate has all input variables and their complements connected to its inputs so that its output is at logic 0 and this will have no effect upon the output of a gate in the OR gate array. The OR array is not programmable and fixed connections in it are indicated by a blob.

To simplify further the drawing of PLD connections, diagrams following the convention shown in Fig. 13.4 are often employed.

Fig. 13.4 *Further simplification of PAL drawing*

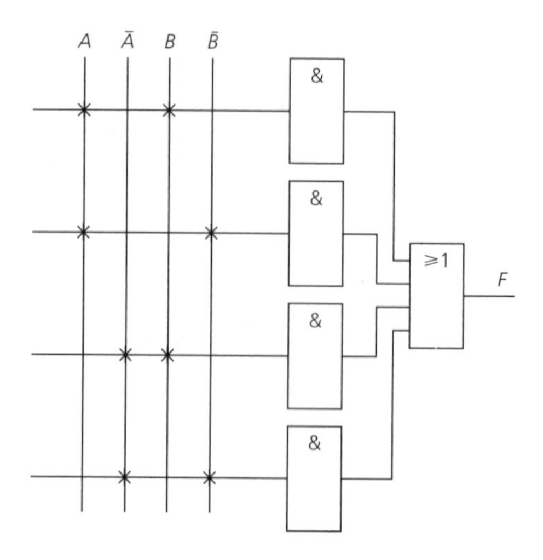

EXAMPLE 13.1

The 2 × 4 PAL shown in Fig. 13.3 is to implement the logical function $F = AB + \bar{A}\bar{B}$. At which intersections should the fuses be blown?

Solution

The connections shown for the top and bottom AND gates are required. At these points the fuses should be left intact. At all other points the fuses should be blown. Figure 13.5 shows the required circuit using the convention shown in Fig. 13.4 (*Ans.*)

Fig. 13.5

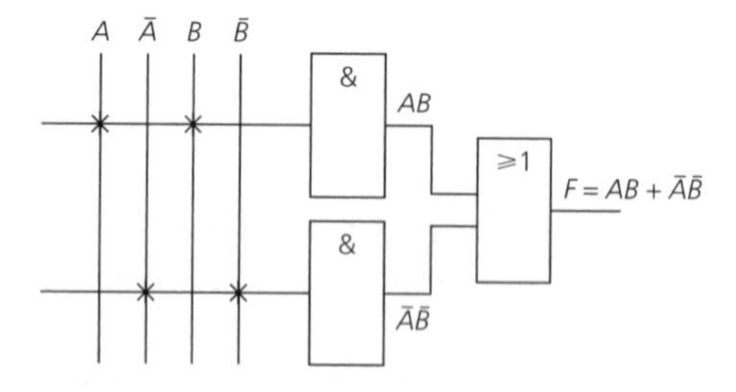

The AND matrix of a 4 × 4 PAL is shown in Fig. 13.6. The A, $\bar{A}$, B and $\bar{B}$ etc. lines can be left connected to the input of an AND gate by not blowing the fuse, or it can be disconnected by blowing the fuse. This matrix has been programmed to give the outputs

Fig. 13.6 *A 4 × 4 AND array*

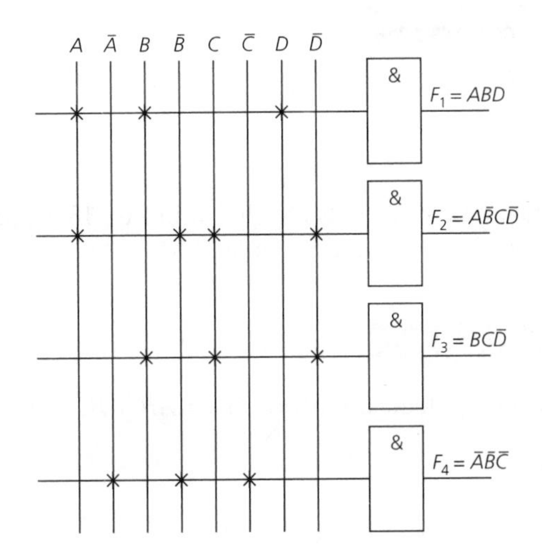

Fig. 13.7

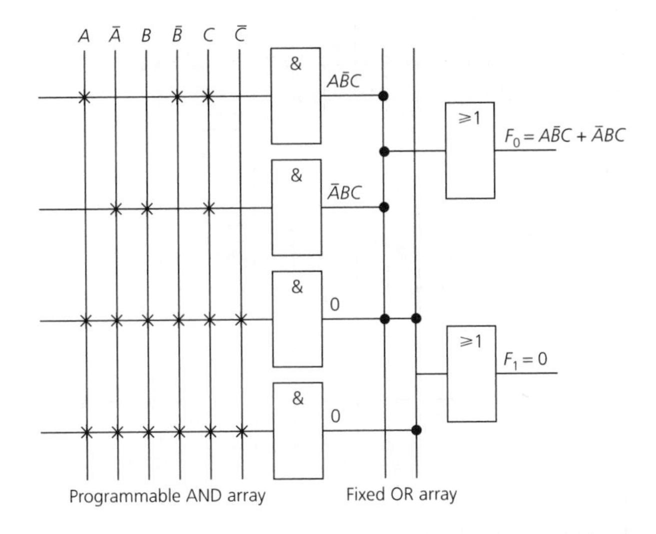

shown. The labelling of a PAL device indicates its performance, e.g. 10H8 indicates 10 inputs and 8 active-HIGH outputs, and 12L6 indicates 12 inputs and 6 active-LOW outputs.

To make a PLD logic diagram as easy to read as possible, figures are drawn in the way illustrated in Fig. 13.7. Each intersection contains a fused link; this fuse has been blown where there is no cross, and left intact if there is a cross. AND array outputs are known as the *product terms*. To form a sum-of-products expression the product terms must be fed into an OR gate. In a PAL device, inputs to the OR gates are fixed, but in a PLA they are also programmable.

EXAMPLE 13.2

Implement the logical function $F = A\bar{B}C + \bar{A}BC$ on a 3×4 PAL.

Solution

The required connections are shown in Fig. 13.7 (*Ans.*)
[Note that if all connections to an AND gate remain made the output is 0.]

EXAMPLE 13.3

Implement the logical functions $F_0 = AB + \bar{B}\bar{C}$ and $F_1 = \bar{A}C + A\bar{B}C$, using a PAL.

Solution

The required circuit is shown in Fig. 13.8 (*Ans.*)

Fig. 13.8

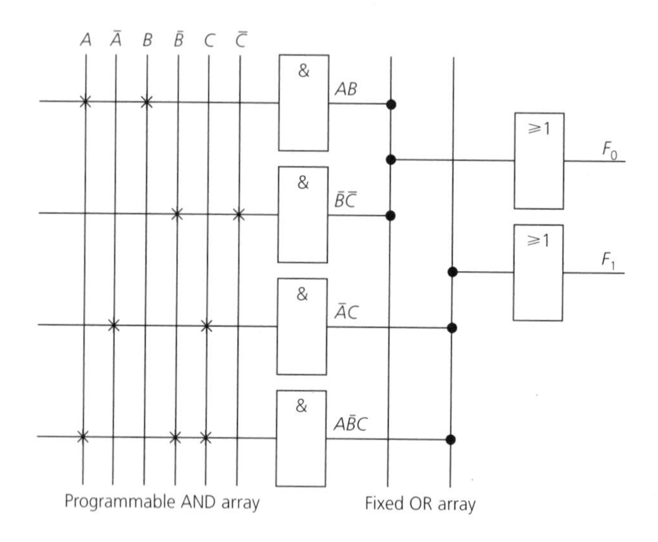

EXAMPLE 13.4

Determine the logical expression that has been implemented by the PAL circuit shown in Fig. 13.9.

Solution

The product terms produced by the AND array are: ABC, $\bar{B}C$, $\bar{A}\bar{B}\bar{C}$, $A\bar{B}C$, $\bar{A}\bar{C}$ and BC. Hence, the output expressions are: $F_0 = ABC + \bar{B}C$, $F_1 = \bar{A}\bar{B}\bar{C} + A\bar{B}C$, and $F_2 = \bar{A}\bar{C} + BC$ (*Ans.*)

Fig. 13.9 *Basic architecture of a PLA*

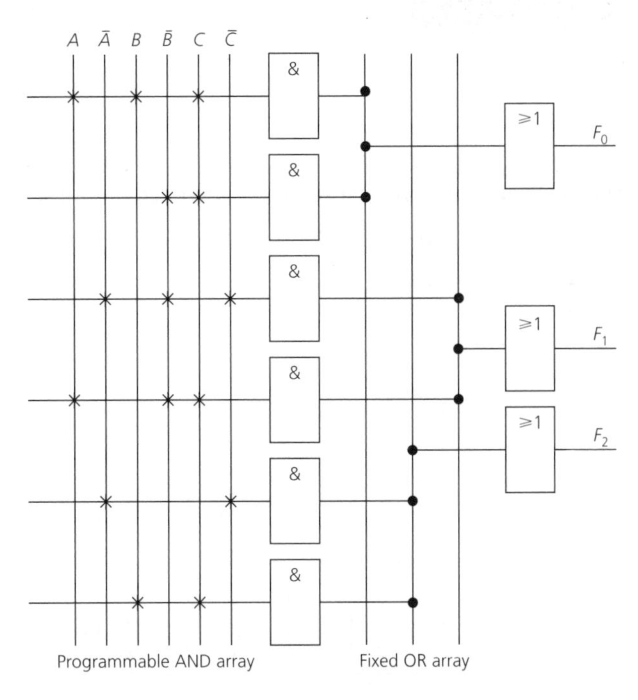

Fig. 13.10 *Further simplification of the circuit*

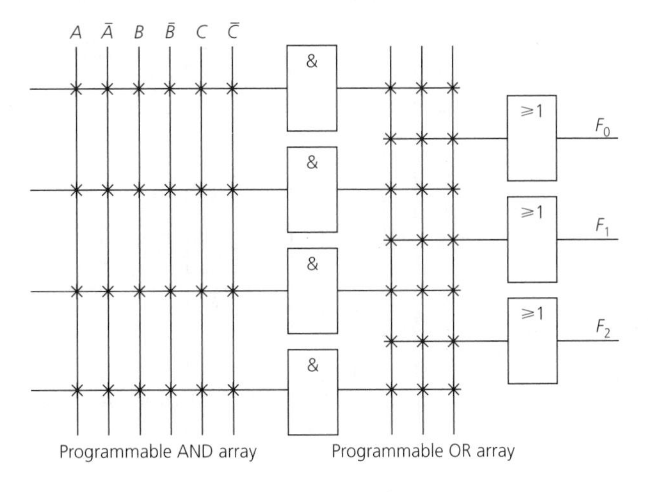

Programmable logic array

A *programmable logic array* (PLA) also contains both an AND gate array and an OR gate array but now both arrays are fully programmable. Figure 13.10 shows the basic architecture of a PLA. Using the same convention as with the PAL.

EXAMPLE 13.5

Implement a full-adder using a PLA.

Solution

The Boolean equations that represent the operation of a full-adder are:

sum $S = ABC_{in} + \bar{A}\bar{B}C_{in} + A\bar{B}\bar{C}_{in} + \bar{A}B\bar{C}_{in}$, and carry $C_{out} = AB + A\bar{B}C_{in} + \bar{A}BC_{in}$

The necessary PLA configuration is shown in Fig. 13.11 (*Ans.*)

Fig. 13.11 *Full-adder*

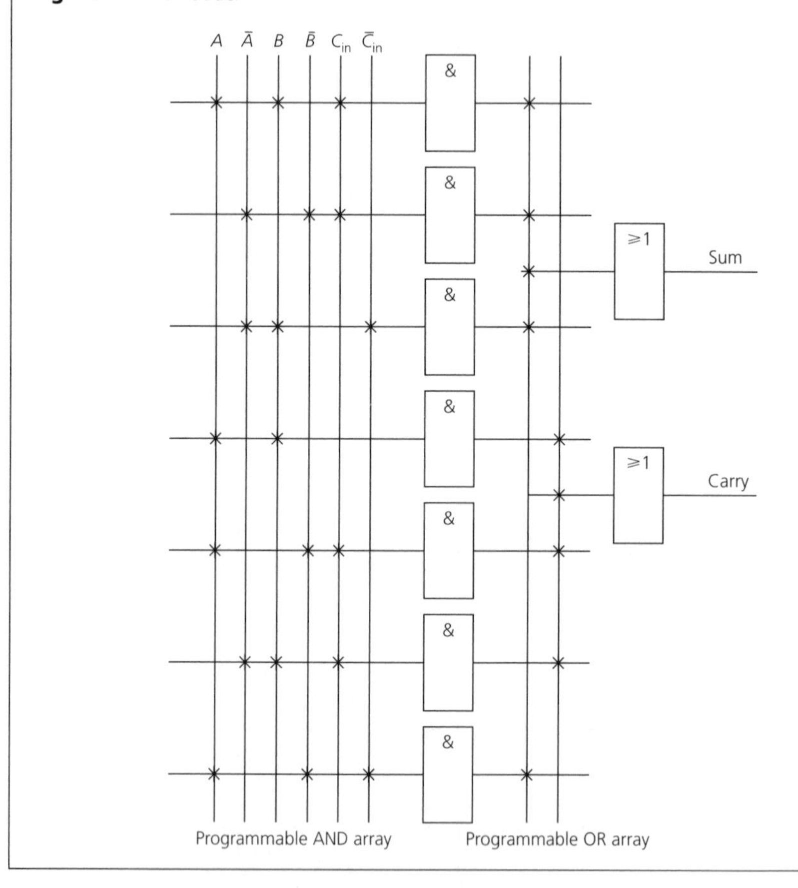

Using PALs and GALs

There are many standard configurations available in bipolar PAL devices and samples of these are shown in Fig. 13.12. Each device is numbered as *m*H*n*, *m*L*n* or *m*R*n*, where *m* indicates the number of inputs, *n* the number of outputs, H indicates active-HIGH and

Fig. 13.12

PAL	No. of data inputs	No. of outputs and configurations	
10H8	10	8 ×	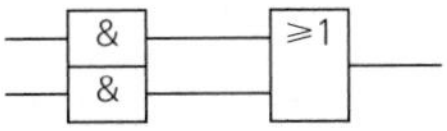
10L8	10	8 ×	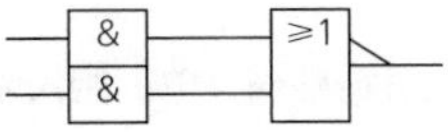
12H6	12	4 ×	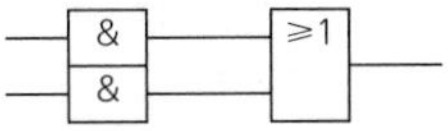 2 × 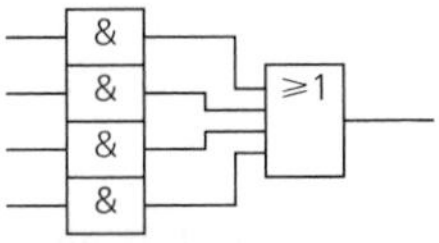
12L6	12	4 ×	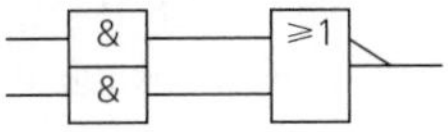 2 × 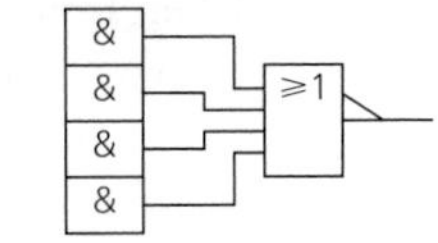
14H4	14	4 ×	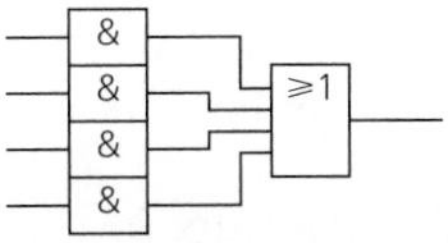
14L4	14	4 ×	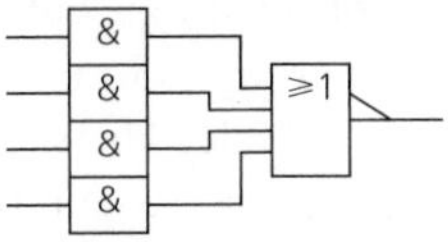
14L8	14	6 ×	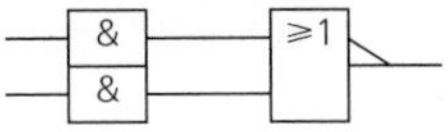 2 × 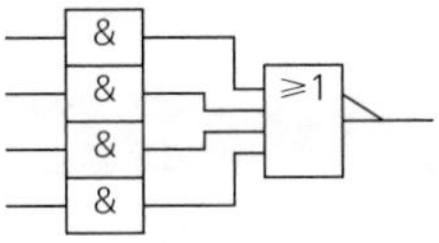
16L2	16	2 ×	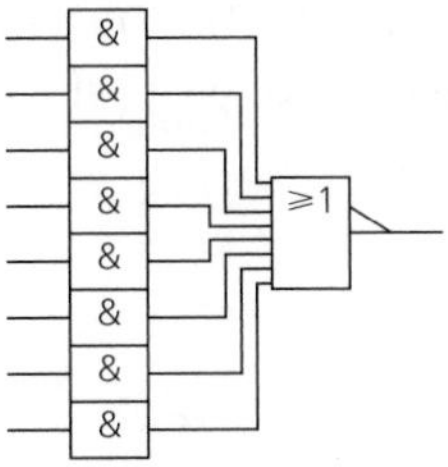
16L6	16	2 ×	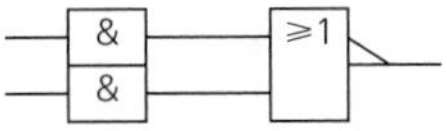 4 × 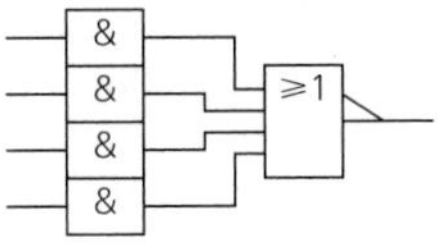

Table 13.1 ***PALs emulated by 16V8 GAL***

10L8, 10H8, 12L6, 12L8, 14L4, 14H4, 16L2, 16H2, 16R8, 16L8 and 16H8

Table 13.2 ***PALs emulated by 20V8 GAL***

14L8, 14H8, 16L6, 16H6, 18L2, 18H2, 20L2, 20H2, 20L8 and 20H8

L indicates active-LOW outputs. The letter R indicates a registered output, i.e. one that is taken from a D flip-flop. The 14L8 PAL, for example, has 14 inputs and 8 active-LOW outputs. Any of the 20-pin IC package devices can be emulated by a 16V8 GAL and any of the 24-pin package devices may be emulated by a 20V8 GAL. The PAL devices that may be emulated are shown in Tables 13.1 and 13.2.

A logic pattern in a PAL can be copied into a GAL with 100% accuracy so that the GAL can directly replace the PAL in a circuit. The programming of GAL devices is beyond the scope of this book (see *Applied Digital Electronics*, Addison Wesley Longman, 1999).

The logic diagram of the 14L8 PAL is shown in Fig. 13.13. The device has 14 input pins each of which is connected to a buffer amplifier. The buffer has two outputs, one inverting and one non-inverting, so that both the true and complemented forms of an input variable are made available. There are eight active-LOW outputs, two of which are from 4-input NOR gates and the remainder are from 2-input NOR gates.

EXAMPLE 13.6

The expressions $F = \bar{A}BC\bar{D} + ABC\bar{D} + A\bar{B}C\bar{D} + \bar{A}\bar{B}C\bar{D} + \bar{A}B\bar{C}D + AB\bar{C}D + A\bar{B}\bar{C}D + \bar{A}\bar{B}\bar{C}D$, and $G = AB\bar{C}\bar{D} + \bar{A}B\bar{C}\bar{D} + A\bar{B}C\bar{D} + \bar{A}\bar{B}C\bar{D} + ABCD + \bar{A}BCD + \bar{A}\bar{B}\bar{C}D + \bar{A}B\bar{C}D$ are to be implemented on a 14L8 PAL. Determine the necessary programming.

Solution

The mapping for F is

CD \ AB	00	01	11	10
00	0	0	0	0
01	1	1	1	1
11	0	0	0	0
10	1	1	1	1

Fig. 13.13

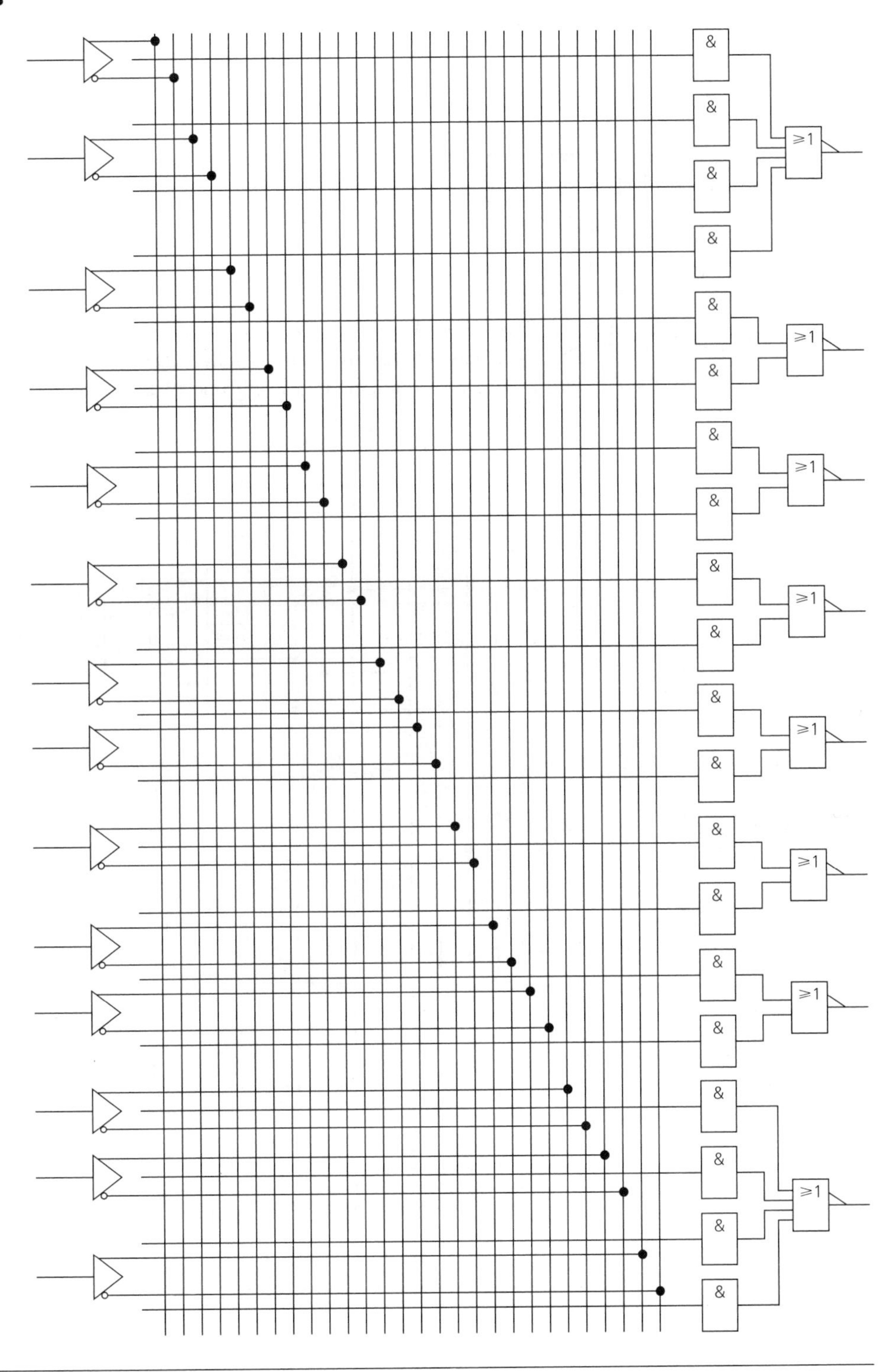

From the map, $F = C\bar{D} + \bar{C}D$ and $\bar{F} = \bar{C}\bar{D} + CD$.

The mapping for G is

CD \ AB	00	01	11	10
00	0	1	1	0
01	1	1	0	0
11	0	1	1	0
10	1	0	0	1

From the map, $G = B\bar{C}\bar{D} + \bar{A}\bar{C}D + \bar{B}C\bar{D} + BCD$, and $\bar{G} = \bar{B}\bar{C}\bar{D} + \bar{B}CD + A\bar{C}D + BC\bar{D}$. The 14L8 has active-LOW outputs taken from NOR gates. Hence, $F = \overline{\bar{C}\bar{D} + CD}$ and this can be implemented using one of the 2-input NOR gates (see Fig. 13.14(a)) (*Ans.*)

Fig. 13.14

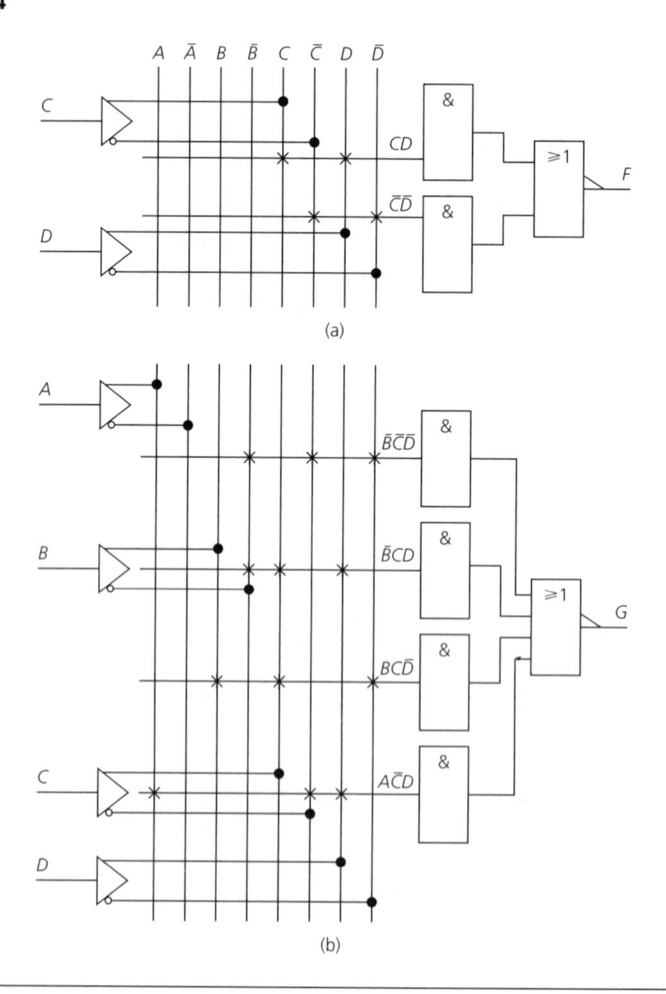

Fig. 13.15

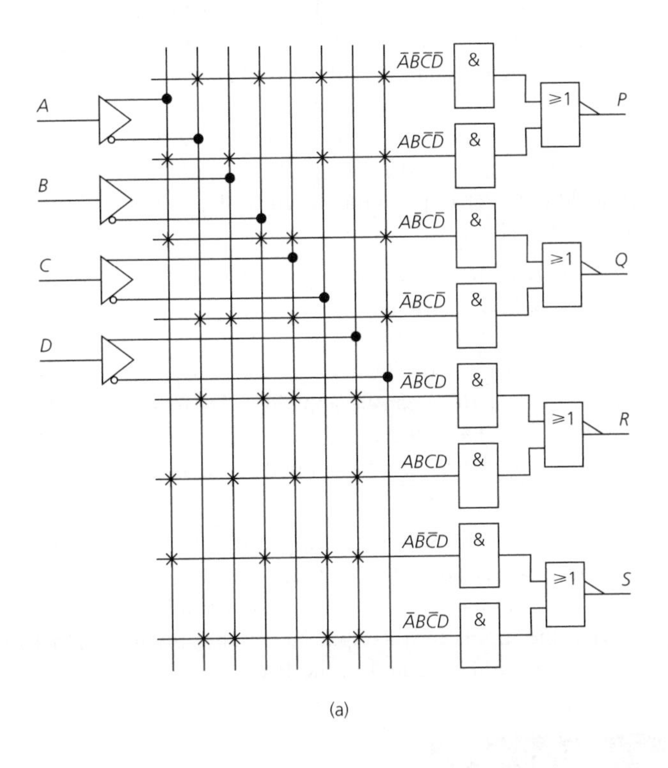
A
B
C
D
$\bar{A}\bar{B}\bar{C}\bar{D}$
$AB\bar{C}\bar{D}$
$A\bar{B}C\bar{D}$
$\bar{A}BC\bar{D}$
$\bar{A}\bar{B}CD$
$ABCD$
$A\bar{B}\bar{C}D$
$\bar{A}B\bar{C}D$
&
$\geqslant 1$
P
Q
R
S

(a)

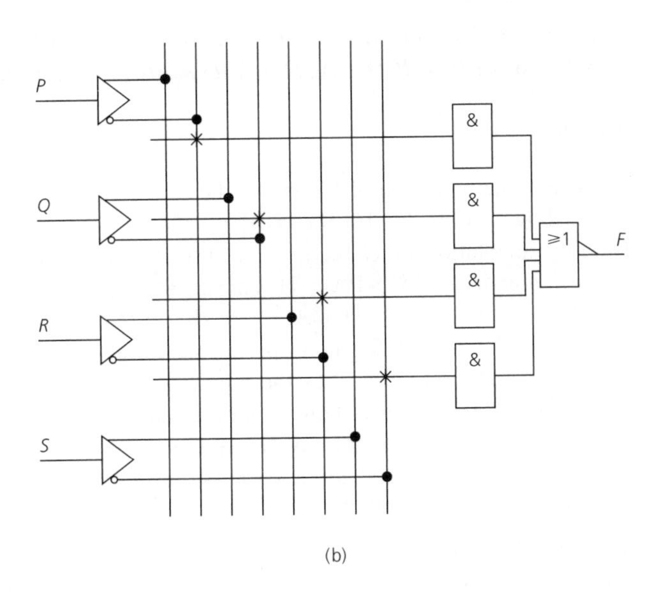
P
Q
R
S
&
$\geqslant 1$
F

(b)

Output G has four terms and so it can be taken from one of the 4-input NOR gates. $G = \overline{\bar{B}\bar{C}\bar{D} + \bar{B}CD + A\bar{C}D + BC\bar{D}}$, and is shown implemented in Fig. 13.14(b) (*Ans.*)

If a Boolean equation with more than four variables is to be implemented using the 14L8 PAL it will be necessary to combine some terms so that they can be applied to a 2-input NOR gate. Consider the function $F = \bar{A}\bar{B}\bar{C}\bar{D} + AB\bar{C}\bar{D} + A\bar{B}C\bar{D} + \bar{A}BC\bar{D} + \bar{A}\bar{B}CD + ABCD + A\bar{B}\bar{C}D + \bar{A}B\bar{C}D$. If the terms of this function are grouped together in pairs then each of the paired terms can be applied to a 2-input NOR gate. Thus:

(a) $\bar{A}\bar{B}\bar{C}\bar{D} + AB\bar{C}\bar{D}$ can be applied to one 2-input NOR gate to give $P = \overline{\bar{A}\bar{B}\bar{C}\bar{D} + AB\bar{C}\bar{D}}$.

(b) $A\bar{B}C\bar{D} + \bar{A}BC\bar{D}$ can be applied to the next 2-input NOR gate to give $Q = \overline{A\bar{B}C\bar{D} + \bar{A}BC\bar{D}}$.

(c) $\bar{A}\bar{B}CD + ABCD$ can be applied to another 2-input NOR gate to give $R = \overline{\bar{A}\bar{B}CD + ABCD}$.

(d) $A\bar{B}\bar{C}D + \bar{A}B\bar{C}D$ can be applied to the last 2-input NOR gate to give $S = \overline{A\bar{B}\bar{C}D + \bar{A}B\bar{C}D}$.

Figure 13.15(a) shows the implementation of these four pairs of variables.

The four variables P, Q, R and S can then be applied to four input terminals and, since the terms are inverted, the complemented output line will be used to provide the inputs to a 4-input NOR gate. Figure 13.15(b) shows how the paired variables are implemented in the PAL.

EXAMPLE 13.7

Implement the functions (a) $\bar{F} = \bar{B}\bar{C}D + A\bar{C}D + A\bar{B}\bar{D} + \bar{A}BC\bar{D}$; (b) $\bar{G} = AC\bar{D} + A\bar{C}D + \bar{A}BD$, and (c) $\bar{H} = BC\bar{D} + \bar{A}C\bar{D} + \bar{A}B\bar{D} + A\bar{B}CD + A\bar{B}\bar{C}\bar{D}$, using a 14L8 PAL.

Solution

(a) The implementation of $\bar{F}$ requires the use of one of the 4-input NOR gates and is shown in Fig. 13.16(a) (*Ans.*)

(b) Implementation of $\bar{G}$ requires the use of a 3-input NOR gate which is not available in the 14L8 PAL. This means that the expression must be rearranged. One way is $G = (AC\bar{D} + A\bar{C}D) + \bar{A}BD$. $(AC\bar{D} + A\bar{C}D)$ can be implemented using a 2-input NOR gate to produce a variable $P = (\overline{AC\bar{D} + A\bar{C}D})$, and then P and $\bar{A}BD$ can be used as the input variables to another 2-input NOR gate. Figure 13.16(b) shows how G is obtained (*Ans.*)

(c) The direct implementation of $\bar{H}$ requires the use of a 5-input NOR gate that is also not available and so the expression for H must also be re-arranged. One possibility is: $H = (BC\bar{D} + \bar{A}C\bar{D}) + (\bar{A}B\bar{D} + A\bar{B}CD) + A\bar{B}\bar{C}\bar{D}$.

In Fig. 13.16(c); $(BC\bar{D} + \bar{A}C\bar{D})$ have been combined to give an output Q, and $(\bar{A}B\bar{D} + A\bar{B}CD)$ have been combined to give output R. $\bar{R}$ has then been combined with $A\bar{B}\bar{C}\bar{D}$ to give another output S. Lastly, outputs Q and S have been applied as the inputs to another 2-input NOR gate to obtain the wanted implementation of $\bar{H}$ (*Ans.*)

Fig. 13.16

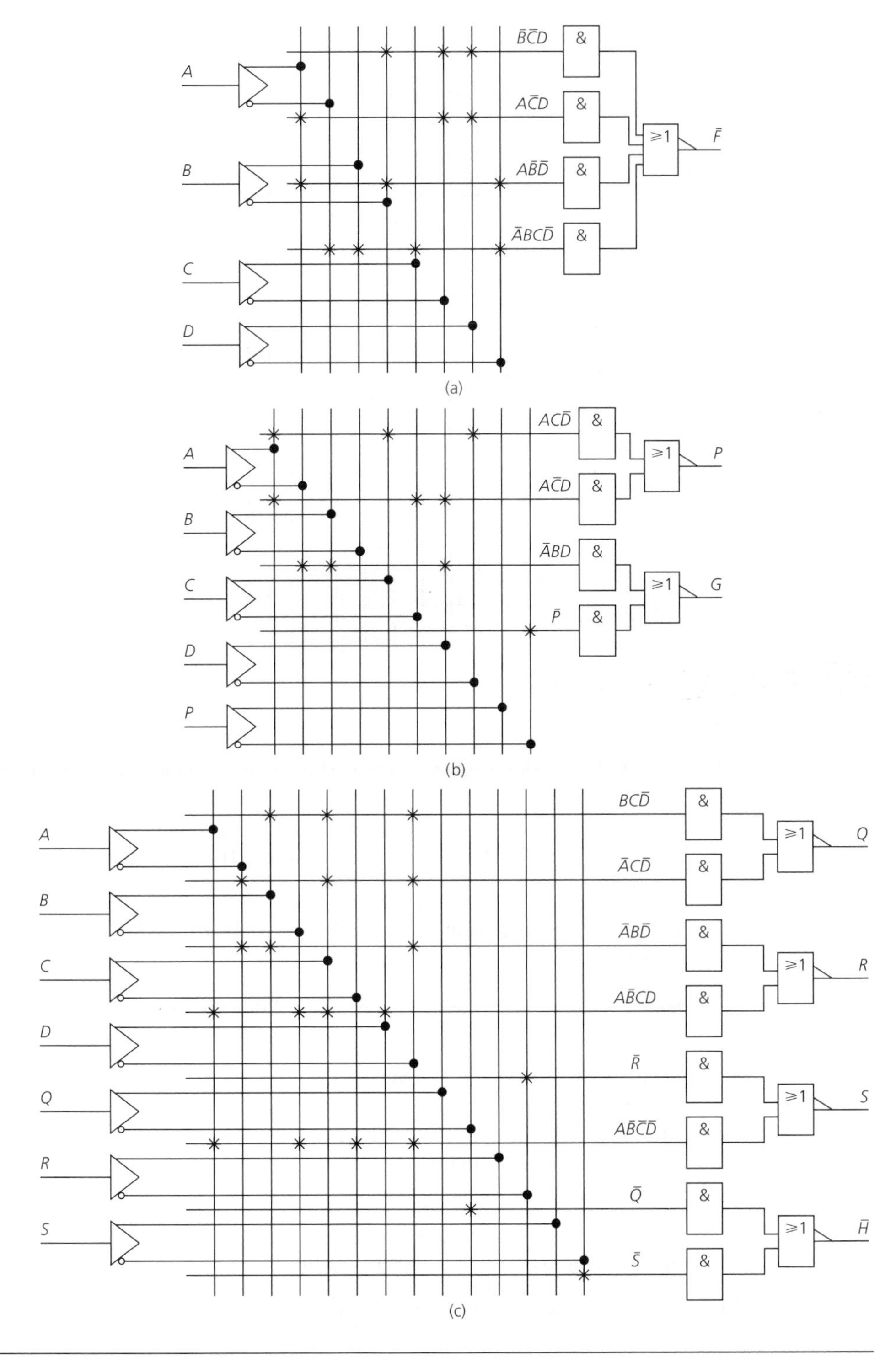
A
B
C
D
$\bar{B}\bar{C}D$
$A\bar{C}D$
$A\bar{B}\bar{D}$
$\bar{A}BC\bar{D}$
&
≥1
$\bar{F}$
(a)
A
B
C
D
P
$AC\bar{D}$
$A\bar{C}D$
$\bar{A}BD$
$\bar{P}$
P
G
(b)
A
B
C
D
Q
R
S
$BC\bar{D}$
$\bar{A}C\bar{D}$
$\bar{A}B\bar{D}$
$A\bar{B}CD$
$\bar{R}$
$A\bar{B}\bar{C}\bar{D}$
$\bar{Q}$
$\bar{S}$
Q
R
S
$\bar{H}$
(c)

Fig. 13.17

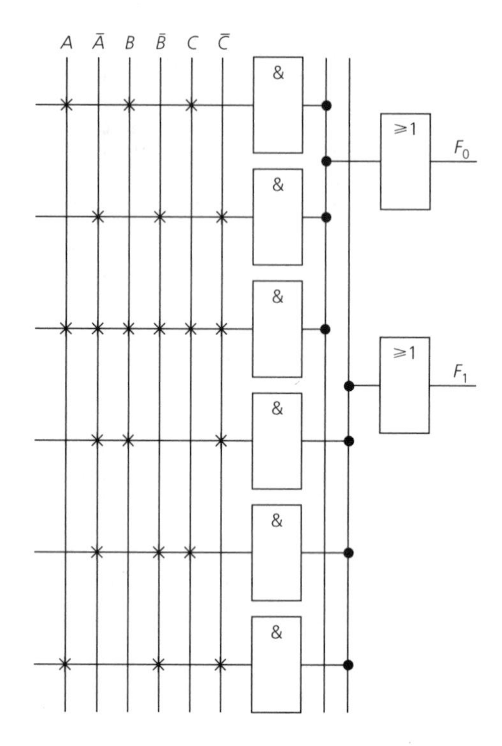

EXERCISES

13.1 (a) A PAL is described as 16 × 8. How many (i) inputs, and (ii) product terms outputs does it have?

(b) A PAL is required to produce the logic expression $F = A + B + C + D + E + F + G + H$. The device available is a 12H4. Explain how the function can be implemented.

13.2 Use a PAL to implement the logical expressions $F_0 = ABC + \bar{B}\bar{C}$ and $F_1 = A\bar{B}\bar{C} + \bar{A}B\bar{C} + \bar{A}C$.

13.3 Determine the logic function which is implemented by the PAL shown in Fig. 13.17.

13.4 Figure 13.18 shows a PAL. State

(a) The number of outputs.

(b) How many product terms there are in each output.

(c) The number of inputs.

(d) The labelling of the device.

13.5 Design a PAL to implement the logical functions $F_0 = ABCD + \bar{A}B\bar{C}D + A\bar{B}C\bar{D} + \bar{A}\bar{B}\bar{C}\bar{D}$, and $F_1 = AB\bar{C}\bar{D} + \bar{A}\bar{B}CD$.

13.6 Implement, using a PLA, the logical function $F = AB + \bar{B}\bar{C} + BC$.

Fig. 13.18

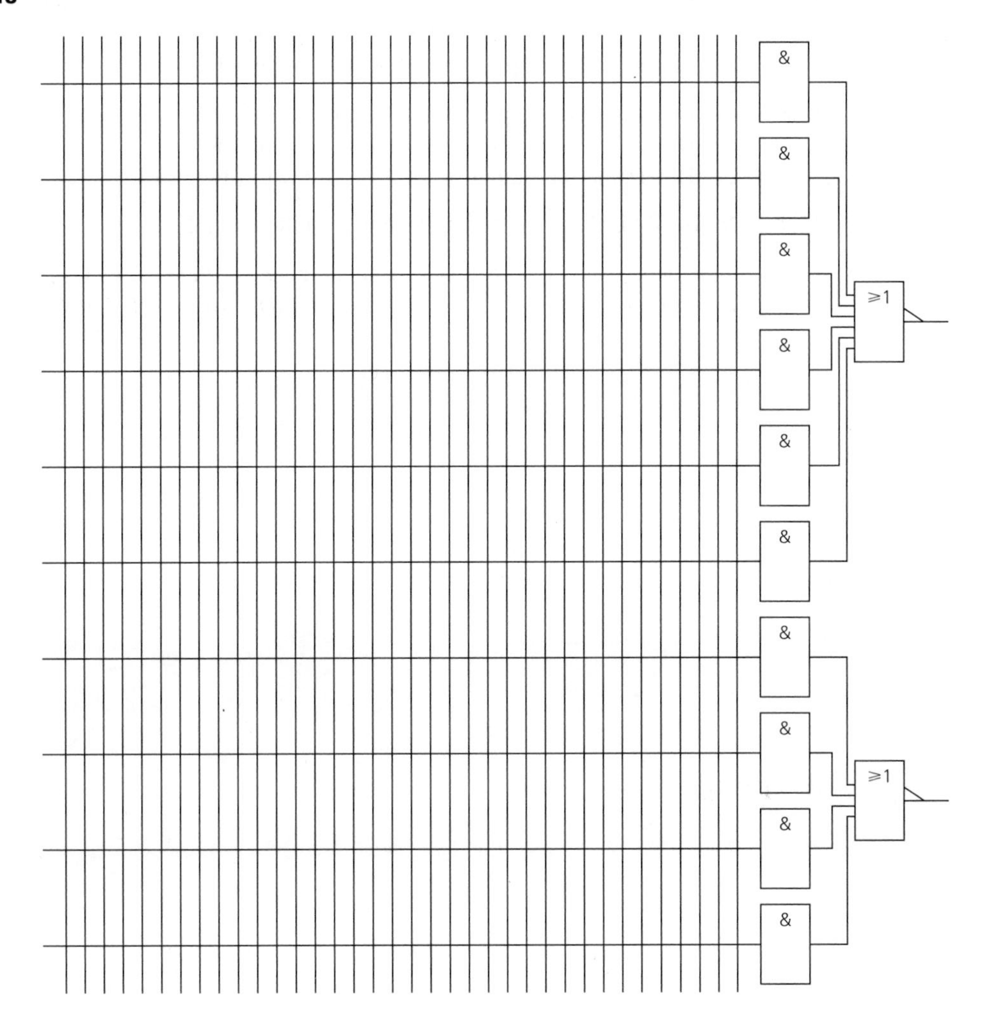

Fig. 13.19

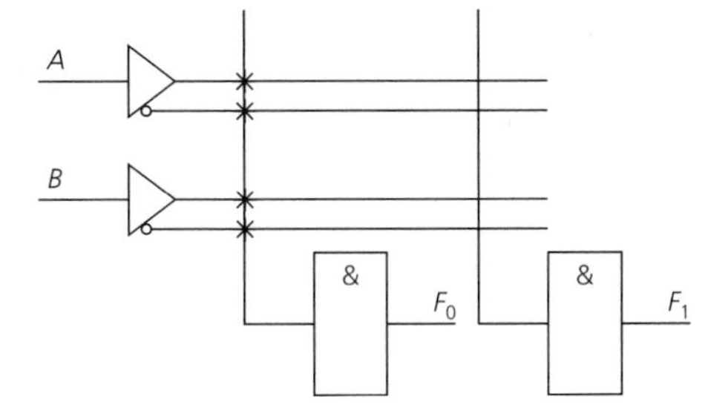

13.7 (a) List some advantages gained by using a PLD to implement a digital design rather than a number of SSI/MSI devices.
(b) List the advantages of E^2PLDs over bipolar fuse-link PLDs.

13.8 (a) Draw the PLD representation of a 2-input AND gate.
(b) Draw the PLD representation of a 2-input OR gate.
(c) Determine the output of the PLA shown in Fig. 13.19.

13.9 Draw a PAL that provides the exclusive-OR function.

13.10 Implement, using a 14L8 PAL, the equations $\bar{F} = A\bar{B}\bar{D} + \bar{A}BC$, $\bar{G} = AC\bar{D} + B\bar{C}D$, $\bar{H} = AB\bar{C}$, $\bar{I} = AD + \bar{A}C\bar{D} + A\bar{B}\bar{C}$, and $\bar{J} = AD + BC + \bar{A}B + \bar{A}C\bar{D} + A\bar{B}\bar{C}$.

14 Visual displays

After reading this chapter you should be able to:

(a) Employ an LED as an indicator in a digital circuit.
(b) Understand the operation of an LCD.
(c) Design a decoder for a 7-segment display.
(d) Use an IC decoder/driver to operate a 7-segment display.
(e) Understand the principles of operation of 16-segment and dot matrix displays.

Visual displays are employed in electronic equipment such as digital watches and clocks, calculators and laptop/notebook computers. A variety of display devices are available but, in this book, the discussion will be limited to just two; namely, the light-emitting diode or LED, and the liquid crystal display or LCD. The circuitry needed to drive the display can be constructed using SSI gates, but MSI decoder/driver ICs are commonly employed, sometimes in the same package as the LEDs.

The light-emitting diode

The majority of *light-emitting diodes* (LEDs) are either gallium phosphide (GaP) or gallium-arsenide-phosphide (GaAsP) devices. An LED radiates energy in the visible part of the electromagnetic spectrum when the forward bias voltage applied across the diode exceeds the voltage that turns it ON. This voltage varies with the type of LED and the colour of its emitted light. Forward voltages at 10 mA current, and the colour of the emitted light, are shown in Table 14.1, for the five most commonly employed semiconductor materials.

Fig. 14.1 *LED symbol*

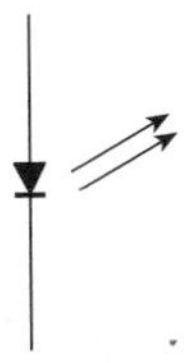

Table 14.1 ***LED types***

Colour	*Material*	*Wavelength (peak radiation) (nm)*	*Forward voltage at 10 mA current (V)*
Red	GaAsP	650	1.6
Green	GaP	565	2.1
Yellow	GaAsP	590	2.0
Orange	GaAsP	625	1.8
Blue	SiC	480	3.0

SiC is silicon carbide.

Fig. 14.2 *Use of LEDs as indicators*

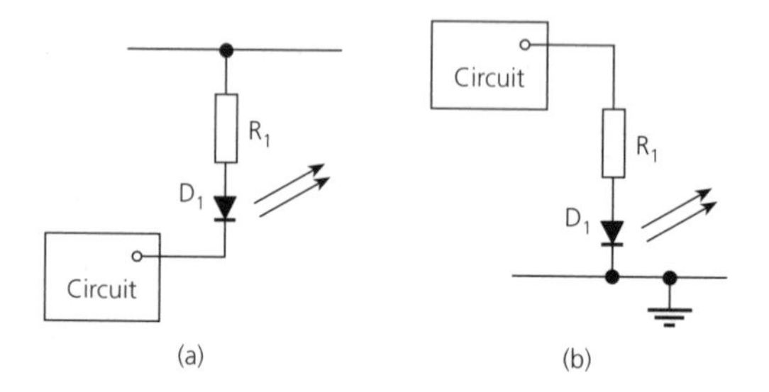

The current flowing in an LED must not be allowed to exceed a safe figure, generally some 20–60 mA and, if necessary, a resistor of suitable value must be connected in series with the diode to limit the current.

Often, an LED is connected between one of the outputs of a TTL or CMOS device and either earth or +5 V depending upon when the LED is required to glow visibly. If, for example, an LED is expected to glow when the output to which it is connected is LOW, the device should be connected as shown in Fig. 14.2(a). Suppose the low voltage to be 0.4 V and the sink current to be 10 mA. Then, if the LED voltage drop is 1.6 V, the value of the series resistor will be

$$(5 - 1.6 - 0.4)/(10 \times 10^{-3}) = 300\ \Omega$$

When the output of the device is high ($\simeq$5 V), no current flows and the LED remains dark. When the LED is to glow to indicate the HIGH output condition, the circuit shown in Fig. 14.2(b) must be used. Now

$$R_1 = (5 - 1.6)/(10 \times 10^{-3}) = 340\ \Omega$$

When an LED is reverse biased it acts very much like a zener diode with a low breakdown voltage ($\simeq$4 V).

Light-emitting diodes are commonly used because they are cheap, reliable, easy to interface, and are readily available from a number of sources. Their main disadvantage is that their luminous efficiency is low, typically 1.5 lumens/watt.

Since light is emitted from the p–n junction of an LED, the maximum junction area must be exposed to the outside world. Most LEDs are manufactured using a ring, or coaxial, form of construction in which the inner ring is made from p-type material while the outer employs n-type material. The physical area of the emitting surface is very small but the package design is such that the emitted light diverges on exit and so gives the impression that it originates from a much larger light source.

Liquid crystal displays

A solid crystal is a material in which the molecules are arranged in a rigid lattice structure. If the temperature of the material is increased above its melting point, the liquid that is formed will tend to retain much of the orderly molecular structure. The material is then said to be in its *liquid crystalline phase*. There are two classes of liquid crystal known, respectively, as *nematic* and *smetic* but only the former is used for display devices.

Nematic means that the liquid crystal has different optical properties which are determined by the orientation of its molecules. The liquid crystal has the fluidic characteristics of a fluid and the molecular orientation characteristics of a solid. The molecules are rod-shaped and possess the ability to re-align themselves in the direction of the electric field whenever a voltage is applied across the crystal. If a liquid crystal is squeezed in between two glass plates, which have fine grooves etched on their surfaces, the molecules near those surfaces will become aligned with the grooves. If the grooves on the two glass plates are orientated to be mutually at right angles then an overall twist will be superimposed on the molecule chain. An LCD with a 90° twist is known as a *twisted nematic* (TN) LCD and an LCD with a twist of between 180° and 270° is known as a *super-twisted nematic* (SNT) LCD. Both NT and SNT LCDs possess the property that the plane of polarization of polarized light passing through the crystal will be rotated through an angle that is equal to the angle of twist. This means that if two polarizing layers, or *polarizers*, are fixed mutually at right angles on either side of an NT liquid crystal, light will pass straight through.

If a voltage is applied across the liquid crystal to set up an electric field, the molecules of the crystal will be aligned with the field. This new alignment will undo the twist in the molecular chain and then the passing polarized light will not be rotated. Consequently, the non-rotated light cannot pass through the rear polarizer. This means that there is then zero light transmission through the crystal over the area in which the electric field exists.

An NT LCD does not give a good contrast between light and dark because the twist angle and the polarizer's relative angle are not exactly equal. Hence, some light leaks through the crystal when there ought to be zero transmission. An SNT LCD is much better in this respect.

The construction of a *liquid crystal* cell is shown in Fig. 14.3. A layer of a liquid crystal material is placed in between two glass plates that have transparent metal film electrodes deposited on to their interior faces. A reflective surface, or mirror, is situated on the outer side of the lower glass plate (it may be deposited on its surface). The conductive material is generally either tin oxide or a tin oxide/indium oxide mixture and it will transmit light with about 90% efficiency. The incident light upon the upper

Fig. 14.3 *Liquid crystal cell*

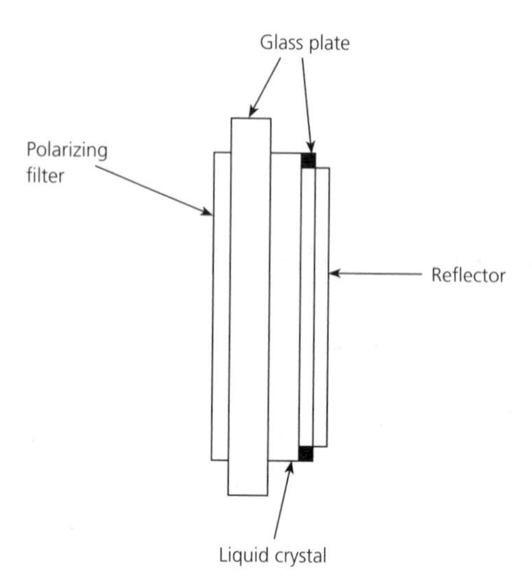

Fig. 14.4 *Types of LCD: (a) reflective; (b) transreflective and (c) transmissive*

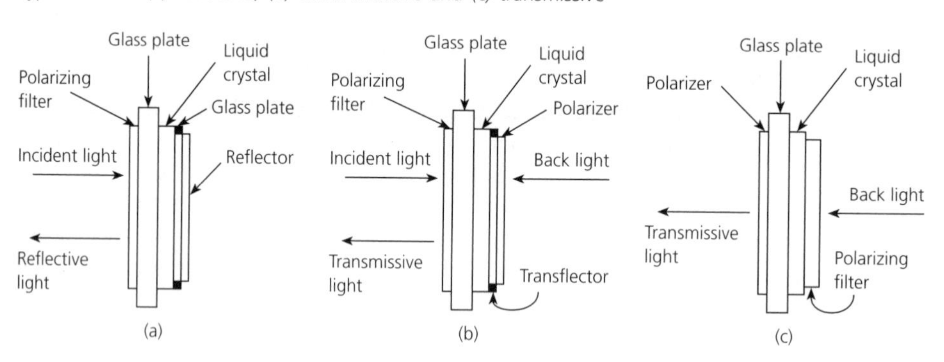

glass plate is polarized in such a way that, if there is *zero* electric field between the plates, the light is able to pass right through and arrive at the reflective surface. Here it is reflected back and the reflected light travels through the cell and emerges from the upper plate. If a voltage is applied across the plates the polarization of the light entering the cell is altered and it is then no longer able to propagate as far as the reflective surface. Thus, no light returns from the upper surface of the cell and the display appears to be dark. Because the LCD does not emit light it dissipates very little power.

There are three ways in which LCDs are used: (a) the reflective type shown in Fig. 14.4(a) for which some ambient light is necessary, (b) the transflective type shown in Fig. 14.4(b); for this type, ambient light is used during the day when it is light and a back light is used when it is dark, and (c) the transmissive type, Fig. 14.4(c) which always requires the use of a back light.

Fig. 14.5 *Action of an LCD in (a) passing light and (b) not passing light*

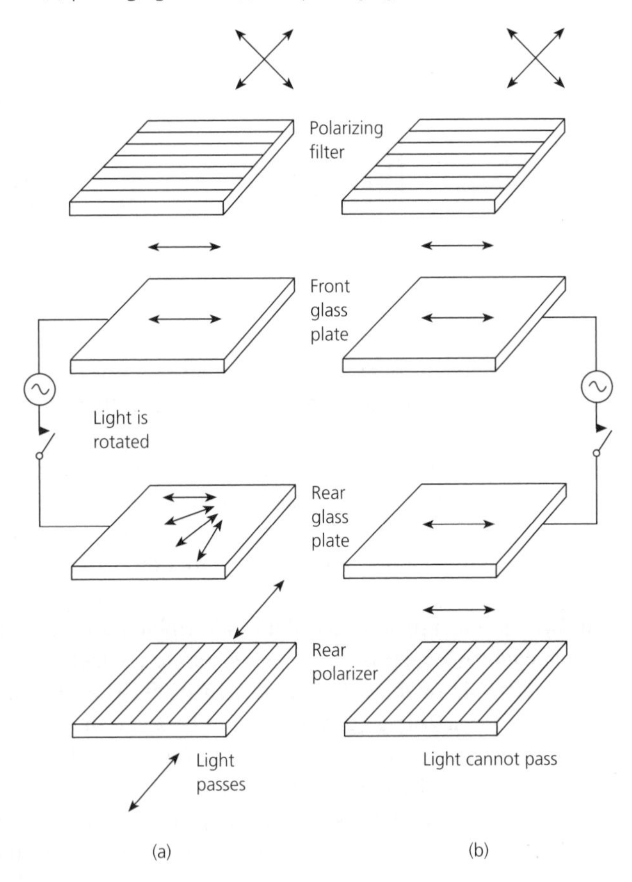

Operation of an LCD

Normally, the molecules of a liquid crystal are twisted through 90° (for a NT device) and possess the property of rotating passing polarized light through the same angle. If a polarizing filter is fitted at the front surface of the crystal it will allow only one plane of polarization of the incident light to pass through the front glass plate and into the liquid crystal. The transmitted polarized light has its plane of polarization rotated through 90° by the liquid crystal molecules before it passes through the rear glass plate to arrive at the rear polarizer. The rear polarizer is fitted at right angles to the plane of the polarizing filter and so it will allow the polarized light to pass through. The action of the twisted nematic liquid crystal is shown in Fig. 14.5(a). When a voltage is applied between the conducting surfaces of the two glass plates the molecules of the liquid crystal become aligned perpendicular to the plates, as shown in Fig. 14.5(b). Now, the polarized light travelling in the crystal is no longer rotated through 90° by the time it reaches the rear glass plate and this means that it is unable to pass through the rear polarizer. Hence, the application of a voltage across the LCD stops the transmission of light through the device.

Fig. 14.6 *LCD 7-segment display*

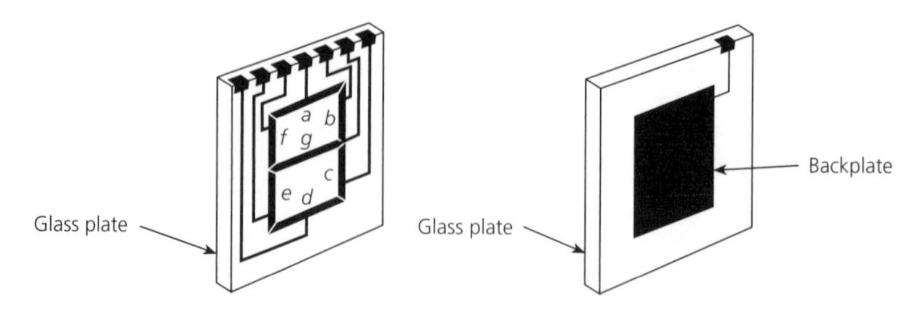

In an alternative method of construction, the polarizing filter and the rear polarizer are fitted in alignment with one other. In an LCD of this type, light will be transmitted through the device when a voltage is applied and there will be zero transmission when the applied voltage is zero.

If a steady d.c. voltage is applied across an LCD, the lifetime of the crystal will be greatly reduced. For this reason, the applied voltage is usually either an interrupted d.c. voltage or an a.c. voltage of frequency somewhere in the range 40–400 Hz.

Liquid crystal displays, unlike LEDs, are not available as single units and are generally manufactured in the form of a 7-segment display. The metal oxide film electrode on the surface of the upper glass plate is formed into the shape of the required seven segments, each of which is taken to a separate contact, and the lower glass plate has a common electrode or *backplate* deposited on it. The idea is shown in Fig. 14.6. With this arrangement, a voltage can be applied between the backplate and any one, or more, of the seven segments to make that, or those, particular segment(s) appear to be dark and thereby display the required number.

Nematic liquid crystal displays possess a number of advantages which have led to their widespread use in battery-operated equipment. First, their power consumption is very small, about 1 μW per segment (much less than the LED); second, their visibility is not affected by bright incident light (such as sunlight); and, third, they are compatible with low-power CMOS circuitry.

The disadvantages of using an LCD for display purposes are: (a) there must be some ambient light, and (b) the response time is slow compared with that of an LED.

Screen displays

Passive screens

A large number of LCDs connected in a matrix can form a screen upon which visible images can be seen. For a mono display, LCDs can be used in a *passive matrix* screen. A typical construction is shown in Fig. 14.7(a). To produce pixels on an LCD screen, two arrays of transparent conductors are arranged at right angles to one another. The column array of lines is positioned below, and the row array of lines is positioned above, the liquid crystals as shown in Fig. 14.7(b). Voltages are applied to the appropriate column and row conductors to select the target pixels. This produces an electric field between the conductors at that point. This electric field causes the liquid crystal at that point to block light and so a black dot is displayed on the screen. The screen is

Fig. 14.7 *Passive LCD screen*

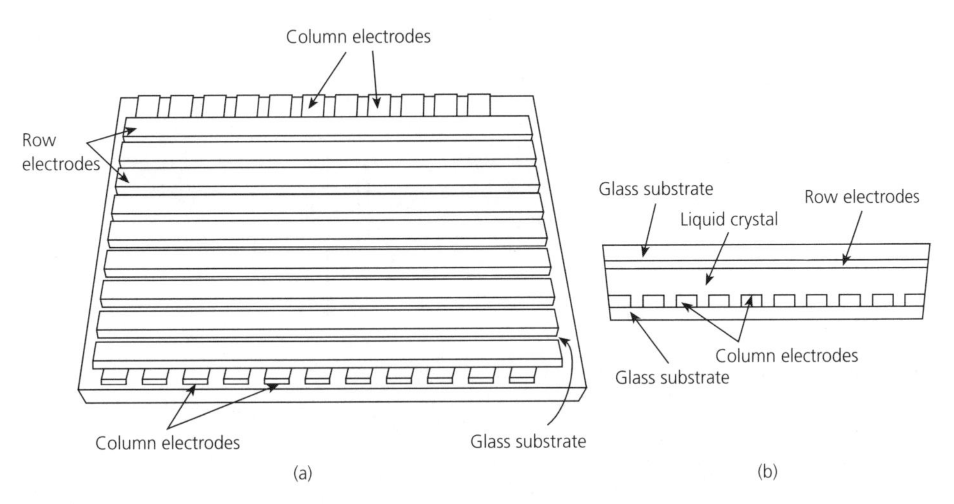

refreshed, at approximately 200 ms intervals, to produce a steady picture. Shades of grey are produced by varying the times for which some pixels are black. The mono display is rather slow to respond to changes in picture content and is not very bright; to overcome these disadvantages, and also to provide a colour display, an *active matrix* screen is often employed.

To increase the speed of operation, some passive matrix screens employ *dual-scan* technology. A dual-scan display splits the screen into two halves and, instead of scanning right across each row in turn, the screen is scanned from the left-hand side to the centre, and from the centre to the right-hand side simultaneously. This technique very nearly doubles the scanning rate. In addition, a double-scan screen provides both better contrast and a wider viewing angle.

Active screens

Each colour pixel consists of three dyed liquid crystals, one red, one green, and the third blue. By controlling the degree to which each dyed liquid crystal is activated any other colour can be obtained. Each part of each colour pixel has an associated *thin-film transistor* (TFT) mounted beneath it, i.e. there are three TFTs per pixel. A TFT is turned ON and OFF by the voltage applied to a control line and the TFT, in turn, drives the pixel. Because there are three connections to every pixel on the screen only the pixels the software wants to alter need to be up-dated and this means that the response time is very fast. The obvious disadvantage is the very large number of transistors that are needed; for a 640 × 480 pixel screen, for example, there are 921 600 TFTs.

Each TFT is connected to its associated liquid crystal in the manner shown in Fig. 14.8. A TFT LCD active screen has a viewing angle of 40° in the vertical plane and 90° in the horizontal direction. A *super TFT* LCD screen increases the viewing angle to about 140° in both directions. The structure of a TFT LCD screen is shown in Fig. 14.9. In some active matrix screens, two *thin-film* diodes connected back-to-back replace each TFT.

Fig. 14.8 *A TFT connected to an LCD cell*

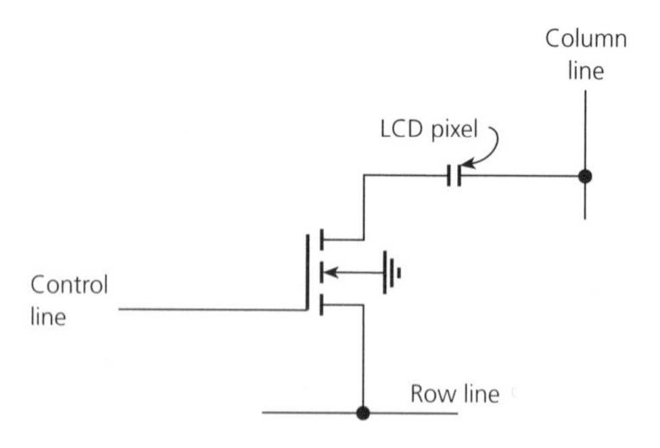

Fig. 14.9 *Active LCD screen*

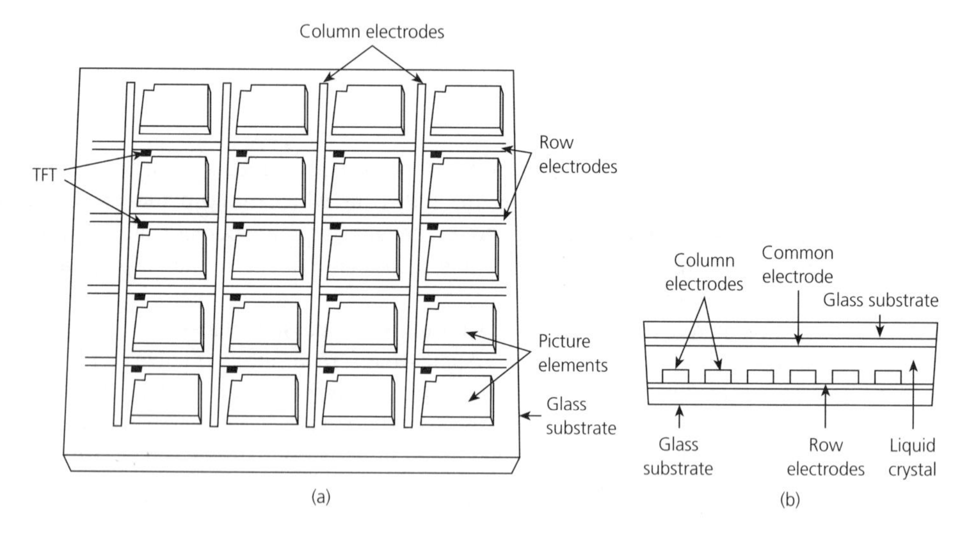

Seven-segment displays

Seven-segment displays are generally employed as indicators of decimal numbers and consist of a number of LEDs arranged in seven segments as shown in Fig. 14.10(a). Any number between 0 and 9 can be indicated by lighting the appropriate segments. This is shown in Fig. 14.10(b).

Clearly, the 7-segment display needs a 7-bit input signal and so a *decoder* is required to convert the digital signal to be displayed into the corresponding 7-segment signal. Decoder/driver circuits *can* be made using SSI devices but, more usually, a ROM or a custom-built IC would be used. Figure 14.11 shows one arrangement, in which the BCD output of a decade counter is converted to a 7-segment signal by a decoder.

When a count in excess of 9 is required, a second counter must be used and be connected in the manner shown in Fig. 14.12. The tens counter is connected to the output of the final flip-flop of the units counter in the same way as the flip-flops inside the counters are connected.

Fig. 14.10 *A 7-segment display: (a) arrangement of LEDs and (b) indication of numbers 0 to 9*

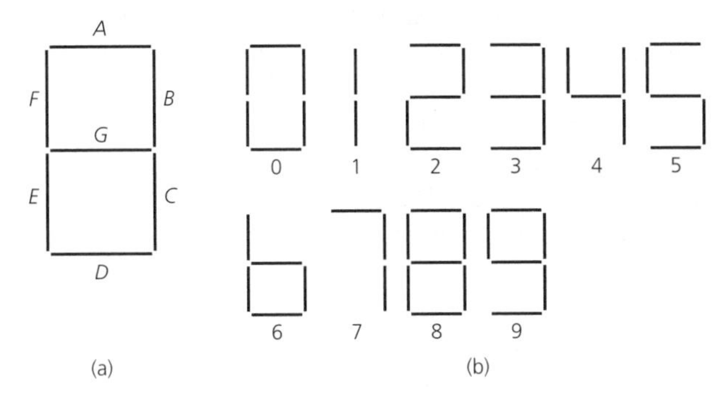

Fig. 14.11 *A BCD decade counter with a 7-segment display*

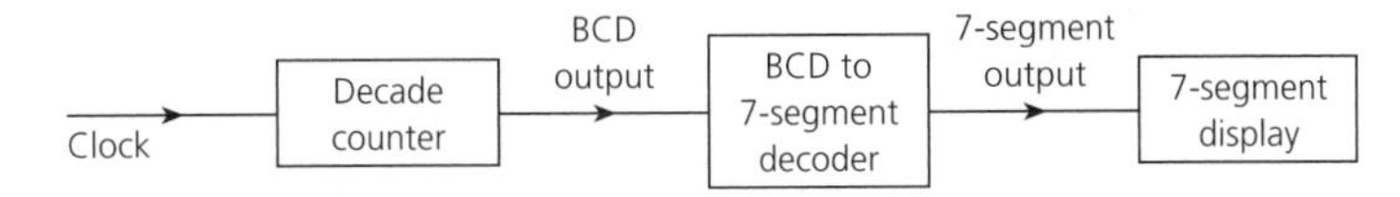

Fig. 14.12 *Decade counters arranged to give a count in excess of 9*

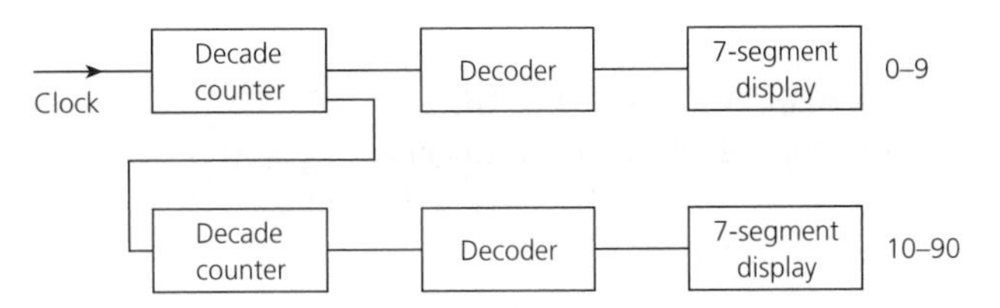

The design of a 7-segment display starts with the truth table for the circuit. This is given in Table 14.2; note that numbers greater than 9 do not appear in the display and so they are 'don't cares'.

Table 14.2 ***7-segment display***

Decimal number displayed	*Inputs*				*Outputs*						
	D	*C*	*B*	*A*	*a*	*b*	*c*	*d*	*e*	*f*	*g*
0	0	0	0	0	1	1	1	1	1	1	0
1	0	0	0	1	0	1	1	0	0	0	0
2	0	0	1	0	1	1	0	1	1	0	1
3	0	0	1	1	1	1	1	1	0	0	1
4	0	1	0	0	0	1	1	0	0	1	1
5	0	1	0	1	1	0	1	1	0	1	1
6	0	1	1	0	0	0	1	1	1	1	1
7	0	1	1	1	1	1	1	0	0	0	0
8	1	0	0	0	1	1	1	1	1	1	1
9	1	0	0	1	1	1	1	0	0	1	1

From the truth table a number of segment maps can be obtained, each of which maps the inputs which must be HIGH for a segment to be illuminated.

Thus, segment a is ON when the input decimal number is 0, 2, 3, 5, 7, 8 or 9. Hence, from the truth table, $a = \bar{A}\bar{B}\bar{C}\bar{D} + \bar{A}B\bar{C}\bar{D} + AB\bar{C}\bar{D} + A\bar{B}C\bar{D} + ABC\bar{D} + \bar{A}\bar{B}\bar{C}D + A\bar{B}\bar{C}D$. The mapping is shown in (a).

From the map, $a = D + AC + AB + \bar{A}\bar{C}$. $b = \bar{A}\bar{B}\bar{C}\bar{D} + A\bar{B}\bar{C}\bar{D} + \bar{A}B\bar{C}\bar{D} + AB\bar{C}\bar{D} + \bar{A}\bar{B}C\bar{D} + ABC\bar{D} + \bar{A}\bar{B}\bar{C}D + A\bar{B}\bar{C}D$.

The mapping is shown in (b).

CD \ AB	00	01	11	10
00	1	1	1	0
01	1	×	×	1
11	×	×	×	×
10	0	0	1	1

(a)

CD \ AB	00	01	11	10
00	1	1	1	1
01	1	×	×	1
11	×	×	×	×
10	1	0	1	0

(b)

From the map, $b = \bar{C} + AB + \bar{A}\bar{B}$.

Similarly, for the other segments, $c = A + \bar{B} + C$, $d = \bar{A}B + \bar{A}\bar{C} + B\bar{C} + A\bar{B}C$, $e = \bar{A}B + \bar{A}\bar{C}$, $f = \bar{A}\bar{B} + \bar{B}C + \bar{A}C + D$, and $g = \bar{A}B + \bar{B}C + B\bar{C} + D$.

Common-cathode or common-anode

Each of the seven LEDs in a 7-segment display has a positive terminal, which is called the anode, and a negative terminal, known as the cathode. This means that there could be as many as 14 connections to be made. To reduce this number to eight, *either* the cathodes *or* the anodes are connected together or commoned. The two methods of connection are shown in Fig. 14.13. With the common-anode system (Fig. 14.13(a)), current will flow in an LED, turning it ON, when its cathode is taken LOW. Conversely, when the anode of a common-cathode LED (Fig. 14.13(b)) is taken HIGH, that LED is turned ON and glows visibly. A resistor is usually connected in series with each LED to limit the current that flows when the device is conducting.

Fig. 14.13 *(a) Common-anode and (b) Common-cathode connected LEDs*

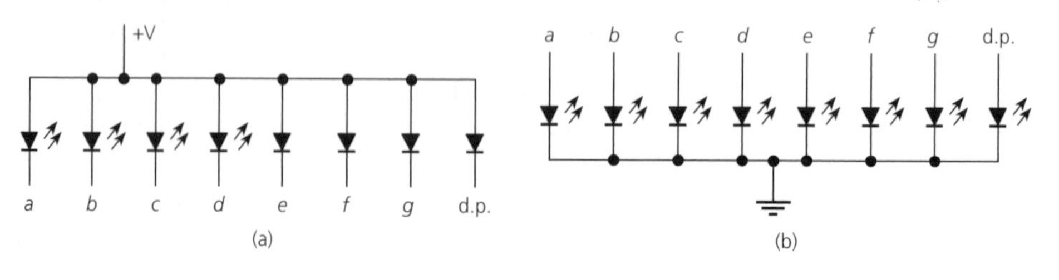

Fig. 14.14 *A 74LS47 7-segment decoder/driver: (a) pinout and (b) logic symbol*

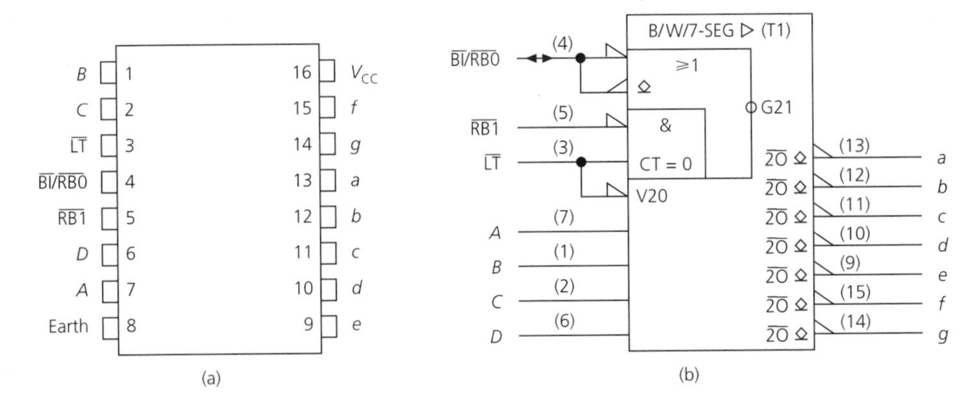

MSI circuits

A commonly employed MSI circuit is the 74LS47 open-collector BCD-to-7-segment decoder/driver. The IC has active-LOW outputs and it is intended for use with common-anode displays. This IC has four input pins, *A*, *B*, *C* and *D* to which the BCD input signal is applied and seven output pins, labelled as *a* through to *g*. When an output is HIGH, the segment to which it is connected lights. If, for example, the input signal is 0101, outputs *a*, *f*, *g*, *c* and *d* go HIGH so that the decimal number 5 is illuminated.

The IC includes a facility known as *remote blanking*. Two other pins, labelled as RB_{in} and RB_{out}, are provided. If the RB_{in} pin of the most significant 7-segment display is earthed and inputs *A*, *B*, *C* and *D* are all LOW, then its RB_{out} pin will be LOW, and so are all segment outputs. The RB_{out} pin of each display is connected to the RB_{in} pin of the next most significant 7-segment display. This connection ensures that leading 0s in a displayed decimal number are not visible, e.g. for a 4-bit display, 617 would be displayed and not 0617.

The pinout of the 74LS47 is shown in Fig. 14.14(a) and its logic symbol is shown in Fig. 14.14(b). The other BCD-to-7-segment decoder/driver available in the TTL logic family, is the 74LS247. This IC incorporates open-collector outputs that can drive LEDs directly, and so it does not require external current-limiting resistors. The 74LS247 IC has the same pinout and the same electrical/functional behaviour as the 74LS47 but its display uses a different fount, i.e. 6 and 9 instead of 6 and 9.

The LT (lamp test) pin is used to test the display. If LT = 0 all segments should be ON, regardless of the logical states of the BCD inputs. The RB0 and RB1 inputs are, respectively, used for the automatic blanking of leading-edge zeros (0 to the left of the highest non-zero digit), or for automatic blanking of trailing-edge zeros (0 to the right of the decimal point).

Figure 14.15 shows how the 74LS247 BCD-to-7-segment decoder may be connected to drive a 7-segment display. The series resistors may be employed to limit the current that flows through an ON LED.

The 74LS48 and 49 are active-HIGH decoder/drivers for driving common-cathode LEDs. Typically, each segment might take a current of about 10 mA. The number 8

Fig. 14.15 *A 74LS247 BCD-to-7-segment decoder*

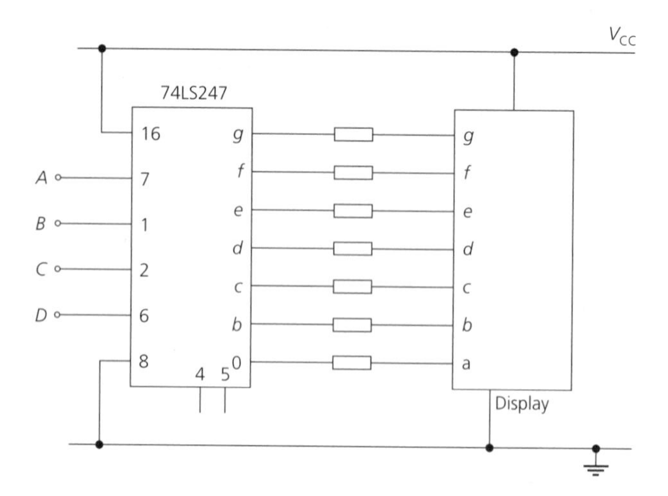

Fig. 14.16 *Operation of an LCD 7-segment display*

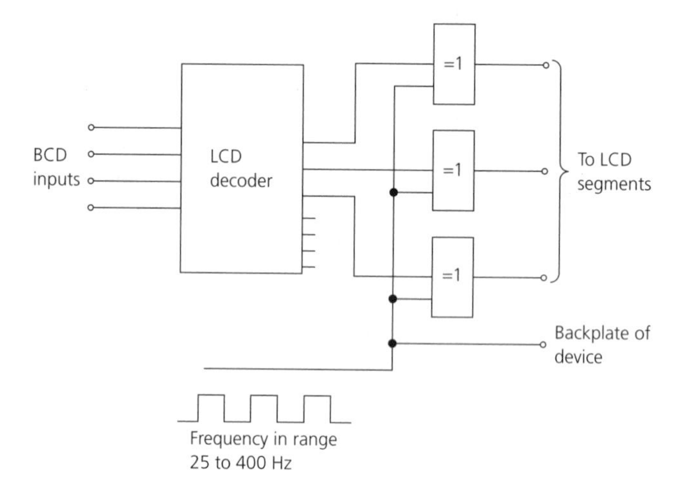

uses all eight segments so that it takes about 80 mA. This means that a 4-digit display could take as much as 320 mA. To reduce the current taken by the display, a form of time division multiplexing is employed. Each digit is energized for part of the time only, for a 4-digit display for one quarter of the periodic time of a signal in the frequency range 50–250 Hz.

LCD decoder/drivers

The voltages applied to an LCD must be alternating with a zero d.c. component to prevent any electrolytic plating action taking place that would damage the device. The basic arrangement used is shown in Fig. 14.16. Suitable LCD decoder/drivers are the Harris ICM7211, ICM7211 and 7224.

PRACTICAL EXERCISE 14.1

To investigate the performance of a 7-segment display.

Components and equipment: one 74LS47 decoder/driver. One 74LS93 ripple counter. One 7-segment common-cathode LED display. Eight 270 Ω resistors. Breadboard. Power supply. Low-frequency pulse generator.

Procedure:

(a) Build the circuit shown in Fig. 14.17. Both the blanking input $\overline{\text{BI}}$ and the ripple blanking input $\overline{\text{RBI}}$ should be held at logic 1. Connect $\overline{\text{LT}}$ to 1 also.

(b) Connect, in turn, each of the binary inputs corresponding to decimal 0 through to decimal 15. For each input note the display. From each displayed number/symbol determine which segments are ON and, hence, complete the truth table started in Table 14.3.

(c) Determine the functions of the other pins on the 74LS47. Apply logic 0 to the $\overline{\text{LT}}$ pin. State what happens to the display. Then remove the 0 from $\overline{\text{LT}}$, restoring $\overline{\text{LT}}$ to logic 1, and apply the 0 to the $\overline{\text{BI}}$ pin. Now what happens?

Fig. 14.17

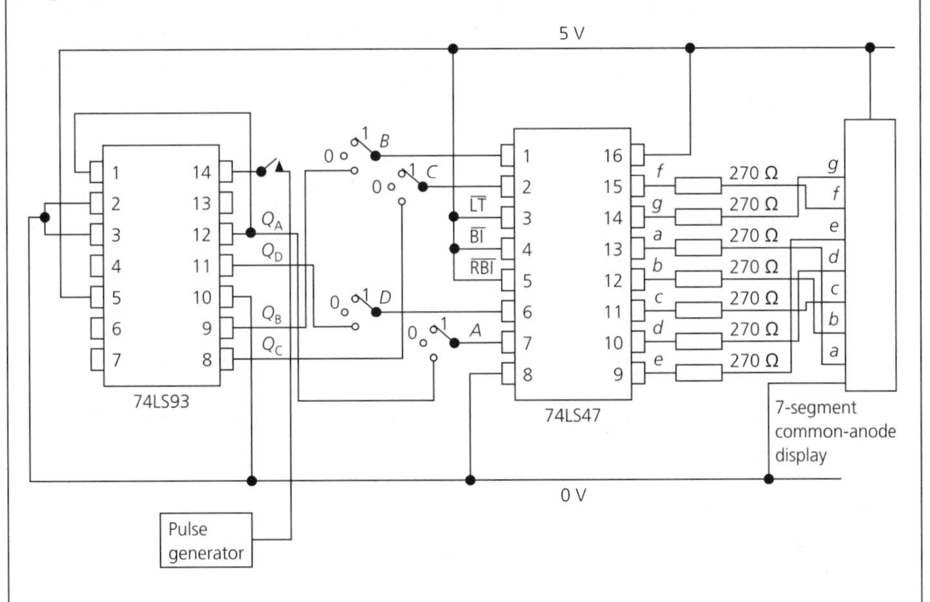

Table 14.3

Decimal number	*D*	*C*	*B*	*A*	*a*	*b*	*c*	*d*	*e*	*f*	*g*
0	0	0	0	0	1	1	1	1	1	1	0
1	0	0	0	1	0	1	1	0	0	0	0

(d) Restore the logic 1 voltage to the $\overline{LT}$, $\overline{BI}$ and $\overline{RBI}$ pins and connect the 74LS93 4-bit ripple counter to the 74LS47 decoder/driver. For the counter, connect the R_{01} and R_{02} pins to logic 0, pin 12 to pin 1, and then connect the counter outputs Q_A, Q_B, Q_C and Q_D to the BCD inputs A, B, C and D of the decoder. Set the frequency of the pulse generator to a low value, say 5 Hz, and connect the generator to the counter. Observe the display. Now modify the connections of the counter so that it resets once its count has reached decimal 9. Apply the pulse generator and again observe the display. Comment on the difference.

Sixteen-segment and dot matrix displays

If alphanumeric characters are to be displayed, either a 16-segment, or a dot matrix, display is required. A 16-segment display, sometimes called a *starburst*, uses 16 LEDs or LCDs arranged in the manner shown in Fig. 14.18. Applying voltages to the display to illuminate the appropriate segments will cause the characters shown in Fig. 14.19 to become visible. A number of 16-segment display driver modules are available, one example being the TSM7752B.

With a dot matrix display, each alphanumeric character is indicated by illuminating a number of dots in a 5 × 7 dot matrix. To allow for lower-case letters and for spaces in between adjacent rows and columns, each character font is allocated a 6 × 12 space. Figure 14.20(a) shows a 6 × 12 dot matrix. Every location in the dot matrix has an LED connected, as shown in Fig. 14.20(b), for the top two rows of the matrix only. All the cathodes of the LEDs in one row, and all of the anodes in one column are connected together. By addressing the appropriate locations in the diode matrix, and making the LEDs at those points to glow visibly, any number or character in the set can be illuminated. Some examples are given in Fig. 14.21.

The circuitry required to drive a dot matrix display is too complex to be implemented using SSI devices. One 3-chip LSI dot matrix display controller, the Rockwell 10939, 10942 and 10943, is a general-purpose controller which is able to interface with other kinds of dot matrix as well as an LED type. The controller can drive up to 46 dots and up to 20 characters selected out of the full 96 character ASCII code.

Fig. 14.18 *A 16-segment display*

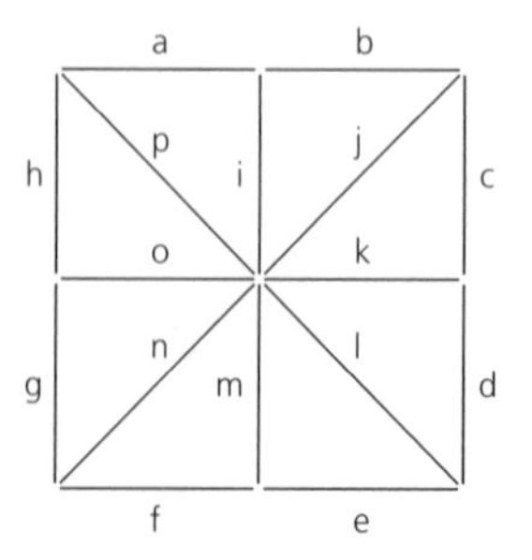

Fig. 14.19 *Sixteen-segment characters*

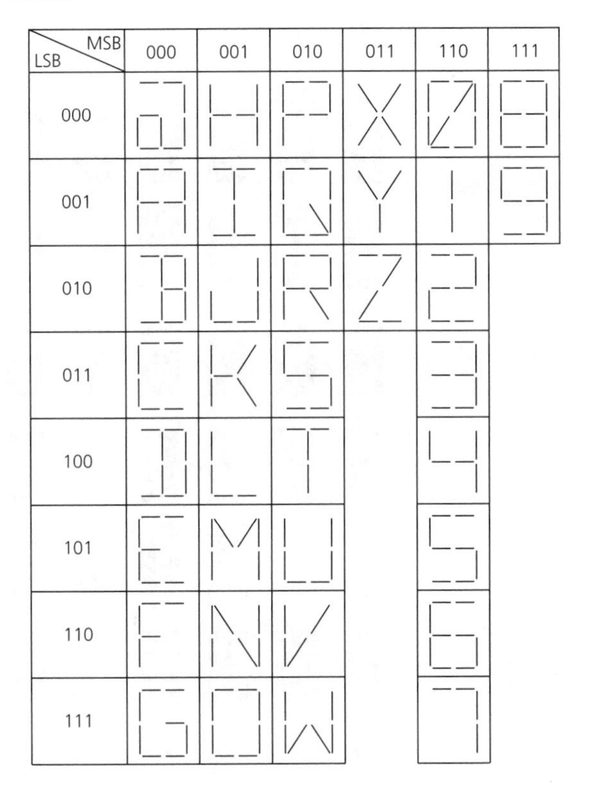

Fig. 14.20 *A 6 × 12 dot matrix*

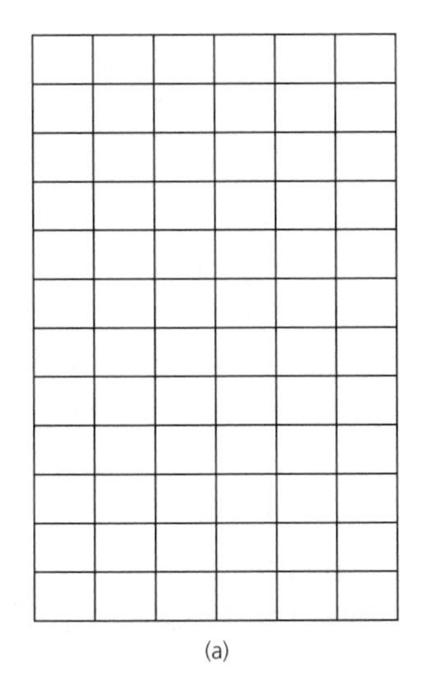

(a)

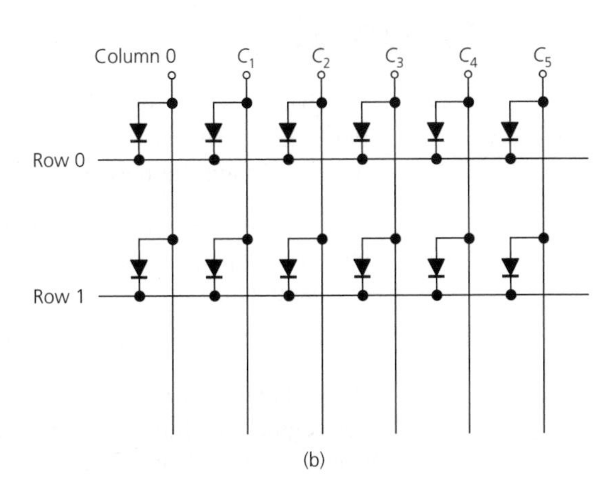

(b)

Fig. 14.21 *Dot matrix characters*

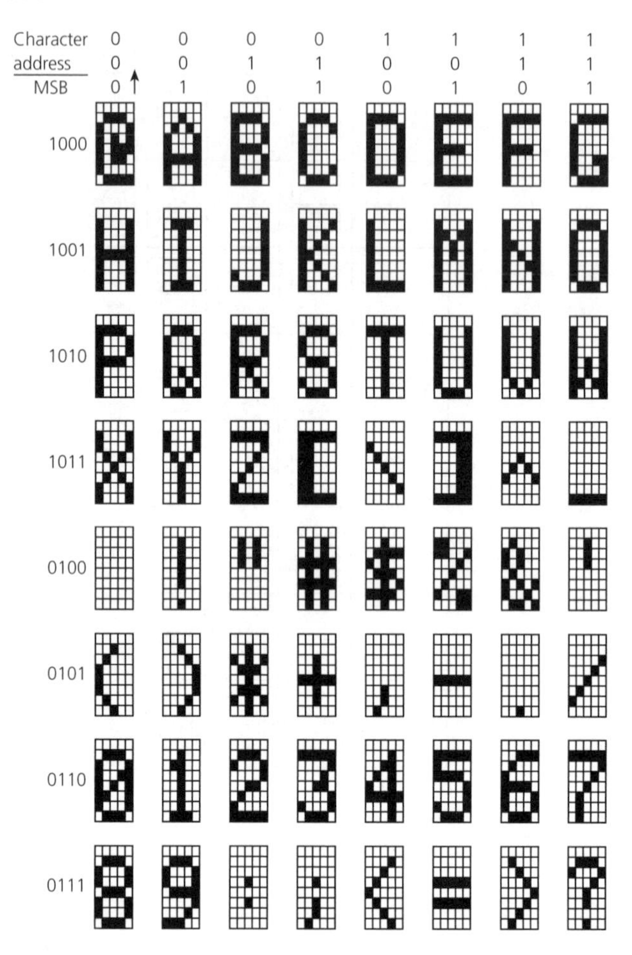

EXERCISES

14.1 Numbers 6 and 9 in a 7-segment display may be 6, 9 instead of 6, 9, as in Fig. 14.5. Write down the truth table for a display that uses these alternative figures. Obtain the new expressions for the segments *a* and *d*.

14.2 Four 74LS47 7-segment decoder/driver ICs are cascaded to give a 4-digit read-out on a display. Show how the RB1 and RB0 pins are connected to blank the left-hand zero.

14.3 Complete Table 14.4 by inserting × in the appropriate columns.

14.4 Write down the truth table for a BCD-to-7-segment decoder, assuming a luminous segment is illuminated by the logic 1 state. Use it to obtain Boolean expressions for the decimal numbers 5 and 9.

Table 14.4

	LED	*LCD*	*TFT LCD*
Bright display			
Ambient light required			
High operating current			
a.c. voltage driven			
Slow response time			
Colour			

14.5 Answer the following questions relating to an LCD display:
(a) Can the display be read in the dark?
(b) Will the display work in very low ambient temperatures?
(c) Is the display affected if it is situated in very bright sunlight?

Give reasons for your answers.

14.6 What is meant by multiplexing when applied to a 6-digit LED display? State one disadvantage and one advantage of using multiplexing.

14.7 An LED is to be connected between the output of a TTL device and +5 V. Determine the value of the series current-limiting resistor required. The LED ON voltage is 1.6 V and the LOW output voltage of the TTL circuit is 0.2 V. Find also the power dissipated in the series resistance. Assume the safe LED current to be 20 mA.

14.8 Draw a 16-segment display and state which segments must be lit for the display to indicate (a) 4, (b) A, (c) R, and (d) T.

14.9 Explain the principle of operation of a dot matrix alphanumeric display, using the letter S as an example.

14.10 A dot matrix liquid crystal display module contains an LCD display and a CMOS LSI drive unit. If the module is operated in conjunction with an LSI controller with on-chip character generator all the ASCII characters can be displayed. Discuss the merits of using such a module.

14.11 Briefly explain the action of an LED. List four advantages and two disadvantages of the LED as a display device.

15 Analogue-to-digital and digital-to-analogue converters

After reading this chapter you should be able to:

(a) Explain the operation of binary-weighted and *R*/2*R* DACs.
(b) Connect a DAC to convert a digital word into an analogue voltage.
(c) Explain the operation, and the relative merits, of several different kinds of ADC.
(d) Use an IC DAC.

The function of many electronic systems is to receive some input information, process it in some way, and then pass the processed information to its destination. Nearly always, the information is of continuous, or *analogue*, nature but, increasingly nowadays, the electronic system employs digital techniques. The analogue data must be converted into a digital codeword so that it can be processed by a digital system. At the output of the system it may be necessary to convert the processed digital data back to analogue form. This process is often known as signal conditioning.

The basic block diagram of a signal conditioning system is shown in Fig. 15.1. The analogue signal produced by the transducer is often amplified and/or filtered before it is passed on to the *analogue-to-digital converter* (ADC) which converts the signal into a digital signal. The digital signal is transferred to the digital logic system or computer for processing and storage. The results of the processing may be made available on a digital display or may be applied to a *digital-to-analogue converter* (DAC) for conversion into analogue form. The analogue signal may then be displayed on an analogue display or it may be transmitted to some other point where it is required for some purpose.

Fig. 15.1 *Signal conditioning*

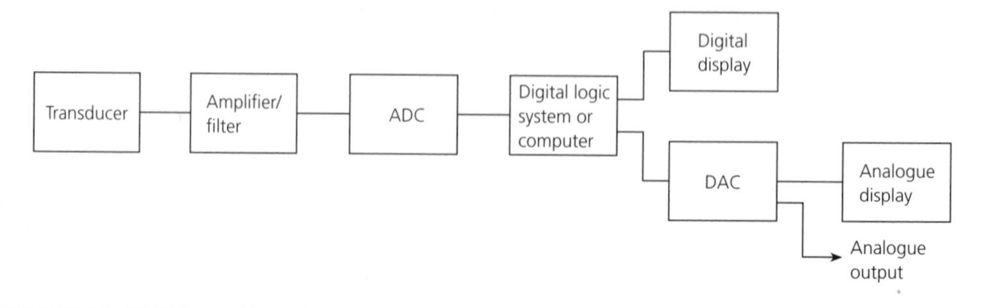

Fig. 15.2 *Block diagram of (a) an ADC converter and (b) a DAC converter*

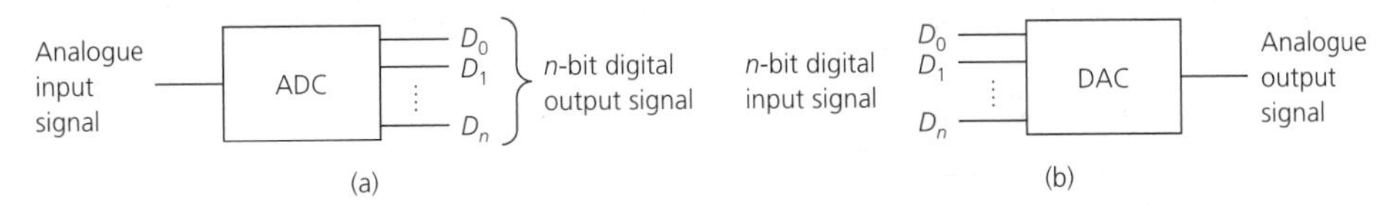

Circuitry is, therefore, required that is able to interface between the analogue world outside the system and the digital system itself. The two interface circuits that are necessary are the ADC and the DAC.

The block diagrams for each of these devices are given in Fig. 15.2(a) and (b). The ADC has an analogue signal applied to its input terminals which is sampled at regular intervals, and an equivalent digital signal is generated to appear at the output terminals. Conversely, the DAC converts an n-bit input digital signal into the corresponding analogue signal.

The main performance criteria for both ADC and DAC are: (a) the resolution, (b) the sampling rate or speed, (c) the conversion time, and (d) linearity.

The *resolution* of an ADC or DAC is the smallest change in the analogue voltage that can be represented by a digital word. The higher the number of bits used the better will be the resolution. The *sampling rate speed*, is the number of times per second the analogue signal is converted into a digital codeword. It must be at least twice the highest frequency contained in the analogue signal. The *conversion time* of a system is the time taken for a conversion to take place. The *linearity* is the accuracy of the analogue or digital conversion over the operating range.

Digital-to-analogue converters

Binary-weighted digital-to-analogue converter

A DAC is used to convert a digital input word into an analogue output voltage or current.

The circuit of a *binary weighted resistor* DAC is shown in Fig. 15.3. An op-amp is connected as a summing amplifier with a number of inputs equal to the number of bits per input digital word. The resistor connected to the most-significant bit (MSB) input D_3 has a resistance of R ohms. The resistor connected to the next MSB has twice that resistance, i.e. $2R$, and so on, for each of the following resistors. If there are n bits in the input digital word, the resistor connected to the LSB input should have a resistance of $2nR$ ohms. If, for example, $R = 2\ \text{k}\Omega$ and $n = 4$, then the least-significant bit (LSB) resistor must have a resistance value of $2^4 = 16\ \text{k}\Omega$.

The voltage at the inverting terminals of the op-amp is very nearly 0 V and this point is known as a *virtual earth*. The value of the feedback resistor R_f is chosen to give the required maximum output voltage.

When a digital signal is applied to the input terminals of the DAC, each input bit that is at the logic 1 voltage level will cause a current to flow in the associated resistor. If an input bit is at the logic 0 level, zero current will flow in that input. The op-amp sums the currents flowing and this total current flows in the feedback resistor R_f to

Fig. 15.3 *A weighted resistor DAC*

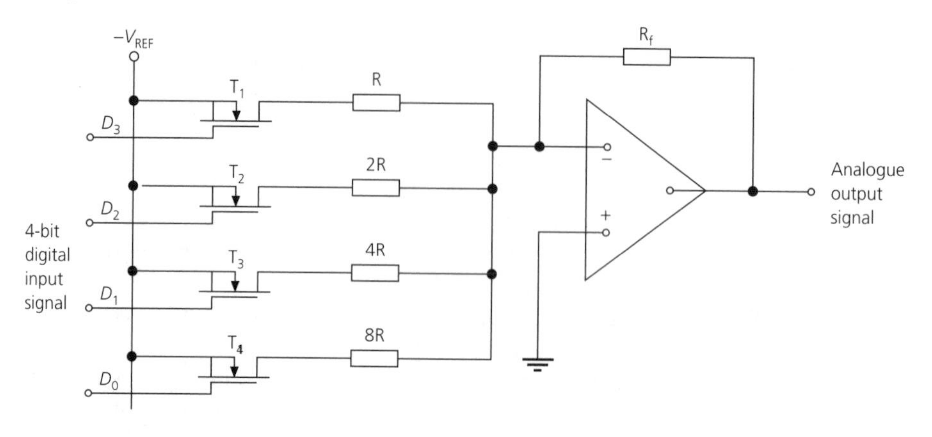

Table 15.1

Input word	Analogue output (V)	Input word	Analogue output (V)
0000	0	1000	2.5
0001	0.3125	1001	2.8125
0010	0.625	1010	3.125
0011	0.9375	1011	3.4375
0100	1.25	1100	3.75
0101	1.5625	1101	4.0625
0110	1.875	1110	4.375
0111	2.1875	1111	4.6875

develop an output voltage. The magnitude of this output voltage is proportional to the total current flowing and, hence, to the digital input word.

If, for example, the digital input word is 1001, $R_f = 5000\ \Omega$, $R = 2R_f = 10\ k\Omega$, and the logic 1 voltage level is +5 V, current will flow in the inputs D_0 and D_3 whose magnitudes are:

$$I_3 = 5/(10 \times 10^3) = 0.5\ \text{mA}$$
$$I_0 = 5/(16 \times 5 \times 10^3) = 5/(80 \times 10^3) = 0.0625\ \text{mA}$$
$$I_1 = I_2 = 0$$

The total current flowing to the op-amp is 0.5625 mA and so the output voltage is equal to $0.5625 \times 10^{-3} \times 5000 = 2.8125$ V. If, now, the input digital word changes to 1101, the currents I_3 and I_0 will be unchanged but there will now also be a current I_2

$$I_2 = 5/(20 \times 10^3) = 0.25\ \text{mA}$$

The total current is now 0.8125 mA and the output voltage is 4.0625 V. Table 15.1 gives the output voltage produced by each of the possible input digital words.

Fig. 15.4 *Transfer function of a DAC*

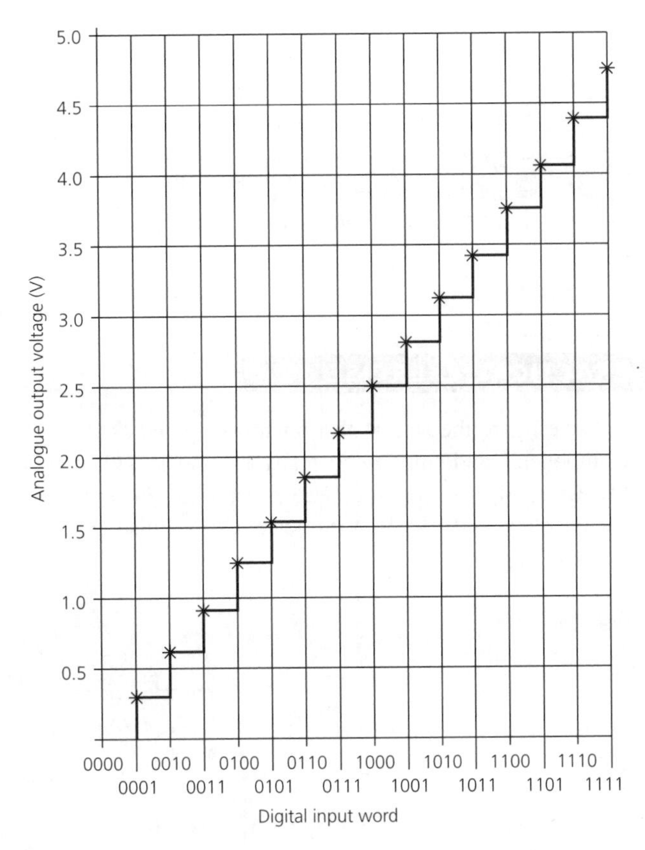

These values are shown plotted in the transfer function of Fig. 15.4 and illustrate that a stepped analogue output voltage signal is obtained. The magnitude of each step could be halved, and resolution improved, if an 8-bit input signal were to be employed.

Although the basic concept is fairly simple its practical implementation is not. The requirement for the input resistors to be *binary weighted* is difficult and, hence, expensive, to satisfy. For this reason the weighted-resistor DAC is employed only where a low resolution is wanted so that a 4-bit input is adequate. It is not practicable to produce an IC version of this type of DAC.

EXAMPLE 15.1

A weighted-resistor DAC has $R_f = 20\ \text{k}\Omega$ and $R = 12.5\ \text{k}\Omega$. Calculate its output voltage when the 4-bit digital input word is (a) 1010, and (b) 0101. Logic 1 voltage = 5 V.

Solution

(a) $I_1 = 5/(50 \times 10^3) = 0.1$ mA.
$I_3 = 5/(12.5 \times 10^3) = 0.4$ mA.
$V_{out} = 0.5 \times 10^{-3} \times 20 \times 10^3 = 10$ V (*Ans.*)

(b) $I_0 = 5/(100 \times 10^3) = 0.05$ mA.
$I_2 = 5/(25 \times 10^3) = 0.2$ mA.
$V_{out} = 0.25 \times 10^{-3} \times 20 \times 10^3 = 5$ V (*Ans.*)

PRACTICAL EXERCISE 15.1

To investigate the action of a binary-weighted DAC.
Components and equipment: resistors: one 10 kΩ, two 20 kΩ, one 39 kΩ, one 82 kΩ. One 741 op-amp. One 74LS93 (or 74HC393) ripple counter. Power supply(ies) giving ±5 V and ±15 V. Voltmeter. CRO. Pulse generator.
Procedure:

Fig. 15.5

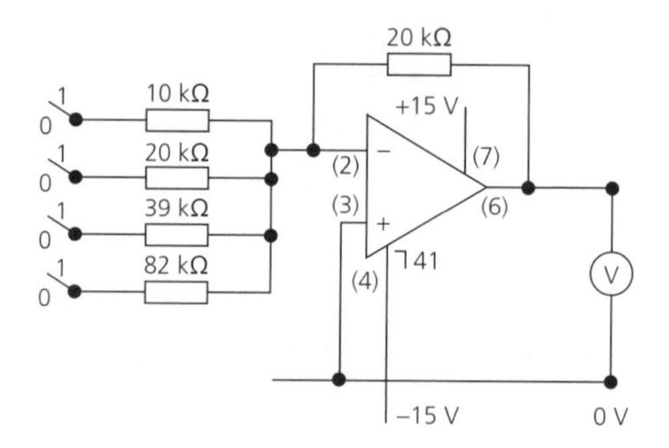

(a) Build the circuit shown in Fig. 15.5.
(b) Connect, in turn, the binary equivalents of decimal 0 through to decimal 15 to the digital inputs of the circuit. Each time measure the output voltage.
(c) Draw the transfer function of the DAC.
(d) Determine the resolution of the circuit, (i) by measurement, and (ii) by calculation.
(e) The values of the four input resistors are not exactly in the required *R*/2*R*/4*R*/8*R* ratio. Calculate the expected output voltage for digital words of (i) 0101, (ii) 1010, and (iii) 1100 and, for each, determine the percentage error introduced.
(f) Connect the inputs of the circuit to the outputs of the 74LS93 ripple counter. Connect the counter to act as a divide-by-16 circuit and then connect the pulse generator to its input. Set the frequency of the pulse generator to 1 kHz. Monitor the output of the circuit on the CRO and draw the displayed waveform.

Fig. 15.6 *A 4-bit R/2R DAC*

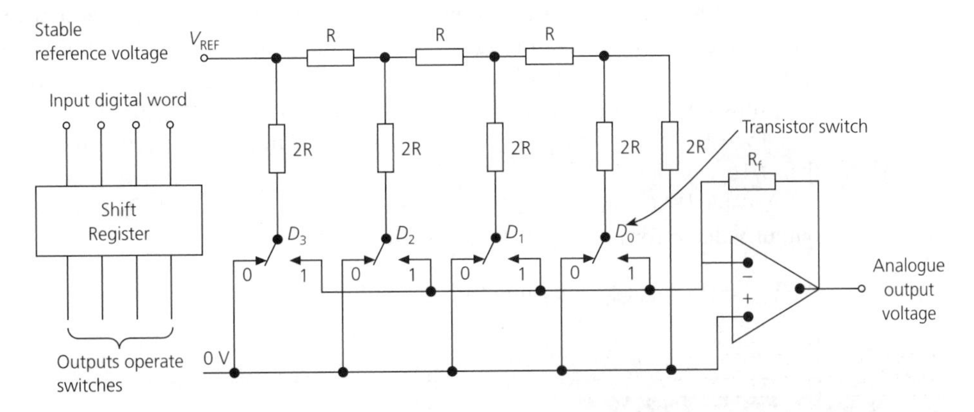

Fig. 15.7 *How an R/2R resistor network produces binary-weighted currents*

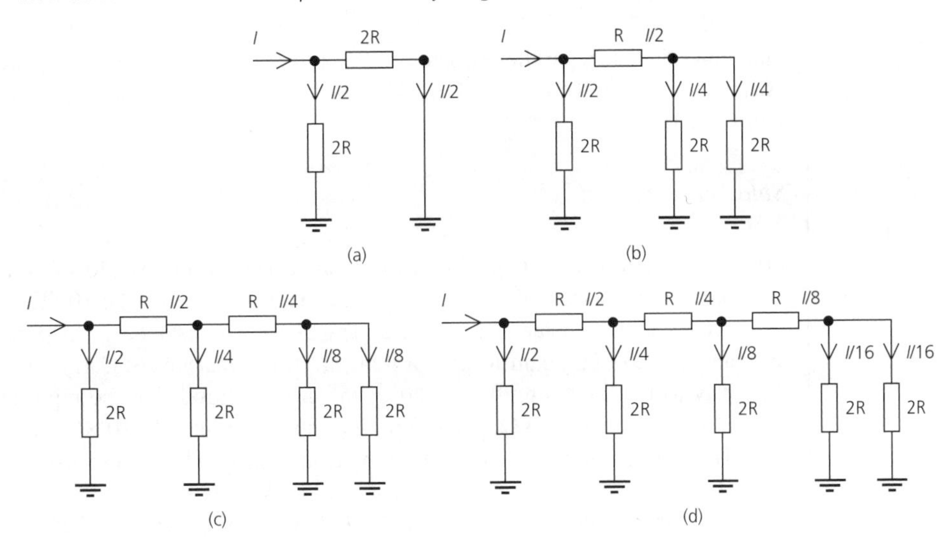

R/2R *digital-to-analogue converter*

The conversion method that is most often employed in an IC digital-to-analogue converter is the *R/2R* circuit shown in Fig. 15.6. The circuit uses resistors of only two different values, one of which is twice the other. The wanted analogue voltage is obtained by switching resistors either to earth or to the op-amp. The total current that is switched is passed through the feedback resistor of the op-amp to generate the output voltage. The switching of the resistors is controlled by a shift register to which the input digital word is applied.

To produce a DAC with a higher resolution, more *R/2R* resistors and switches to the left of D_3 need to be added to the circuit. Eight-bit and 12-bit IC DACs are made in this way.

The operation of the *R/2R* resistor network to generate the required analogue voltage is demonstrated in Fig. 15.7.

When a transistor switch is in its logic 1 position a current flows through both the associated resistor $2R$ and the summing resistor R_f. This current is proportional to the binary value of the switch. In Fig. 15.7(a), for the D_0 part of the circuit, the input current splits into two equal parts of $I/2$. Figure 15.7(b) shows the D_0 and D_1 parts of the circuit; now currents of $I/2$ and $I/4$ are obtained. In Fig. 15.7(c), for D_0, D_1 and D_2, and Fig. 15.7(d), for all four parts of the circuit, the currents splits even more into $I/8$ and then $I/16$.

A 4-bit DAC can have $2^4 = 16$ different combinations of D_0 through to D_3. The output voltage for any input digital word is given by

$$V_{out} = [V_{REF} \times (\text{decimal input number})]/8 \qquad (15.1)$$

EXAMPLE 15.2

An $R/2R$ DAC network has $R = 10$ kΩ and $R_f = 20$ kΩ. (a) Calculate the current that flows in each parallel resistor if the reference voltage is +5 V and all the switches are at logical 1. (b) Calculate the output voltage when the input digital word is 1010.

Solution

(a) Starting from the D_0 position, the total resistance of two 20 kΩ resistors in parallel is 10 kΩ. This 10 kΩ resistance is in series with the 10 kΩ resistor to give a total resistance of 20 kΩ. This 20 kΩ resistance is in parallel with another 20 kΩ, again to give a total of 10 kΩ resistance. Carrying on in this way to the most significant end of the circuit shows that the input resistance of the circuit is 10 kΩ. Therefore, the input current = $5/(10 \times 10^3) = 0.5$ mA. The current in resistor $D_3 = 0.25$ mA, in resistor $D_2 = 0.125$ mA, in resistor $D_1 = 62.5$ μA, and in resistor $D_0 = 31.25$ μA (*Ans.*)

(b) The current in feedback resistor $R_f = 0.25$ mA + 62.5 μA = 0.3125 mA. Output voltage = $0.3125 \times 10^{-3} \times 20 \times 20 \times 10^3 = 6.25$ V (*Ans.*)

EXAMPLE 15.3

Calculate the output voltage of a 4-bit DAC when the input digital word has a decimal equivalent of (a) 3, (b) 9, and (c) 15. $V_{REF} = 5$ V.

Solution

(a) $V_{out} = 5 \times 3/8 = 1.875$ V (*Ans.*)
(b) $V_{out} = 5 \times 9/8 = 5.625$ V (*Ans.*)
(c) $V_{out} = 5 \times 15/8 = 9.375$ V (*Ans.*)

Fig. 15.8 *An AD557 8-bit DAC pinout*

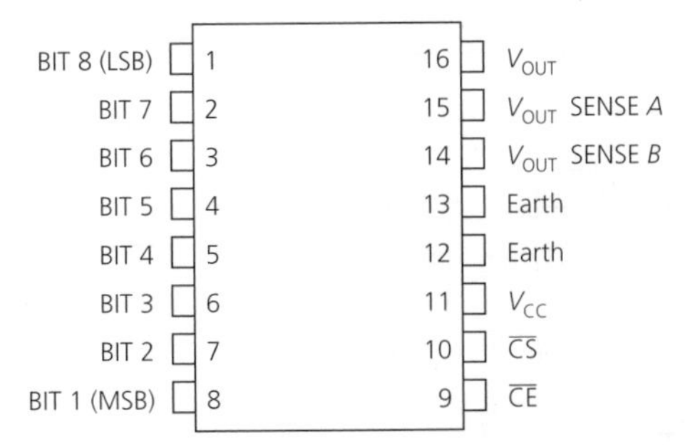

Integrated circuit digital-to-analogue converters

A wide variety of DACs are offered by several manufacturers. Some examples are: NSC 0808 8-bit, AD75756 8-bit, AD573 10-bit, AD574 12-bit. Fourteen-bit, 16-bit and 24-bit devices are also available.

The AD557 8-bit digital-to-analogue converter

The pinout of the AD557 8-bit DAC is shown in Fig. 15.8. The circuit operates from a 5 V supply and it produces an analogue output voltage in the range 0–2.56 V. The actual output voltage is equal to (decimal equivalent of input digital word × 0.1) V. The IC employs input latches that are controlled by the chip select $\overline{CS}$ and chip enable $\overline{CE}$ inputs. If the input latches are not wanted they are disabled by holding $\overline{CS}$ and $\overline{CE}$ LOW. If the inputs are to be latched, the data is inputted at the leading edge of either the $\overline{CS}$ or $\overline{CE}$ pulse and held until both $\overline{CS}$ and $\overline{CE}$ go LOW. Then the data is transferred into the DAC proper for conversion. Latched inputs are employed when the circuit is connected directly to the data bus of a microprocessor. The three V_{out} pins are normally tied together.

PRACTICAL EXERCISE 15.2

To test the performance of the DAC 0808.
Components and equipment: one DAC 0808. Two 74LS93 (or 74HC393) binary counters. One 741 op-amp. Three 5.1 kΩ resistors. One 220 pF capacitor. Power supply(ies) for +5 V, +10 V and −15 V. Pulse generator. CRO.
Procedure:

(a) Build the circuit shown in Fig. 15.9.
(b) Set the pulse generator to 1 kHz and observe the display on the CRO. The displayed analogue voltage ought to start from 0 V and increase in $2^8 = 256$ steps to very nearly 10 V. The time for each step is $1/(1 \times 10^3) = 1$ ms.
(c) Alter the frequency of the pulse generator to 100 Hz and repeat (b).
(d) Remove the pulse generator and the two ripple counters from the circuit. Apply logic 1 and 0 voltages to the input pins A_1 through to A_8 to input the digital word 10110101. Use the CRO to measure the output voltage. Repeat for digital words of (i) 11010000, and (ii) 00011100.

Fig. 15.9

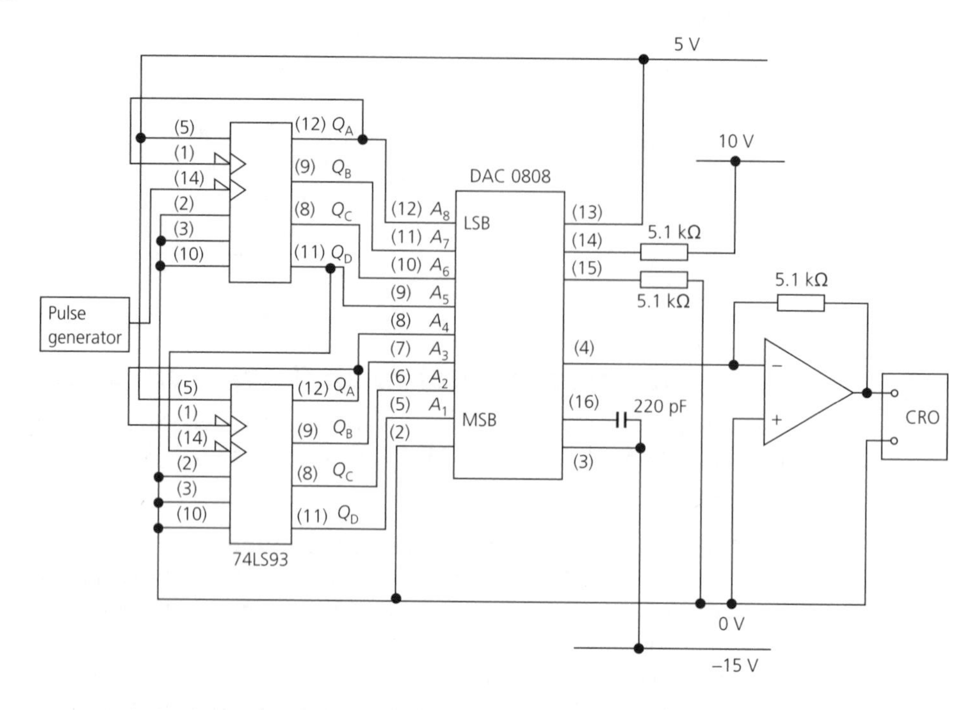

Analogue-to-digital converters

The function of an ADC is to convert an input analogue voltage into a digital number. The digital number represents the input voltage in discrete steps having a finite resolution. The resolution is determined by the number of bits in each digital word and it is equal to one part in 2^n. The ADC samples the analogue signal present at its input terminals at regular intervals and, for each sample, outputs a digital word which indicates the sampled voltage. An ADC may be arranged to output its data either in parallel or in serial form.

EXAMPLE 15.4

(a) Calculate the resolution of a 12-bit ADC. (b) What is this equal to for a 10 V input voltage range?

Solution

(a) Resolution = 1 part in 2^{12} = 1 part in 4096 (*Ans.*)
(b) Resolution = 10/4096 = 2.44 mV (*Ans.*)

Fig. 15.10 *Transfer function of a 3-bit ADC*

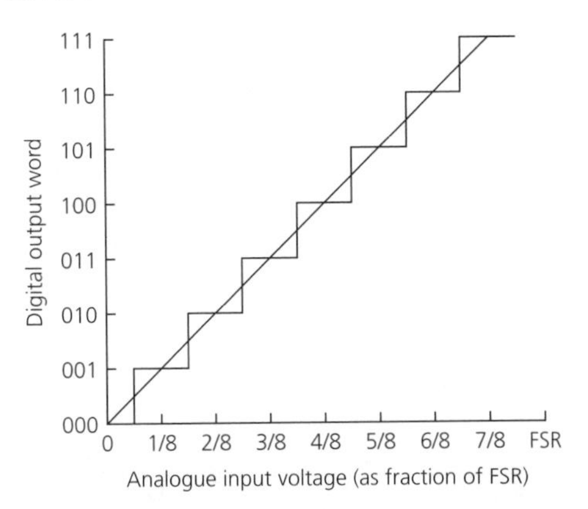

The transfer function of an ADC is shown in Fig. 15.10. The function is stepped because input voltages ± one LSB either side of a *quantization level* have the same digital word. The number of quantization levels is equal to 2^n, where n is the number of bits.

Ramp ADC

The basic circuit of a ramp-type ADC is given in Fig.15.11. At the start of a conversion the counter is cleared so that it has zero count. The clock then makes the counter go through its counting sequence. The output of the counter is applied to the input to the DAC and is also made available to the output terminals of the circuit. The DAC output then increments in steps, each of which corresponds to one LSB and is applied to the inverting input of the op-amp. The op-amp is operated as a voltage comparator; hence, when the DAC output voltage becomes more positive than the analogue input voltage the comparator switches and its output goes LOW. This low voltage acts as an *end-of-conversion* signal which disables the counter. The count then reached indicates the digital equivalent of the analogue voltage. The digital output word can be latched by using a quad D latch that will be clocked by the end-of-conversion signal. When the circuit is reset the output word will be held until the next end-of-conversion signal arrives. The counter must then be reset to zero for a new count to begin and another conversion to start. The circuit is able to produce only one conversion for every 2^n clock pulses and so it is suitable for use only when the analogue signal changes fairly slowly.

EXAMPLE 15.5

An 8-bit ramp ADC uses a 100 kHz clock. Calculate the maximum conversion time.

Solution

Maximum count = $2^8 = 256$. The circuit will take $256 \times 1/(100 \times 10^3)$ seconds to count from 0 to 255. Therefore, the maximum conversion time is 2.56 ms (*Ans.*)

Fig. 15.11 *Ramp-type ADC*

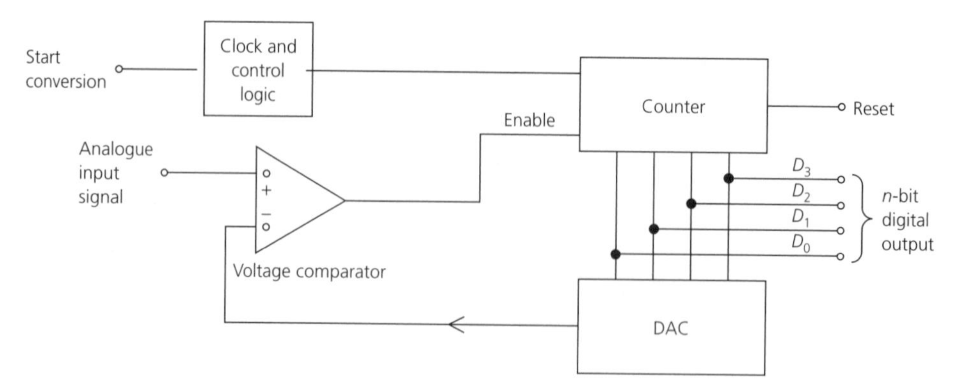

Fig. 15.12 *Tracking ADC*

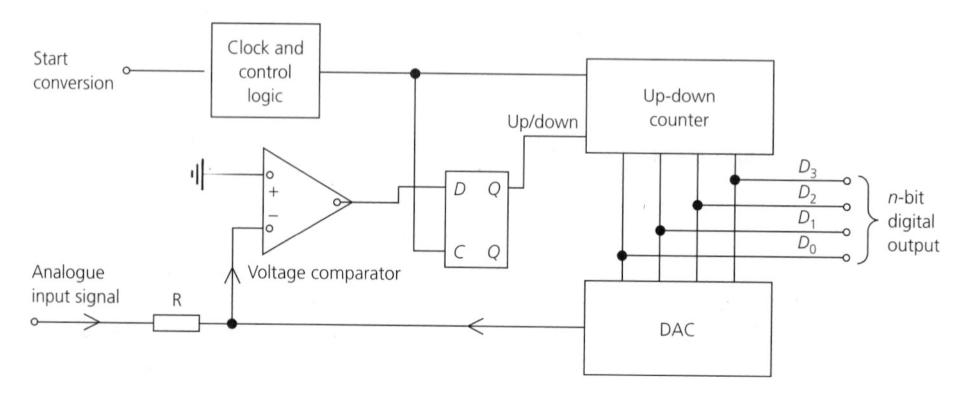

Tracking ADC

A much improved performance can be obtained at very little extra cost if the counter in Fig. 15.11 is replaced by an up-down counter. The circuit of a tracking ADC is given in Fig. 15.12.

The voltage applied to the inverting input of the comparator-connected op-amp is the difference between the analogue signal voltage and the output voltage of the DAC. If this difference is positive the output of the comparator will be LOW and the counter will cause the DAC output to increment by one LSB. This happens repeatedly until the difference voltage becomes negative, i.e. the output voltage of the DAC is more positive than is the analogue input voltage. The comparator will then switch and its output will go HIGH. This makes the counter commence down-counting. At all instants in time the comparator will switch the counter to count in the direction (up or down) which tends to reduce the difference voltage.

Once an initial lock has been established, the ADC will *track* any changes in the analogue voltage so long as they are not too rapid. Any break in the signal will mean that the lock is lost. Continuously present at the counter output is the binary-coded equivalent of the analogue voltage. The purpose of the D flip-flop is to ensure that there is an adequate time period between any change in the comparator output and the next count.

Fig. 15.13 *Successive approximation ADC*

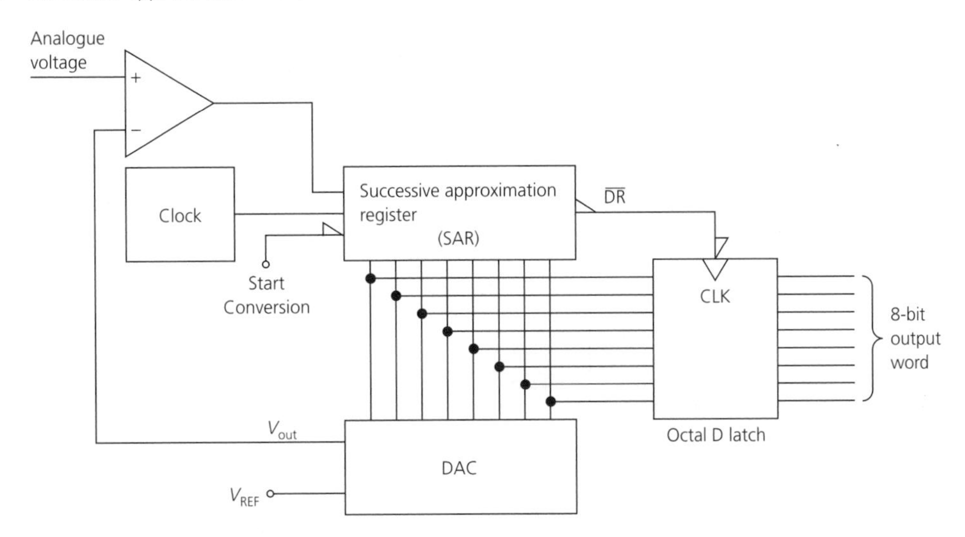

Successive approximation ADC

The conversion technique that is employed in most ICs is known as *successive approximation*. The block diagram of a successive approximation ADC is given in Fig. 15.13. The circuit operates by repeatedly comparing the analogue signal voltage with a number of approximate voltages which are generated at the DAC output.

Initially, the shift register is cleared and then the DAC output is zero. Conversion is initiated by taking the *start conversion* line LOW. The first clock pulse then applies the MSB of the register to the DAC. The output of the DAC is then one-half of its full-scale voltage range (FSR). If the analogue voltage is greater than FSR/2 the MSB is retained (stored by a latch). If it is less than FSR/2 the MSB is taken LOW by the successive approximation register (SAR) and is lost. The next clock pulse applies the next lower MSB to the DAC, producing a DAC output of FSR/4. If the MSB has been retained, the total DAC output voltage is now 3FSR/4; if the MSB had been lost the output of the DAC is now FSR/4. In either case the analogue and DAC voltages are again compared. If the analogue voltage is the larger of the two, the second MSB is retained (latched); if it is not, the second MSB is lost.

A succession of similar trials are carried out and after each the shift register output bit is either retained by a latch or it is not. Once $n + 1$ clock pulses have been supplied to the register, the conversion has been completed and the register output gives the digital word that represents the analogue voltage.

The $\overline{DR}$ (data ready) end-of-conversion pin is taken LOW to indicate that the conversion is complete and the data are ready to be used. The trailing edge of the $\overline{DR}$ pulse clocks the data into the D latch, making it available at the output. If continuous conversions are wanted, the $\overline{DR}$ line is connected to the start conversion line. The latched outputs will always show the results of the previous conversion.

The approximation process is illustrated in Fig. 15.14 for a 4-bit ADC in which the analogue voltage is equal to 5/16 FSR.

Fig. 15.14 *Action of a successive approximation ADC*

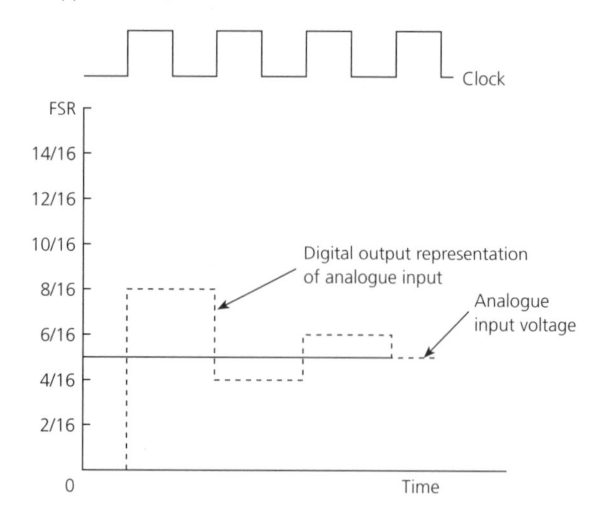

The first approximation is 8/16 FSR. This is too big and so the next clock pulse turns the MSB OFF and the second approximation is 4/16 FSR. This value is less than the input voltage and so this bit is retained. The third clock pulse turns ON the next register bit and the DAC output is incremented by 2/16 FSR to 6/16 FSR. Now the approximation is again too large and so the next clock pulse turns the third bit OFF and the LSB ON. The DAC output is then (4/16 FSR + 1/16 FSR) or 5/16 FSR, and this is equal to the analogue input voltage. The codeword produced by the DAC is then 0101.

EXAMPLE 15.6

Show the timing waveforms in a successive approximation ADC when an analogue voltage of 3.8 V is converted to a 4-bit binary word. The full-scale input voltage is 10 V.

Solution

The waveforms are shown in Fig. 15.15. The conversion starts when the start conversion input goes LOW and the $\overline{DR}$ input goes HIGH. At the end of four clock pulses (for a 4-bit device), $\overline{DR}$ goes LOW and the converted voltage is outputted (*Ans.*) [Fourteen-bit, 16-bit and 17-bit devices are also offered by manufacturers.]

Flash ADC

The circuit of a *flash* ADC is shown in Fig. 15.16. A reference voltage, whose value is equal to the full-scale analogue input voltage, is applied across a voltage divider that consists of $2^n + 1$ resistors connected in series. The divider splits the analogue voltage

Fig. 15.15

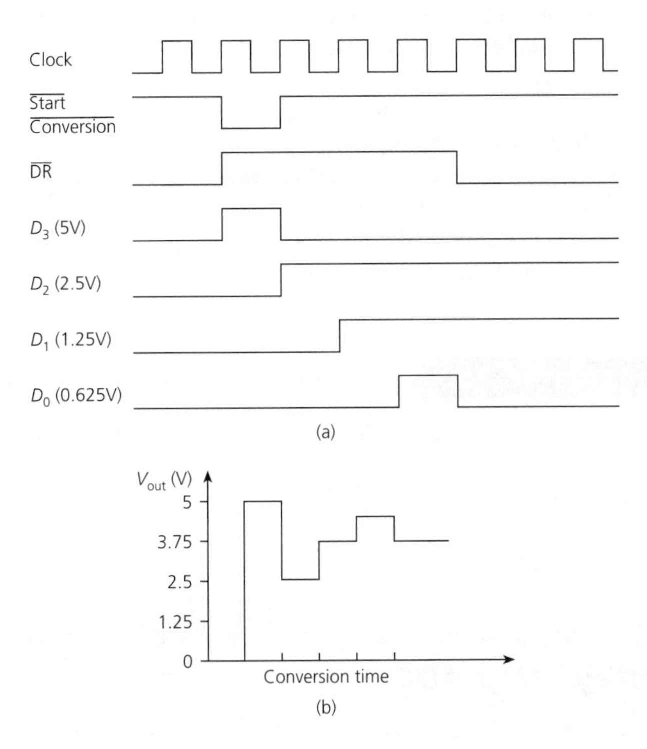

Fig. 15.16 *Flash ADC*

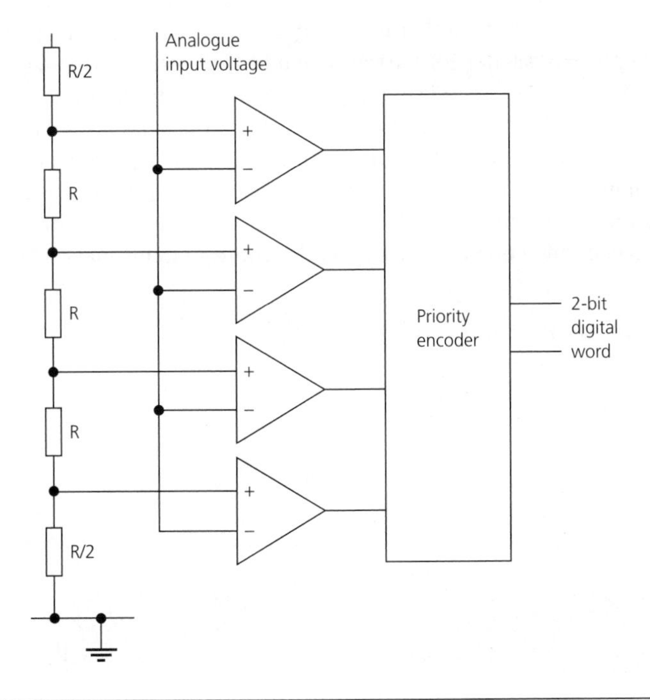

into $2^n - 1$ parts. The value of the input voltage is determined by the highest voltage comparator to have a HIGH output. For any given value of input voltage this comparator plus all those below it will have a HIGH output. All comparator outputs are connected to a priority encoder (see p. 193). This circuit produces an output that corresponds to the input that represents the largest voltage which is equal to, or less than, the input voltage. Then the binary output represents the voltage which is the nearest to the input voltage.

Flash converters are very fast to operate and can employ clock frequencies of up to 500 MHz, because all bits are determined in parallel. Because one voltage comparator is required per bit, flash ADCs generally employ no more than eight bits. The flash ADC is often used in digital CROs.

EXAMPLE 15.7

Determine how many voltage comparators an 8-bit flash ADC needs.

Solution

$2^n - 1 = 255$ (*Ans.*)

Integrating ADC

An integrating ADC measures the time taken for a capacitor to charge to determine the input analogue voltage and then converts this time into a digital word. In the *dual-slope technique*, a current that is directly proportional to the analogue input voltage is used to charge a capacitor for a fixed period of time. The average input analogue voltage is then determined by measuring the time taken to discharge the capacitor using a constant current. The block diagram of an integrating ADC is shown in Fig. 15.17. A positive input voltage generates a negative-going ramp voltage and vice versa. Suppose that the input voltage is negative so that a positive ramp is produced; the ramp continues for a fixed time determined by the counter reaching a specific count. When this count is reached the counter is reset and the control circuit makes the input switch change over

Fig. 15.17 *Dual-slope integrating ADC*

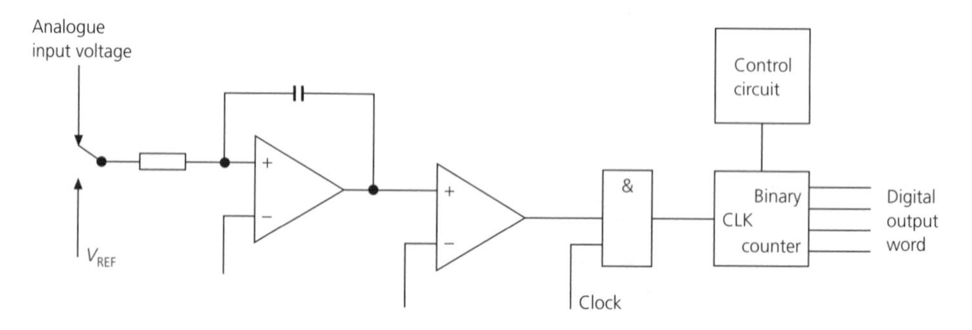

from the input voltage to the reference voltage. This reference voltage always has the opposite polarity to the input voltage and, hence, it causes the integrator to produce a negative-going ramp voltage that starts from the positive value reached when the counter reset. When the negative ramp reaches zero volts the voltage comparator switches state to have a LOW output and this disables the AND gate. No more clock pulses are applied to the counter and so it stops counting. The count it contains is directly proportional to the time it took the negative ramp to reach zero volts and this is proportional to the time it took the positive ramp to reach its final value. This means that the count is a measure of the input analogue voltage. The binary output of the counter is read to give an output digital word. The integration process reduces the effects of any noise that is picked up by the circuit and so this technique is often used for digital multimeters. The disadvantage of the integrating ADC is that it has the slow conversion rate of a few hundred hertz.

EXERCISES

15.1 (a) The DAC circuit shown in Fig. 15.3 has a largest value input resistor of 12 kΩ. Calculate the value of the smallest value input resistor.
(b) If the feedback resistor is 2 kΩ, calculate the maximum output voltage.
(c) Calculate the resolution of the circuit.

15.2 (a) A 4-bit ADC has a resolution of one count per 200 mV. An analogue voltage of peak value 2.1 V is applied to the circuit. What will be the output digital word?
(b) An ADC has a resolution of 100 mV per count. Determine the analogue voltage that is represented by the third significant bit at the output.

15.3 (a) What is the function of the voltage comparator in an ADC?
(b) An ADC is operated from a 5-bit counter. It has a reference voltage of 16 V. Calculate its resolution.

15.4 An 8-bit ADC has an input voltage of range of ±5 V. Determine the digital word that represents
(a) +2.5 V.
(b) −2.5 V.

15.5 An ADC has a sampling frequency of 10 kHz.
(a) Calculate the maximum input analogue signal frequency.
(b) Calculate the maximum allowable conversion time.

15.6 Figure 15.18 shows an alternative form of $R/2R$ DAC. The op-amp is connected as a voltage follower and has a very high input impedance. Determine the output voltage when the input digital word is 1000 and logic 1 = 5 V.

15.7 (a) Calculate the resolution of a 16-bit ADC.
(b) Determine its value for a 12 V range.

Fig. 15.18

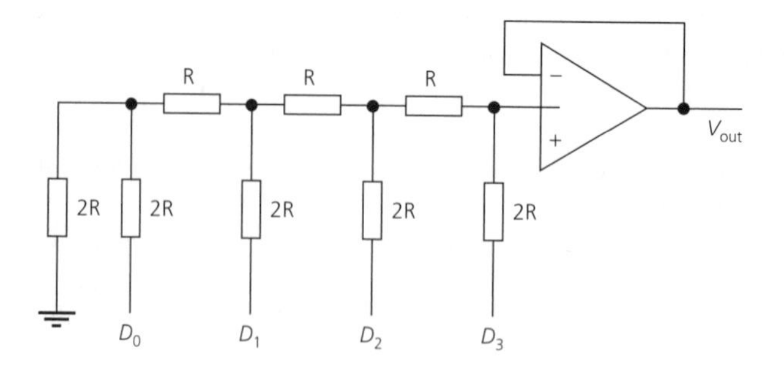

Fig. 15.19

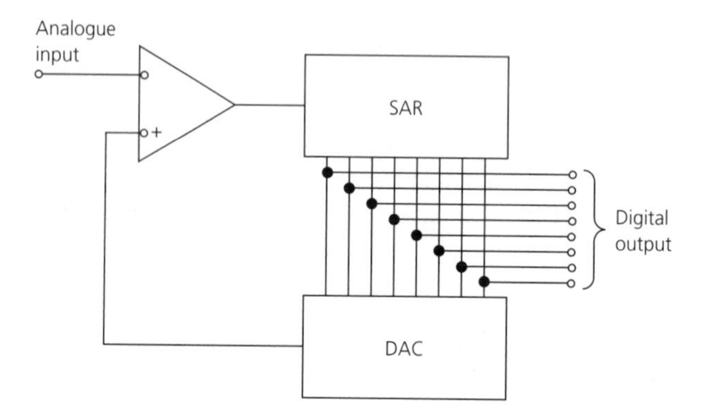

15.8 Explain, with the aid of a block diagram, the difference between a digital-to-analogue converter and an analogue-to-digital converter.

15.9 Most DACs are of either the binary weighted resistor, or the *R*/2*R* type. What are the disadvantages of the former and when might it be employed?

15.10 Figure 15.19 is the basic circuit of an 8-bit successive approximation ADC. Explain its action.

15.11 The maximum input voltage to a successive approximation ADC is 5 V. Calculate the voltage corresponding to the LSB if (a) 8 bits and (b) 16 bits are used. Hence, compare the circuits on the basis of both speed and noise.

15.12 An 8-bit DAC of the binary-weighted type uses a lowest resistor value of 1 kΩ. Calculate the values required for the other resistors. Comment on your answer.

15.13 The internal connections of the ZN588E DAC are shown in Fig. 15.20. Explain the operation of the IC.

Fig. 15.20

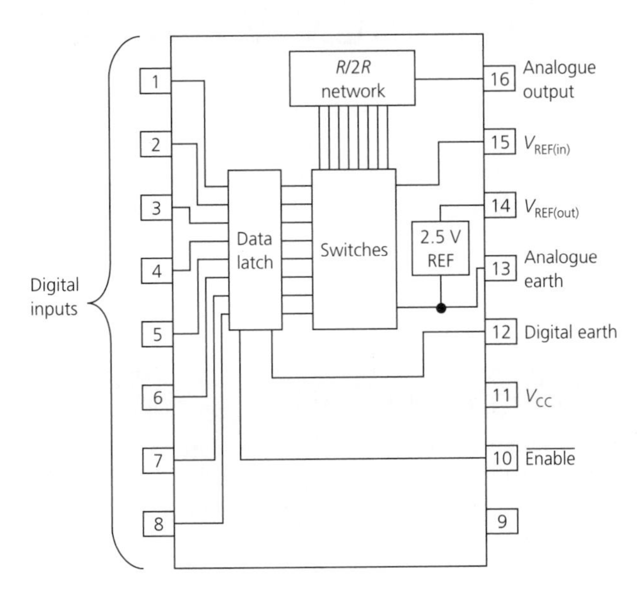

15.14 An analogue signal of peak voltage +5 V is applied to a successive approximation ADC. Sketch the output voltage of the circuit.

15.15 A weighted-resistor type 4-bit DAC uses a voltage of 5 V and the feedback resistor is 1000 Ω. If the output voltage produced by the LSB is to be 0.05 V, determine the required value for R.

16 Electronics Workbench

After working through the exercises in this chapter on your PC you should:

(a) Have gained an appreciation of the use of Electronics Workshop.
(b) Be able to modify and test a circuit that is displayed on the circuit window.
(c) Have improved on your understanding of digital electronic circuits.

Electronics Workbench is an electronic design tool that provides all the components, measuring equipment and voltage/current sources necessary to create designs on a computer and then test their performance. Electronics Workbench includes both analogue and digital devices as well as mixed ICs such as timers, analogue-to-digital converters and digital-to-analogue converters. Electronics Workbench is an extension of the *Simulation Program with Integrated Circuit Emphasis* (SPICE) industry standard set of algorithms for the simulation of analogue circuits, that provides a user-friendly graphical interface. In this chapter each of the practical exercises provided in the previous chapters are the subject of an exercise using Electronics Workbench and have the same number. When a file has been loaded the circuit to be tested will appear and can then be tested in a similar way to the breadboard design met earlier. Most circuits are given in complete form but a few have been left unfinished; this is to allow the user to gain some experience in drawing connections and connecting test instruments.

In the practical exercises an LED is employed to indicate the logical state of an output. An LED can be employed in the circuits created on Electronics Workbench when the symbol arrows highlight to indicate that the device is ON. However, it is not easy to see the highlighting and it is better to replace each LED with a device peculiar to Electronics Workbench, i.e. a LED probe. The probe lights up when it is ON and it does not require a series-connected current-limiting resistor.

When Electronics Workbench is first loaded a blank circuit window is visible. Click **FILE** and when a menu appears, click **OPEN**. Access the floppy disc provided with this book, select the file name, and then click **OPEN**. The circuit to be tested, and perhaps completed, will then appear.

- To select a component click on it.
- To deselect a component right click on it and then click on delete from the menu.
- To move a component click on it when it will turn red, then drag it, using the mouse, to its new position.

- To change the default value of a component click on it twice and enter the required value in the appropriate box.
- To wire components together, (a) point at the component's terminal until it highlights and then press and hold the mouse button and drag. A wire will appear. (b) Drag the wire to another component, to an instrument or to another wire. (c) When the terminal on the destination component, or instrument, highlights, release the mouse button when the connection will be made. The first few connections are neatly made and routed at right angles, but later ones sometimes take unexpected routes. If the routing looks confusing, wires can be given different colours. Click the wire twice and choose the schematic options tab; this will bring up a colour button, click this and choose the required colour.

The circuits given in the following Electronic Workbench exercises use LS devices, since it was thought that (probably) the practical exercises would also employ them. If it is desired to replace an LS IC by its HC equivalent, it is merely necessary to double click on the IC to display the properties menu. Next, click on CMOS in the Library menu and then click HC in the Model menu. Should it be desired to substitute a device from one of the other CMOS logic families it will be necessary to edit the HC parameters; to do this click EDIT and then insert the correct values for the various parameters for the wanted device.

Exercises using Electronics Workbench

EXERCISE EWB1

(a) Open file Ex.1.1 and activate it.
(b) Use the CRO to measure (i) the frequency and (ii) the duty cycle, of the pulse waveform. Compare the values obtained with those calculated using equation (1.1).
(c) Alter the values of R_1 and R_2 to 7.5 kΩ each and again measure both the frequency and the duty cycle of the displayed pulse waveform.
(d) Calculate values for R_1 and R_2 (with R_2 approximately 10 times larger than R_1) for a frequency of 100 Hz. Alter R_1 and R_2 to these new values and measure the pulse waveform using the CRO. Does the circuit work and at what frequency and duty cycle? If the circuit doesn't work, try a larger capacitance value and re-calculate the values of R_1 and R_2.

EXERCISE EWB2

(a) Open file Ex.3.1a and activate the circuit.
(b) Apply each of the four possible combinations of logical 0 and logical 1 voltages to inputs *1A* and *1B* of the AND gate IC. This can best be done by double clicking each d.c. source and altering its voltage. Each time note the logical state of the red probe (lit = 1, dim = 0).
(c) Close file Ex.3.la and open file Ex.3.1b.
(d) Repeat procedure (b).

EXERCISE EWB3

(a) Open file Ex.3.2a and activate the circuit.
(b) Note whether the red probe glows visibly.
(c) Try, in turn, each of the other combinations of logical 1 and 0, applied to inputs *1A* and *1B* and each time note the logical state of the probe. Write down the truth table for the gate.
(d) Close the file and then open file Ex.3.2b. The circuit that appears is not finished and requires some more connections to be made. Refer to Fig. 3.25(b) and complete the circuit. Note that the LED and the series 270 Ω resistor are replaced by the red probe. A connection is required only from the output *4F* to the probe terminal.
(e) Repeat (b) and (c) above.
(f) Open the logic converter. Connect the left-hand terminal of the logic converter to inputs *1A/2A* of the circuit. Connect its next-door terminal to inputs *1B/3B* of the circuit. Connect the output *4F* of the circuit to the right-hand terminal of the logic converter. The logic converter is able to calculate and display the truth table of the circuit (but note that it has *A* as the most-significant bit and *B* as the least-significant). Click the Circuit to Truth Table button. The truth table should then appear in the converter's display. Then click the Truth Table to Boolean Expression button, when the Boolean expression describing the operation of the circuit will appear at the bottom of the converter.

EXERCISE EWB4

(a) Open file Ex.3.3a and activate the circuit.
(b) Apply the four possible combinations of logic 1 and 0 voltages to the inputs *1A* and *1B*. Each time note the logical state of the LED probe.
(c) Write down the truth table of the gate.
(d) De-select the voltage sources connected to *1A* and *1B*. Select the logic converter. Connect the left-hand converter terminal to *1A* and the next terminal to *1B*. Then connect the right-hand terminal to *1Y*. Activate the circuit.
(e) Click the Circuit to Truth Table button and the truth table of the gate will be displayed. Then click the Truth Table to Boolean expression button and the expression $\bar{A} + \bar{B}$ is displayed. This is equivalent to $\overline{AB}$ (see p. 84).
(f) Close file Ex.3.3a and open file Ex.3.3b.
(g) Activate the circuit and repeat (b) and (c) above.
(h) The inputs *2A* and *2B* are connected together so that gate 2 acts as an inverter. A voltage source is connected to the common input. Vary the voltage of this source between 0 and 5 V and each time check the logical state of the probe. Did the gate act as an inverter?
(i) De-select the voltage source connected to input *1A*. Select the function generator and set it to pulse waveform with a frequency of 100 Hz, a duty cycle of 50% and a voltage of 5 V. Connect the function generator to input *1A*.
(j) Select the CRO and connect it to output *1Y*. Leave the probe connected. Activate the circuit. Vary the voltage applied to input *1B* between 0 and 5 V and each time

check the waveform displayed by the CRO. The waveform will be displayed when the NOR gate is enabled and absent when the NOR gate is disabled. Which polarity voltage applied to input *1B* enabled the gate?

EXERCISE EWB5

(a) Load file Ex.4.1a and activate the circuit.

(b) Vary the voltage of the two d.c. voltage sources to apply each of the four possible combinations of logic 1 and logic 0 voltages to the inputs of the circuit.

(c) Each time note the logical state of the LED probes (lit = 1, dim = 0) and, hence, write down the truth tables for the two parts of the circuit.

(d) What do you notice about them? Which of De Morgan's rules is confirmed?

(e) Close file Ex.4.1a and open file Ex.4.1b. Activate the circuit and repeat steps (b), (c) and (d) above.

(f) Remove the connections between the two d.c. voltage sources and the inputs to the circuit. Select the logic converter and connect its left-hand terminal to input *1A* of the 72LS32 IC (without losing the connection to inputs *1A*/*1B* of the 74LS00 IC). Connect the next logic converter terminal to input *1B* (and inputs *2A*/*2B*), and the right-hand terminal to the output of the 74LS08 IC.

(g) Click the Circuit to Truth Table button and note the truth table that is then displayed. Next, click the Truth Table to Boolean Expression button and note the displayed equation. Compare this with the equation deduced from the truth table obtained in (e).

(h) Remove the connection from the logic converter to pin 3 of the 74LS08 and move it to pin 8 of the 74LS00. Repeat procedure (g).

(i) Delete the logic converter and select both the word generator and the logic analyser. Connect the right-hand terminal of the word generator to input *1A*/*1B* and the next terminal to input *3A*/*3B*. Open the controls by clicking the icon. Since the circuit has only two inputs, a bit pattern must be created for the first five rows of the display. Ensure that the current address is 0000 (if it is not, click pattern and on the pop-up menu click clear buffer and then accept). Place the cursor in the second word of the hex field and click. Then move the cursor to the extreme right, or least-significant 0 in the binary field and alter it to 1 (delete 0 first). Now move the cursor to the third hex word and click; move the cursor to the binary field and alter the second 0 from the right to 1. Last, select the fourth hex word and then alter both the right-hand binary bits to 1.

(j) Move the cursor to the tenth hex word and click breakpoint. Click burst to send the digital words in sequence once.

(k) Connect the top terminal of the logic analyser to *1A*/*1B* and the next terminal to *3A*/*3B*. Connect the third terminal down to the output *2Y* of the circuit. Open the logic analyser controls by clicking its icon twice.

(l) Set the clock frequencies of the two instruments to the same value and observe the waveforms displayed on the logic analyser. The rate at which patterns are drawn can be altered using the clocks per division setting. The top pattern is the input waveform to *1A/1B*, the second waveform is the input to *3A/3B*, and the bottom pattern is the output waveform. Deduce from the waveforms the logic function performed by the circuit.

EXERCISE EWB6

(a) Open file Ex.5.1a. It shows the required circuit. Make a note of it.

(b) Close file Ex.5.1a and then open file Ex.5.1b. Activate the circuit.

(c) Apply each of the four combinations of logical 1 and 0 voltage to the inputs of the circuit. Each time note the logical state of the LED probe.

(d) Check that the design requirements of the circuit are met.

(e) Close file Ex.5.1b and open file Ex.5.1c and activate the circuit. Repeat (c) and (d) above.

(f) Delete the connections between the d.c. voltage sources and the inputs to the circuit. Select the logic converter. Connect (i) its left-hand terminal to input *1A* of the inverter, (ii) the adjacent terminal to input *2A* of the inverter, (iii) the next terminal to the commoned *2A/2B* terminals of the NOR gate and, (iv) the fourth terminal to the *2C* terminal of the NOR gate. Connect the right-hand terminal to the output of the circuit.

(g) Double click the word generator icon to access its controls. Click the Circuit to Truth Table button and note the result.

(h) Click the Truth Table to Boolean Equation button and note the equation for the circuit. Then press the button below to obtain the simplified expression describing the circuit.

(i) If you have the time, replace the logic converter with the word generator and connect the logic analyser to the output of the circuit. Set up the truth table of the circuit on the generator, set the breakpoint to, say, 20H, and press burst. Then double click the logic analyser icon and display the circuit waveforms.

EXERCISE EWB7

(a) Open file Ex.6.1a and activate the circuit.

(b) Apply each of the four combinations of logic 1 and 0 voltage to the two inputs and each time note the logical state of the LED probe. Write down the truth table for the circuit. What logic function has been performed?

(c) Delete the connections between the voltage sources *A* and *B* and select the logic converter. Connect its two LSB terminals to inputs *1A* and *1B* and its right-hand terminal to output *3Y*. Double click the converter's icon to access its controls.

(d) Click the Circuit to Truth Table button to display the truth table of the circuit. What logic function does it represent? Then click the Truth Table to Boolean Expression button to obtain the equation describing the circuit. What is it?

(e) Now close file Ex.6.1a and open file Ex.6.1b. Activate the circuit and then repeat (b); (c) and (d) above.

EXERCISE EWB8

(a) Open file Ex.6.2a and activate the circuit.
(b) Confirm that the circuit performs the exclusive-OR logical function (i) by varying the voltages of the two d.c. voltage sources, and (ii) by using the logic converter.
(c) Delete the circuit and then build the circuit using two 74LS00 and 2-input NAND gates. (i) Click the IC button D when the digital IC menu will be displayed. (ii) Drag the left-hand icon on to the circuit window. Select 7400 and click accept. (iii) Double click the IC package and select ttl and LS on the pop-up menu. Press OK. (iv) Press the sources button to display the sources menu. Drag one 5 V, two battery, and three earth icons to the circuit window. (v) To obtain whatever connection you require, click the basic button and drag connectors as required to the circuit window.
(d) Build the circuit shown in file Ex.6.2a, activate it, and then repeat (b) above.
(e) Close file Ex.6.2a and open file Ex.6.2b. Activate the circuit and then repeat (b) and (c) above (except that when building the circuit use the 74LS02 IC).

EXERCISE EWB9

(a) Open file Ex.7.1 and activate the circuit. Set the voltage of the d.c. voltage source to 0 V. Increase its voltage in a number of steps and each time note the reading of the voltmeter (allow enough time for the voltage to reach its steady value).
(b) State (i) the output voltage when the input voltage was 0 V, and (ii) the input voltage at which the output voltage suddenly fell to a low value. (iii) What was this low value? Compare with the values quoted in Fig. 7.11.

[Note that Electronics Workshop allows the voltages that represent logic 1 and logic 0 at the inputs and output of the selected IC to be set to any desired value.]

EXERCISE EWB10

(a) Open file Ex.7.2a and activate the circuit. With at least one input set to logical 0 measure, and note, the output voltage at pin *1Y*. The high-value resistor shown would not be used in a practical digital circuit but it is required here so that Electronics Workbench can simulate the output drive characteristics.
(b) Close file Ex.7.2a and open file Ex.7.2b. The circuit shown is the gate in the previous file with its output connected to both inputs of a second gate. Activate the circuit and measure the output voltage.
(c) Close file Ex.7.2b and open file Ex.7.2c. Now the output of the first gate is connected to inputs *2A*/*2B* and *3A*/*3B* to give a fan-out of 4. Activate the circuit and measure the output voltage.
(d) Close file Ex.7.2c and open file Ex.7.2d. The commoned inputs of the fourth gate have now also been connected to the output of the first gate. This means that the fan-out is now 6. Activate the circuit and measure the output voltage.

(e) Modify the circuit by connecting the commoned *1A*/*1B* terminals of the second NAND gate IC to the output. Now the fan-out is 8. Activate the circuit and measure the output voltage.

(f) Carry on using the other three gates in this second IC to obtain, in turn, fan-outs of 10, 12 and 14. Each time activate the circuit and measure the output voltage.

(g) Draw a graph of output voltage plotted against fan-out and mark on it the maximum voltage that will be taken as logic 0. Estimate the maximum fan-out for the 74LS00 NAND gate IC.

EXERCISE EWB11

(a) Open file Ex.8.1 and activate the circuit.

(b) Set the select inputs *A* and *B* to logical 0. Then determine which combinations of the data inputs $1C_0$ through to $1C_3$ (D_0 through to D_3 in chapter 8) give a HIGH output. Compare the results with the truth table for a 4-to-1 multiplexer given in Table 8.1 on p. 178.

(c) Delete the earth connection to $1\overline{G}$ (pin 1) and select the function generator. Connect the function generator between $1\overline{G}$ and earth.

(d) Double click the function generator icon to access its controls. Set it to pulse waveform, frequency 1 Hz, duty cycle 50% and amplitude 4 V.

(e) Set the data input to 0101 and activate the circuit. Vary the select input voltages (on *A* and *B*) so that each data input is selected in turn. Each time note the logical state of the LED probe.

(f) Explain how this arrangement allows the device to be employed to multiplex four channels.

EXERCISE EWB12

(a) Open file Ex.8.2a and activate the circuit. The circuit ought to implement the logic function given in example (8.7).

(b) Write down the truth table for four inputs *A*, *B*, *C* and *D*. Then work through the table by making *A*, etc. equal to either 1 or 0 as required. For each input word note the logical state of the LED probe (lit = 1, dim = 0).

(c) In the output *F* column of the truth table enter either 1 or 0 as indicated by the probe. Write down the Boolean equation obtained from the truth table and then compare it with the one given in equations (8.6) and (8.7).

(d) Now close file Ex.8.2a and open file Ex.8.2b. This gives the same circuit as before, but the voltage sources applied to the inputs have been replaced by the logic converter. The output of the circuit is also connected to the logic converter.

(e) Activate the circuit and double click the logic converter icon. Click the Circuit to Truth Table button and wait while the truth table is calculated and displayed. Compare it with the one obtained in part (c). Now click the Truth Table to Boolean Expression button and note the expression that is displayed at the bottom of the converter. Compare this with the equation the circuit was designed to implement.

EXERCISE EWB13

(a) Open file Ex.8.3a and activate the circuit.
(b) Connect 5 V to the $1\overline{G}$ input pin and then apply all combinations of 1 and 0 logic voltage to the *1A* and *1B* inputs. Confirm that the voltages applied to *1A* and *1B* have zero effect on the output. Why?
(c) Connect 0 V to the $1\overline{G}$ input. Apply, in turn, each of the four possible combinations of logical 1 and 0 voltage to inputs *1A* and *1B*. Note which LED turns OFF. What circuit function is the circuit performing? Are the outputs active-HIGH or active-LOW?
(d) Close file Ex.8.3a and open file Ex.8.3b. Double click the function generator icon to access its controls. Set it for a rectangular 5 V output at a frequency of 1 Hz and with 20% duty cycle.
(e) Activate the circuit and confirm that the input data (the 20% duty cycle waveform) appear at the output that has been selected by the voltages applied to *1A* and *1B*. What function is the circuit now performing?

EXERCISE EWB14

(a) Open file Ex.8.4 and activate the circuit.
(b) Apply each of the four combinations of logical 1 and 0 voltage to the inputs of the circuit. Each time note the logical states of the four LED probes.
(c) Write down the truth table of the circuit. Compare it with the expected truth table and state what error has occurred.
(d) Find the fault in the circuit, correct it, and then check that the correct truth table is then obtained.

EXERCISE EWB15

(a) Open file Ex.8.5 and activate the circuit. An HCMOS device is used because the Electronics Workbench library does not include a 74LS series full-adder.
(b) Apply different 4-bit numbers to the A_0–A_3 and B_0–B_3 input terminals with carry-in = 0. Note the sum and carry-out indicated by the LED probes. Now make the carry-in 1 and note the indicated sum and carry-out.
(c) Repeat this for several other input binary numbers and check whether the circuit always gives the correct sum and carry-out.

EXERCISE EWB16

(a) Open file Ex.9.1a and activate the circuit.
(b) Determine the truth table of the $\bar{S}$-$\bar{R}$ latch. Compare it with the expected truth table and state how it differs from it. Why does this circuit act differently in one respect?
(c) Close file Ex.9.1a and open file Ex.9.1b. The displayed circuit uses the 74LS279 quad $\bar{S}$-$\bar{R}$ latches. The $\bar{S}$-$\bar{R}$ inputs are normally held HIGH. When the $\bar{S}$ input is taken LOW, the Q output goes HIGH. When $\bar{R}$ is taken LOW, the Q output goes LOW.

The S-R inputs should not be both taken LOW at the same time since the Q output will then be unpredictable. Latches 1 and 3 have two S inputs, labelled as $1\bar{S}_1$ and $1\bar{S}_2$ or $3\bar{S}_1$ and $3\bar{S}_2$; both latches have both $\bar{S}$ inputs applied to the same NAND gate as the output from the other NAND gate.

(d) Determine the truth table for latch 4 and compare it with the truth table of a S-R latch.

(e) Determine the truth table for latch 1. Suggest a reason for using this latch rather than either latch 2 or 4.

EXERCISE EWB17

(a) Open file Ex.9.2 and activate the circuit.

(b) Set both $1\overline{CLR}$ and $1\overline{PRE}$ to 1. Set $J = K = 1$. Double click the function generator icon and set it to produce a rectangular wave with 50% duty cycle, 1 Hz frequency, and 5 V amplitude.

(c) Double click the CRO icon and set the timebase to 0.5 s/div and both channel A and channel B to 1 V/div. Note the displayed waveforms. It will probably be best if some offset is applied to both the X and the Y positions.

(d) Set, first $J = 1$ $K = 0$, and, second, $J = 0$ $K = 1$, and note the display.

(e) Minimize the CRO and then confirm that the circuit follows the truth table given in Table 9.4.

EXERCISE EWB18

(a) Open file Ex.9.3a and activate the circuit.

(b) Set $\overline{PRE} = \overline{CLR} = 1$.

(c) Set $D = 1$ and then vary the clock input from 0 to 1 and back again. Notice that the LED probes did not change their logical states. Why not?

(d) Remove the voltage source connected to the CLK input and replace it with the function generator. Double click the function generator to access its controls. Set its output voltage to rectangular waveform, with 50% duty cycle, 1 Hz frequency, and 5 V amplitude.

(e) Check that the flip-flop now works in the expected manner.

(f) Close file Ex.9.3a and open file Ex.9.3b. The 74LS75 is a quad D latch.

(g) With the clock held HIGH, check that the Q output takes up the same logical state as the D input, no matter how many times the D input changes state.

(h) See what happens when the clock input is held LOW. Compare with the function table given in Table 9.5.

EXERCISE EWB19

(a) Open file Ex.10.1 and activate the circuit.

(b) Set the function generator to give a rectangular waveform with 50% duty cycle, 1 Hz frequency, and 5 V amplitude.

(c) Note the logical states of the $\overline{PRE}$ and $\overline{CLR}$ inputs and of the three LED probes.

(d) Take the $\overline{\text{CLR}}$ inputs HIGH and the $\overline{\text{PRE}}$ inputs LOW. What now are the logical states of the three probes?
(e) Put $\overline{\text{CLR}}$ and $\overline{\text{PRE}}$ HIGH. Now the circuit ought to start counting from 0 through to 7 repeatedly.
(f) Rearrange the output connections of each flip-flop so that the clock inputs CLKB and CLKC are connected to the preceding Q outputs instead of to Q. Determine the function that is now performed by the circuit.

EXERCISE EWB20

(a) Open file Ex.10.2a and activate the circuit. Inputs R_{01} and R_{02} are held HIGH so that the counter is reset and all four LED probes are OFF.
(b) Double click the function generator icon to access its controls. Set its output to rectangular waveform with 50% duty cycle, 1 Hz frequency, and 5 V amplitude.
(c) Move the connections to R_{01} and R_{02} from +5 V to earth. The circuit should now count from 0 through to 15 repeatedly. Observe the LED probes to check that it does so. The CRO could be employed to observe the input and output waveforms.
(d) Close file Ex.10.2a and open file Ex.10.2b. The 74LS93 has now been connected to act as a decade counter. Activate the circuit and check that the circuits counts repeatedly from 0 through to 9.
(e) Refer to Table 10.2 and connect the circuit to act as (i) a divide-by-9 circuit, and (ii) a divide-by-14 circuit. Each time check that the circuit works correctly.

EXERCISE EWB21

(a) Open file Ex.10.3 and activate the circuit.
(b) Double click the function generator icon and set its controls to square wave, frequency 1 Hz, and amplitude 5 V.
(c) Observe the LED probes and determine the count of the circuit.
(d) Increase the frequency to 1 kHz. Connect the CRO into the circuit, input A to Q_D and input B to Q_A. Observe the displayed waveforms. Measure (i) their frequencies, and (ii) their duty cycles.
(e) Remove the connections between pins 1 and 12 and then connect pin 11 to pin 14 and connect the function generator to pin 1. Observe the displayed waveforms on the CRO and determine (i) their frequencies, and (ii) their duty cycles.
(f) Work out how to convert the circuit into a divide-by-6 circuit. Set up the circuit and check its operation.

EXERCISE EWB22

(a) Open file Ex.10.4 and activate the circuit.
(b) $\overline{\text{CLR}}$ is LOW so all stages in the counter are cleared. The input data word is 1001. To load this data into the counter $\overline{\text{CLR}}$ must be taken HIGH and then $\overline{\text{load}}$ must be taken LOW. Hence, connect $\overline{\text{CLR}}$ to 5 V and connect $\overline{\text{load}}$ to earth.

(c) Double click the icon of the function generator and set it to rectangular wave with 50% duty cycle, amplitude 5 V and some low frequency, say, 1 Hz.
(d) Note, which LED probes are ON and, hence, determine the initial count of the circuit.
(e) Disconnect ENP/ENT from earth and, instead, connect them to 5 V. Then connect $\overline{\text{load}}$ to 5 V instead of to earth; the circuit should then commence to count. Determine the count.
(f) To stop the count take ENP LOW.
(g) Load another data word on to terminals A, B, C and D and then repeat (b) etc. above.
(h) Modify the circuit so that it counts up from decimal 9 through to 15 and 0, and stops when the count reaches 3. To do this drag one, or more, gates on to the circuit window and use it to take ENP LOW when $Q_A = Q_B = 1$ and $Q_C = Q_D = 0$.

EXERCISE EWB23

(a) Open file Ex.10.5. It shows the 74LS191 4-bit synchronous up-down counter. Activate the circuit.
(b) The counter has been pre-set by the input data word 0111. Take $\overline{\text{load}}$ LOW to load the data word into the counter. Note which LED probes are ON and say what the initial count is.
(c) Double click the function generator icon to access its controls. Set the output voltage to square waveform, 1 Hz frequency, and 5 V amplitude.
(d) The direction of the count is set by the level of the voltage applied to the $\overline{U}$/D terminal. This terminal is held LOW so the counter will count up. The count is disabled since the $\overline{\text{CTEN}}$ terminal is held HIGH. To start the count, take $\overline{\text{CTEN}}$ HIGH and then take $\overline{\text{load}}$ HIGH. Note the logical states of the LED probes and, hence, determine the count of the circuit.
(e) The count can be stopped at any point in the cycle by taking $\overline{\text{CTEN}}$ HIGH at that point. Devise a gating arrangement to stop the count when it is at decimal 14. Connect up your circuit and check that it counts from decimal 7 (0111) up to decimal 14. To restart the count cycle take $\overline{\text{CTEN}}$ LOW.
(f) Repeat first (d) and then (e) with the circuit operated as a down counter.

EXERCISE EWB24

(a) Open file Ex.11.1 and activate the circuit. Logic 0 is applied to the $\overline{\text{CLR}}$ line and logic 1 to the $\overline{\text{PR}}$ line to ensure that all stages are cleared. Check that all four LED probes are dim.
(b) Alter the voltages applied to the $\overline{\text{CLR}}$ and $\overline{\text{PRE}}$ lines to 5 V.
(c) Set the clock to 1 MHz, 50% duty cycle, and 5 V amplitude.
(d) Start the circuit running and press the space bar to operate the switch. This will change the input data between 1 and 0 as required to obtain a particular data word. Observe the LED probes and check that the input word passes through the register.

The Q_D LED probe indicates the serial output. All four probes indicate the parallel output.

(e) Clear all stages, either by taking the $\overline{CLR}$ line to 0 or by using the space bar to input four 0s. Then pause the circuit.

(f) Set up the input parallel word 1000. Set the clock to 1 kHz and press Resume. Note the logical states of the LED probes (i) initially, and (ii) after several clock pulses. Now set the input parallel word to 0010 and repeat.

EXERCISE EWB25

(a) Open file Ex.11.2 and activate the circuit. The 8-bit shift register is connected as a ring counter. The circuit is not self-starting.

(b) All stages are intially cleared. Press the space bar; the switch will then operate to disable the clear.

(c) To start the counter it is necessary to load it with a single 1 bit. In a practical circuit this would be achieved by using additional logic and timing circuitry. Here, press the Z key quickly twice. The first press will operate the switch to apply logic 1 to the first stage in the register. The second press, which must be made quickly, will remove the 1 bit. If the switch is not operated rapidly enough, two or more bits will be loaded into the circuit.

(d) The single bit will circulate around the counter. This can be seen by observing the LEDs which will be seen to glow one after the other repeatedly until the space bar is pressed.

(e) Try entering two 1 bits into the counter by keeping the Z key pressed for a little longer. What happens?

(f) If time allows, devise a NAND gate arrangement which can be employed to make the circuit self-starting.

EXERCISE EWB26

(a) Open file Ex.14.1a and activate the circuit.

(b) Press twice, in order, the keys W, X, Y and Z. State what happens.

(c) Try various combinations of W, X, Y and Z that give BCD numbers and state the results.

(d) Now try some key combinations that correspond to numbers greater than 9. Each time note the displayed character.

(e) Close file Ex.14.1a and open file Ex.14.1b. Activate the circuit.

(f) The switches W, X, Y and Z in the previous circuit have been replaced by the outputs Q_A, Q_B, Q_C and Q_D of the 4-bit counter. Resume the circuit and note the displayed numbers.

(g) Obviously, it is undesirable to have symbols displayed for numbers greater than 9. Modify the 74LS93 to act as a decade counter and then observe that the display counts from 0 through to 9 repeatedly.

EXERCISE EWB27

(a) Open file Ex.15.1a and activate the circuit. Operate the four switches by pressing keys W, X, Y and Z to enter the binary equivalents of the decimal numbers 0 through to 15. W is the MSB and Z is the LSB. Each time measure the output voltage (ignore the minus sign).

(b) Complete the truth table shown below.

W	X	Y	Z	Output voltage (V)
0	0	0	0	6×10^{-3}
0	0	0	1	1.2
0	0	1	0	
0	0	1	1	

(c) From the table determine the resolution of the circuit. Compare it with the calculated value and account for any discrepancies.

EXERCISE EWB28

(a) Open file Ex.15.2 which shows the generic voltage-output DAC. Activate the circuit.

(b) Press key 7. This operate switch 7 and applies 5 V to the MSB input of the circuit. Note that the indicated voltage is 2.5 V, i.e. one-half of the applied 5 V. Press key 7 again to remove the 5 V from the MSB input.

(c) Press, in order, key 0, key 1, keys 0 and 1, key 2, keys 0 and 2, keys 1 and 2, keys 0 and 1 and 2, and so on. Each time note the indicated output voltage.

(d) Use the results to draw the transfer function of the DAC.

(e) Determine the resolution of the circuit (i) from the transfer function, and (ii) by calculation.

Appendix A International alphabet 5 or ASCII code

The ASCII (American Standard Code for Information Interchange) codes for numbers, alphabet letters and other common symbols

Decimal numbers	ASCII code		Alphabetical characters	ASCII code	
	in binary	*in hex*		*in binary*	*in hex*
0	0110000	30	@	1000000	40
1	0110001	31	A (a)	1000001	41 (61)
2	0110010	32	B (b)	1000010	42 (62)
3	0110011	33	C (c)	1000011	43 (63)
4	0110100	34	D (d)	1000100	44 (64)
5	0110101	35	E (e)	1000101	45 (65)
6	0110110	36	F (f)	1000110	46 (66)
7	0110111	37	G (g)	1000111	47 (67)
8	0111000	38	H (h)	1001000	48 (68)
9	0111001	39	I (i)	1001001	49 (69)
			J (j)	1001010	4A (6A)
Other			K (k)	1001011	4B (6B)
symbols			L (l)	1001100	4C (6C)
:	0111010	3A	M (m)	1001101	4D (6D)
;	0111011	3B	N (n)	1001110	4E (6E)
<	0111100	3C	O (o)	1001111	4F (6F)
=	0111101	3D	P (p)	1010000	50 (70)
>	0111110	3E	Q (q)	1010001	51 (71)
?	0111111	3F	R (r)	1010010	52 (72)
Space	0100000	20	S (s)	1010011	53 (73)
!	0100001	21	T (t)	1010100	54 (74)
"	0100010	22	U (u)	1010101	55 (75)
#	0100011	23	V (v)	1010110	56 (76)
$	0100100	24	W (w)	1010111	57 (77)
%	0100101	25	X (x)	1011000	58 (78)
&	0100110	26	Y (y)	1011001	59 (79)
'	0100111	27	Z (z)	1011010	5A (7A)
(	0101000	28			

Decimal numbers	ASCII code		Alphabetical characters	ASCII code	
	in binary	in hex		in binary	in hex
)	0101001	29	[	1011011	5B
*	0101010	2A	\	1011100	5C
+	0101011	2B	]	1011101	5D
,	0101100	2C	↑	1011110	5E
-	0101101	2D	←	1011111	5F
.	0101110	2E	a	01100001	61
/	0101111	2F	b	01100010	62
			c	01100011	63
			d	01100100	64
			e	01100101	65
			f	01100110	66
			g	01100111	67
			h	01101000	68
			i	01101001	69
			j	01101010	6A
			k	01101011	6B
			l	01101100	6C
			m	01101101	6D
			n	01101110	6E
			o	01101111	6F
			p	01110000	70
			q	01110001	71
			r	01110010	72
			s	01110011	73
			t	01110100	74
			u	01110101	75
			v	01110110	76
			w	01110111	77
			x	01111000	78
			y	01111001	79
			z	01111010	7A

Appendix B

Gray code

Decimal number	*Binary number*	*Gray code number*
0	0000	0000
1	0001	0001
2	0010	0011
3	0011	0010
4	0100	0110
5	0101	0111
6	0110	0101
7	0111	0100
8	1000	1100
9	1001	1101
10	1010	1111
11	1011	1110
12	1100	1010
13	1101	1011
14	1110	1001
15	1111	1000

Answers to exercises

Chapter 1

1.1 (a) Digital. (b) Analogue. (c) Digital. (d) Analogue. (e) Digital. (f) Analogue. (g) Digital.
1.2 (a) Unipolar (b) Positive.
1.5 (a) Analogue. (b) Digital. (c) Digital. (d) Digital. (e) Digital. (f) Analogue.
1.7 $f = 1/(240 \times 10^{-9}) = 4.167$ MHz.
1.8 (a) Duty cycle = 1/3 = 0.33. (b) Mark–space ratio = 1/2 = 0.5.
1.10 (a) 1/4800 = 208.3 μs. (b) $1/(140 \times 10^6) = 7.14$ ns. (c) $1/(565 \times 10^6) = 1.77$ ns.

Chapter 2

2.1 (a)
$$\begin{array}{r} 11101110 \\ 1010101 \\ +\ 1110011 \\ \hline 110110110 \end{array}$$
(b)
$$\begin{array}{r} 11100101 \\ 11010001 \\ +\ 10110111 \\ \hline 1001101101 \end{array}$$

2.2 (a) (i)
$$\begin{array}{r} 1001001 \\ -101110 \\ \hline 0011011 \end{array} = 27_{10}$$
(ii)
$$\begin{array}{r} 1001001 \\ +\ 010001 \\ \hline 1011010 \end{array}$$
Add the carry to get $011011 = 27_{10}$
(iii)
$$\begin{array}{r} 1001001 \\ +\ 010010 \\ \hline 011011 \end{array} = 27_{10}$$

(b) (i)
$$\begin{array}{r} 100001 \\ -\ 10001 \\ \hline 10000 \end{array} = 16_{10}$$
(ii)
$$\begin{array}{r} 100001 \\ +\ 01110 \\ \hline 101111 \end{array}$$
Add the carry to get $10000 = 16_{10}$
(iii)
$$\begin{array}{r} 100001 \\ +\ 01111 \\ \hline 10000 \end{array} = 16_{10}$$

(c) (i)
$$\begin{array}{r} 1001011 \\ -0111000 \\ \hline 0010011 \end{array}$$
(ii)
$$\begin{array}{r} 0111000 \\ +0110100 \\ \hline 1101100 \end{array}$$
(iii)
$$\begin{array}{r} 0111000 \\ +0110101 \\ \hline 1101101 \end{array}$$

(ii) No carry, so the result is negative
Complementing gives $0010011 = -19_{10}$

(iii) No carry, so the result is negative
The 1s complement is 1101100 and complementing gives $0010011 = -19_{10}$

2.3 (a) 01111111 + 11010010 = 01010001 = $+81_{10}$. (b) 01111111 + 10000000 = 11111111 = -1_{10}. (c) 00000001 + 11111101 = 11111110 = -2_{10}.

2.4 (a) 01111111 + 11010001 = 101010000. Add the carry to the LSB to get 01010001 = $+81_{10}$. (b) $+37_{10}$ = 00100101 and $+91_{10}$ = 01011011. The 1s complements are

 11011010
+10100100
101111110

Adding the carry gives 01111111, which is incorrect because of overflow.

2.5 58–33: 58 = 111010 and 33 = 100001. Ones complement = 011110 and 2s complement = 011111.

 111010
+011111
1011001 There is no carry, so the result = $+25_{10}$.

25 + 14 = 011001
+001110
100111 = 39_{10}.

39 – 45: 39 = 100111 and 45 = 101101. Ones complement = 010010 and 2s complement = 010011.

 100111
+010011
 111010 There is no carry, so the result is negative. Complementing gives 000101 and adding 1 gives 000110 = -6_{10}.

2.6 (a) (i) 212/8 = 26 remainder 4, 26/8 = 3 remainder 2, 3/8 = 0 remainder 3. Therefore, $212_{10} = 324_8$. (ii) 399/8 = 49 remainder = 7, 49/8 = 6 remainder = 1, 6/8 = 0 remainder = 6. Hence, $399_{10} = 667_8$. (iii) 42/8 = 5 remainder = 2, 5/8 = 0 remainder = 5. Therefore, $42_{10} = 52_8$. (iv) 1000/8 = 125 remainder = 0, 125/8 = 15 remainder = 5, 15/8 = 1 remainder = 7, 1/8 = 0 remainder = 1. Therefore, $1000_{10} = 1750_8$. (b) (i) 212_8 = 010001010 = 138_{10}. (ii) 377_8 = 011111111 = 255_{10}. (iii) 42_8 = 100010 = 34_{10}. (iv) 26_8 = 010110 = 22_{10}.

2.7 (a) (i) AF = $10 \times 16 + 15 = 175_{10}$. (ii) 2BC = $2 \times 16^2 + 11 \times 16 + 12 = 700_{10}$. (iii) 77 = $7 \times 16 + 7 = 119_{10}$. (iv) 1EA = $1 \times 16^2 + 14 \times 16 + 10 = 490_{10}$. (b) (i) 100/16 = 6 remainder = 4, 6/16 = 0 remainder = 6. Therefore, 100_{10} = 64H. (ii) 328/16 = 20 remainder = 8, 20/16 = 1 remainder = 4, 1/16 = 0 remainder = 1. Thus, 328_{10} = 148H. (iii) 1000/16 = 62 remainder = 8, 62/16 = 3 remainder = E, 3/16 = 0 remainder = 3. Therefore, 1000_{10} = 3E8H. (iv) 7300/16 = 456 remainder = 4, 456/16 = 28 remainder = 8, 28/16 = 1 remainder = C, 1/16 = 0 remainder = 1. Therefore, 7300_{10} = 1C84H.

2.8 (a)

(i)	(ii)	(iii)	(iv)
10110111	11010101	1111	10110110
+11100011	+ 1111101	+1101	+ 11011
110011010	101010010	11100	110110001

(b)

(i)	(ii)	(iii)	(iv)
47H	BBH	40DH	FFH
+AAH	+172H	+27H	+EAH
F1H	22DH	434H	1E9H

2.9 (a) (i) $\begin{array}{r} 101101 \\ -\quad 111 \\ \hline 100110 \end{array}$ (ii) $\begin{array}{r} 1001 \\ -\quad 11 \\ \hline 110 \end{array}$ (iii) $\begin{array}{r} 10001101 \\ -\quad 100011 \\ \hline 1101010 \end{array}$ (iv) $\begin{array}{r} 110111 \\ -\quad 101 \\ \hline 110010 \end{array}$

(b) (i) $\begin{array}{r} \text{AAH} \\ -\ \text{74H} \\ \hline \text{36H} \end{array}$ (ii) $\begin{array}{r} \text{172H} \\ -\text{BBH} \\ \hline \text{B7H} \end{array}$ (iii) $\begin{array}{r} \text{40DH} \\ -\quad \text{27H} \\ \hline \text{3E6H} \end{array}$ (iv) $\begin{array}{r} \text{FFH} \\ -\text{EAH} \\ \hline \text{15H} \end{array}$

2.10 (a) 7 = 111, 1s complement = 000, 2s complement = 001. (b) 77 = 1001101, 1s complement = 0110010, 2s complement = 0110011. (c) 126 = 11111100, 1s complement = 0000001, 2s complement = 0000010. (d) 200 = 11001000, 1s complement = 00110111, 2s complement = 00111000.

2.11 (a) 16_{10} = 10H, -16_{10} = $\begin{array}{r} \text{0H} \\ -\text{10H} \\ \hline \text{F0H} \end{array}$ (b) 66_{10} = 42H, -66_{10} = $\begin{array}{r} \text{0H} \\ -\text{66H} \\ \hline \text{9AH} \end{array}$

(c) 166_{10} = A6H, -166_{10} = $\begin{array}{r} \text{0H} \\ -\text{A6H} \\ \hline \text{5AH} \end{array}$ (d) 200_{10} = C8H, -200_{10} = $\begin{array}{r} \text{0H} \\ -\text{C8H} \\ \hline \text{38H} \end{array}$

2.12 (a) 10, 4, 10, 8. (b) 10000000000, 11011, 10110100000, 0111111111.

2.13 (a) 11011.01111, 101.010101, 100.011. (b) 45.5, 119.375, 7.3125.

2.14 (a) 010011. (b) 1.010011. (c) 0.110. (d) 0.0001001. (e) 1.01011.

2.15 (a) 100010001. (b) 11010010. (c) 100011110. (d) 101101000. (e) 11000.11. (f) 10.11111.

2.16 (a) 101. (b) 100.001101. (c) 1111.01. (d) 10.1. (e) 1001. (f) 1.000101.

2.17 (a) (i) 19. (ii) 56. (iii) 293. (iv) 31.2. (b) (i) 31. (ii) 126. (iii) 1223. (iv) 61.1.

2.18 (a) (i) ABH = 10101011, 45H = 01000101, (ii) ABH = 171_{10}, 45H = 69_{10}. (iii) ABH = 253_8, 45H = 105_8. (b) (i) DEH = 13 × 16 × 14 = 222_{10}, 5FEH = 5 × 256 + 15 × 16 + 14 = 1534_{10}, 123H = 1 × 256 + 2 × 16 + 3 = 291_{10}. (ii) DEH = 11011110, 5FEH = 010111111110, 123H = 000100100011.

2.19 (a) 10.11. (b) 11011. (c) 101.11.

2.20 (a) 100100.101. (b) 11000110. (c) 100000.011.

2.21 (a) 0000H, FF. (b) 65535. (c) (i) 2227. (ii) 8B3H.

2.22 00010111. 11101001. 00111000. 11001000. 01011010. 10100110.

2.23 (a) (i) 109. (ii) 43. (iii) 95. (iv) 96. (b) As (a).

2.24 (a) (i) 155_8. (ii) 53_8. (iii) 277_8. (iv) 300_8. (b) (i) 100010. (ii) 010100. (iii) 111111. (iv) 001101111010.

2.25 (a) (i) D9CH. (ii) C38H. (iii) C66FH. (b) (i) 011001010. (ii) 10111101. (iii) 000101000010. (iv) 101111000010.

2.26 (a) (i) 11010100 = 324_8. (ii) 11000111 = 617_8. (iii) 101010 = 52_8. (iv) 1111101000 = 1750_8. (b) (i) $218_8 = 2 \times 64 + 1 \times 8 + 2 = 428_{10}$. (ii) $366_8 = 3 \times 64 + 6 \times 8 + 6 = 246_{10}$. (iii) $42_8 = 4 \times 8 + 2 = 34_{10}$. (iv) $26_8 = 2 \times 8 + 6 = 22_{10}$.

2.27 (a) 208. (b) 150. (c) 4011. (d) 169.

2.28 (a) 1011.101 = 8 + 2 + 1 + 0.5 + 0.125 = 11.625. (b) 33.3 = 32 + 1 + 0.25 + 0.3125 + 0.015625 = 100001.010011. (c) 25.4H = 00100101.01.

2.29 (a) 1101010.10001 (b) 110.010011. (c) 1010.1101. (d) 110010.1010011.

Chapter 3

3.1 The truth table is

A	0	1	0	1	0	1	0	1
B	0	0	1	1	0	0	1	1
C	0	0	0	0	1	1	1	1
$A \oplus B$	0	1	1	0	0	1	1	0
$A \oplus B \oplus C$	0	1	1	0	1	0	0	1

3.2 The truth table is

A	0	1	0	1	0	1	0	1
B	0	0	1	1	0	0	1	1
C	0	0	0	0	1	1	1	1
$D = \overline{A + \mathrm{B}}$	1	0	0	0	0	0	0	0
$E = \overline{B + C}$	1	1	0	0	0	0	0	0
$F = \overline{DE}$	0	1	1	1	1	1	1	1

The Boolean equation is $F = A\bar{B}\bar{C} + \bar{A}B\bar{C} + AB\bar{C} + \bar{A}\bar{B}C + A\bar{B}C + \bar{A}BC + ABC$. [Note: this equation can be simplified to $F = A + B + C$ (see chapters 4 and 5).]

3.3 The truth table is

A	0	1	0	1	0	1	0	1
B	0	0	1	1	0	0	1	1
C	0	0	0	0	1	1	1	1
$A + C$	0	1	0	1	1	1	1	1
$\overline{B + C}$	1	1	0	0	0	0	0	0
$A + \overline{B + C}$	1	1	0	1	0	1	0	1
F	0	1	0	1	0	1	0	1

The Boolean equation is $F = A\bar{B}\bar{C} + AB\bar{C} + A\bar{B}C + ABC$. [This can be reduced to $F = A$.]

3.4. (a)

A	0	1	0	1	0	1	0	1	0	1	0	1	0	1	0	1
B	0	0	1	1	0	0	1	1	0	0	1	1	0	0	1	1
C	0	0	0	0	0	0	0	0	1	1	1	1	1	1	1	1
F	0	0	0	0	0	0	0	0	0	0	0	0	0	0	0	1

(b)

A	0	1	0	1	0	1	0	1	0	1	0	1	0	1	0	1
B	0	0	1	1	0	0	1	1	0	0	1	1	0	0	1	1
C	0	0	0	0	0	0	0	0	1	1	1	1	1	1	1	1
F	0	1	1	1	1	1	1	1	1	1	1	1	1	1	1	1

3.5 $F = A\bar{B}\bar{C} + AB\bar{C}$, $G = \bar{A}B\bar{C} + \bar{A}BC + ABC$, $H = \bar{A}B\bar{C} + ABC$.

3.6 (a) When room temperature > set temperature. (b) When pilot is lit. (c) AND.

(d)

Temperature	0	1	0	1
Control	0	0	1	1
Valve	0	0	0	1

3.7 (a) A possible circuit is shown in Fig. 3.50. (b) HIGH.

Fig. 3.50

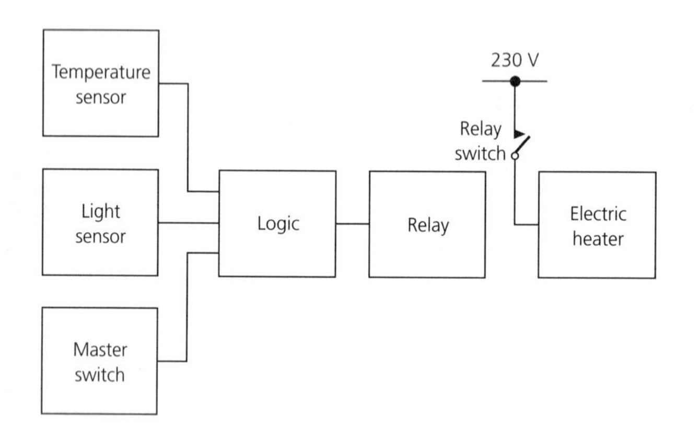

3.8 (a) $\overline{\bar{A} + \bar{B}}$

A	0	1	0	1	
B	0	0	1	1	
$\bar{A}$	1	0	1	0	
$\bar{B}$	1	1	0	0	
F	0	0	0	1	$= AB$

(b)

A	0	1	0	1	
B	0	0	1	1	
$\bar{A}$	1	0	1	0	
$\bar{B}$	1	1	0	0	
F	0	1	1	1	$= A + B$

3.9

A	0	1	0	1	0	1	0	1	
B	0	0	1	1	0	0	1	1	
C	0	0	0	0	1	1	1	1	
$D = \overline{A + B}$	1	0	0	0	1	0	0	0	
$E = \overline{A + C}$	1	0	1	0	0	0	0	0	
$F = \overline{D + E}$	0	1	0	1	0	1	1	1	$F = A + BC$

3.10

A	0	1	0	1	0	1	0	1
B	0	0	1	1	0	0	1	1
C	0	0	0	0	1	1	1	1
$D = \overline{AB}$	1	1	1	0	1	1	1	0
$E = \overline{AC}$	1	1	1	1	1	0	1	0
$F = \overline{DE}$	0	0	0	1	0	1	0	1

$F = AC + AB$

3.11 (a)

A	0	1	0	1	0	1	0	1
B	0	0	1	1	0	0	1	1
C	0	0	0	0	1	1	1	1
F	1	0	1	0	1	0	1	0

(b) $F = \bar{A}\bar{B}\bar{C} + \bar{A}B\bar{C} + \bar{A}\bar{B}C + \bar{A}BC$. [Note that the circuit only requires $A = 0$ for the output to be HIGH. Thus, the equation can be simplified to $F = \bar{A}$.]

3.12 (a)

A	0	1	0	1	0	1	0	1
B	0	0	1	1	0	0	1	1
C	0	0	0	0	1	1	1	1
F	1	0	0	1	0	1	0	0

(b) $F = \bar{A}\bar{B}\bar{C} + AB\bar{C} + A\bar{B}C$.

3.13 The truth table is

A	B	F
0	0	1
1	0	1
0	1	1
1	1	0

From the table, when either input is at 1 (HIGH) the logical state of the output F is the complement of the logical state of the other input.

3.14 (a) OR. (b) AND. (c) exclusive-OR.

3.15 $F = AB + C$.

3.16 $\mathrm{H} = \overline{\mathrm{T}}_1 + \mathrm{T}_2$, $\mathrm{A} = \overline{\mathrm{T}}_2$.

3.17 $P = \bar{P}_1\bar{P}_2\bar{P}_3$.

3.19 (a) $AB + C$.

3.20 $F = \overline{(\bar{A} + B)(\overline{B + C})}$.

3.21 (a) Hold one input at 1 and apply signal to other input. (b) Hold one input LOW.

Chapter 4

4.1 (a) $F = \overline{\bar{A} + \bar{B}} = \bar{\bar{A}}\bar{\bar{B}} = AB$. (b) $F = \overline{\bar{A}\bar{B}} = \bar{\bar{A}} + \bar{\bar{B}} = A + B$. (c) $F = \bar{A}\bar{B} = \overline{A + B}$.

4.2 $F = \overline{(\overline{AB})C + D} = (\overline{\overline{AB}.C}).\bar{D} = (AB + \bar{C})\bar{D} = AB\bar{D} + \bar{C}\bar{D}$.

4.3 Output of first NAND gate = $\overline{AB}$. Output of second NAND gate = $\overline{BC}$. Output of NOR gate = $\overline{\overline{AB} + \overline{BC}} = (AB)(BC) = ABC$.

4.4 (a) $F = (AB)(B + C) = AB + ABC = AB$. (b) $F = \overline{ABC} + B\bar{C} + \overline{\bar{A}C} = \bar{A} + \bar{B} + \bar{C} + B\bar{C} + A + \bar{C} = 1 + \bar{B} + \bar{C} = 1$. (c) $F = \overline{(\bar{A}C + B\bar{C})(A + \bar{B} + D)} = \overline{(\bar{A}C + B\bar{C})} + \overline{(A + \bar{B} + D)} = (\overline{\bar{A}C})(\overline{B\bar{C}}) + \bar{A}B\bar{D} = (A + \bar{C})(\bar{B} + C) + \bar{A}BD = A\bar{B} + AC + \bar{B}\bar{C} + \bar{A}BD$.

4.5 (a) Two AND and one OR gates. $F = A(B + C)$ needs one AND and one OR gate. (b) (i) $ABC + \overline{ABC}$. (ii) $AB + ABC = AB(1 + C) = AB$. (c) $\overline{\overline{ABC}} + A = ABC + A = A(1 + BC) = A$. (d) $BC + \bar{C}D + BD = BC + \bar{C}D$ (rule 14).

4.6 (a) $F = (\overline{AC} + BC)(\overline{A + B + D}) = (\bar{A} + \bar{C} + BC)(\bar{A}\bar{B}\bar{D}) = \bar{A}\bar{B}\bar{D} + \bar{A}\bar{B}\bar{C}\bar{D} + \bar{A}\bar{B}\bar{D} = \bar{A}\bar{B}\bar{D}$. (b) $F = \overline{\overline{\bar{A}B\cdot C}} + \overline{\overline{\bar{B}C}} = \bar{A}BC + \bar{B}C = (\bar{A} + \bar{B})C + \bar{B}C = \bar{A}C + \bar{B}C + \bar{B}C = \bar{A}C + \bar{B}C$.

4.7 (a) $F = \overline{A + \bar{B} + \bar{C} + \bar{D}} + \bar{A}BC\bar{D} = \bar{A}BCD + \bar{A}BC\bar{D} = \bar{A}BC$. (b) $F = (A + B)(\bar{A} + B)(A + \bar{B}) = (AB + \bar{A}B + B)(A + \bar{B}) = B(A + \bar{B}) = AB$. (c) $F = (A + \bar{B} + \bar{C})(\bar{A} + \bar{B} + \bar{C}) = A\bar{B} + A\bar{C} + \bar{A}\bar{B} + \bar{B} + \bar{B}C + \bar{A}\bar{C} + \bar{B}\bar{C} + \bar{C} = \bar{B} + \bar{C}$. (d) $F = \overline{\bar{A}B(CD + \bar{A}\bar{C})} = \overline{\bar{A}B} + \overline{CD + \bar{A}\bar{C}} = A + \bar{B} + \overline{CD}(\overline{\bar{A}\bar{C}}) = A + \bar{B} + (\bar{C} + \bar{D})(A + C) = A + \bar{B} + A\bar{C} + A\bar{D} + C\bar{D} = A + \bar{B} + C\bar{D}$.

4.8 The input combinations are: 010, 101 and 110. Hence, the truth table is

A	0	1	0	1	0	1	0	1
B	0	0	1	1	0	0	1	1
C	0	0	0	0	1	1	1	1
F	1	1	0	0	1	0	1	1

4.9 Example (4.17(a)) is $F = \bar{A}\bar{B} + BC$. $\bar{F} = \overline{\bar{A}\bar{B} + BC} = (\overline{\bar{A}\bar{B}})(\overline{BC}) = (A + B)(\bar{B} + \bar{C}) = A\bar{B} + A\bar{C} + B\bar{C}$. Example (4.17(b)) is $(\bar{A} + B + C)(A + \bar{B} + C)(\bar{A} + \bar{B} + C)(\bar{A} + B + \bar{C}) = (\bar{A}\bar{B} + \bar{A}C + AB + BC + AC + \bar{B}C + C)(\bar{A} + \bar{B} + C)(\bar{A} + B + \bar{C}) = (C + AB + \bar{A}\bar{B})(\bar{A} + \bar{B} + C)(\bar{A} + B + \bar{C}) = (\bar{A}C + \bar{A}\bar{B} + \bar{B}C + \bar{A}\bar{B} + C + \bar{A}\bar{B}C + ABC)(\bar{A} + B + \bar{C}) = (C + \bar{A}\bar{B})(\bar{A} + B + \bar{C}) = \bar{A}C + BC + \bar{A}\bar{B} + \bar{A}\bar{B}\bar{C} = \bar{A}C + BC + \bar{A}\bar{B}$.

4.10 $F = (A + B)(A + \overline{B + C})(\overline{B + D}) = (A + B)(A + \bar{B}\bar{C})(\bar{B}\bar{D}) = (A + A\bar{B}\bar{C} + AB)\bar{B}\bar{D} = A\bar{B}\bar{D}$.

The simplified circuit is shown in Fig. 4.10. Two ICs are needed, one quad 2-input NAND and one triple 3-input AND gates.

4.11 $F = (A + \bar{B} + C)(\bar{A} + \bar{B} + C + D)(\bar{A} + B + \bar{C} + \bar{D})(A + C) = (C + \bar{B} + AD)(\bar{A} + B + \bar{C} + \bar{D})(A + C) = (A\bar{C} + \bar{A} + \bar{B}C + ABC + ABD + A\bar{B}\bar{D} + AC\bar{D} + \bar{A}C + BC + C\bar{D} + A\bar{B}\bar{C} + \bar{B}C\bar{D} + ABCD) = BC + C\bar{D} + \bar{A}C + A\bar{B}\bar{C} + A\bar{B}\bar{D} + ABD + A\bar{C}D = BC + C\bar{D} + \bar{A}C + A\bar{C}D + A\bar{B}\bar{C}$.

4.12 $F = \bar{A}C + BC + \bar{A}\bar{B}(1 + \bar{C}) = \bar{A}C + BC + \bar{A}\bar{B} = \bar{A}\bar{B} + BC$.

Fig. 4.10

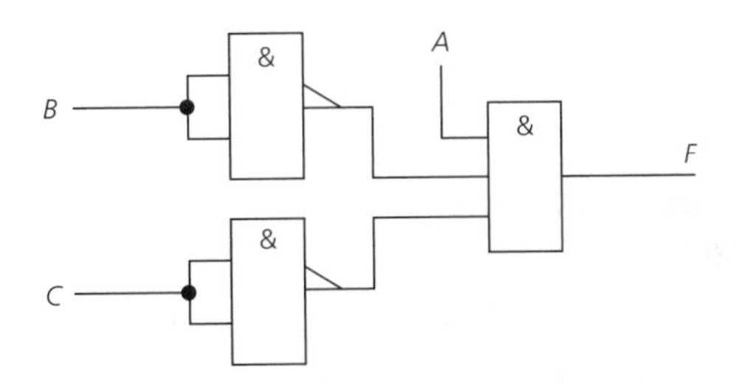

4.13 Output of gate 1 = $\overline{B + C}$. Output of gate 5 = $\overline{A + B}$. Output of gate 6 = $\overline{A + \overline{B + C}} = \bar{A}B + \bar{A}C$. Output of gate 7 = $\overline{\bar{A} + \bar{C}} = AC$. Output of gate 8 = $\overline{\bar{A} + \bar{B}} = AB$. Output of gate 9 = $\overline{\overline{A + \overline{B + C}} + \overline{A + B}} = (A + \overline{B + C})(A + B) = (A + \bar{B}\bar{C})(A + B) = A$. Output of gate 10 = $\bar{A}$.

4.14 (a) $F = AD(B + C)$. (b) $F = (A + B)(C + D)$.

4.15 Output $F = 0$, hence $G = 0$ and output of 3-input NAND gate = 0. Thus, the inputs to the NAND gate are all 1. Therefore, $A = B = 1$ and $E = F = 1$. The output of the NOR gate is 1, so both its inputs are at 0. Hence, $C = D = 0$.

4.16 $F = \overline{(\overline{AB})(\overline{C + D})(\overline{EF})} + G = \overline{AB} + \overline{\overline{C + D}} + \overline{EF} + G = \bar{A} + \bar{B} + C + D + \bar{E} + \bar{F} + G$. An 8-input AND gate and a hex inverter are needed.

4.17 Top half of circuit: $G = (AB)(CD)$. Bottom half of circuit: $H = (\bar{A}\bar{B})(EF)$. Therefore, output = $(ABCD)(\bar{A}\bar{B}EF) = 0$, so no circuit is required!

4.18 $F = \bar{A}BC\bar{D} + \bar{A}BCD + A\bar{B}\bar{C}D + A\bar{B}CD + AB\bar{C}D + ABCD = \bar{A}BC + A\bar{B}D + ABD = \bar{A}BC + AD$.

4.19 (a) $F = (0 + 0)(0 + 0) = 0$. (b) $F = (1 + 1)(1 + 1) = 1$. (c) $F = (1 + 0)(0 + 0) = 0$.

4.20 (a) $F = \overline{A + B + \overline{A + B}} = \bar{A}\bar{B}(A + B) = 0$. (b) $F = \overline{\bar{A}\bar{B} + \overline{A + B}} = (\overline{\bar{A}\bar{B}})(A + B) = (A + B)(A + B) = A + B$. (c) $F = (AB + \bar{C})(A + BC) = AB + ABC + A\bar{C} = AB + A\bar{C}$. (d) $F = (\overline{AC + D})(\overline{AC} + B) = (\overline{ACD})(\bar{A} + B + \bar{C}) = (\bar{A} + \bar{C})\bar{D}(\bar{A} + B + \bar{C}) = (A + \bar{C})(\bar{A}\bar{D} + \bar{C}\bar{D} + B\bar{D}) = \bar{A}\bar{D} + \bar{A}\bar{C}\bar{D} + \bar{A}B\bar{D} + \bar{A}C\bar{D} + BC\bar{D} = C\bar{D}(\bar{A} + B)$.

4.21 (a) $F = (AB + \bar{C}) + (\bar{A} + \bar{B})C = AB + \bar{C} + \bar{A}C + \bar{B}C = AB + \bar{C} + C = 1 + AB = 1$. (b) $F = B(AD + \overline{BC})(\bar{A} + \bar{D})\bar{C} = B(\overline{AD}BC)(\bar{A}C + \bar{C}\bar{D}) = (\bar{A} + \bar{D})BC(\bar{A}C + \bar{C}\bar{D}) = (\bar{A}BC + BC\bar{D})(\bar{A}C + \bar{C}\bar{D}) = \bar{A}BC + \bar{A}BCD = \bar{A}BC$.

4.22 (a) $F = AB\bar{C} + A\bar{B}C + \bar{A}\bar{B}\bar{C}$ $\bar{F} = \overline{AB\bar{C} + A\bar{B}C + \bar{A}\bar{B}\bar{C}} = (\overline{AB\bar{C}})(\overline{A\bar{B}C})(\overline{\bar{A}\bar{B}\bar{C}}) = (\bar{A} + \bar{B} + C)(\bar{A} + B + \bar{C})(A + B + C) = \bar{A}B + \bar{A}C + A\bar{B}\bar{C} + ABC + BC + BC = BC + \bar{A}B + \bar{A}C + A\bar{B}\bar{C}$ $F = \overline{BC + \bar{A}B + \bar{A}C + A\bar{B}\bar{C}} = (\overline{BC})(\overline{\bar{A}B})(\overline{\bar{A}C})(\overline{A\bar{B}\bar{C}}) = (\bar{B} + \bar{C})(A + \bar{B})(A + \bar{C})(\bar{A} + B + C)$. (b) $F = (A + B + \bar{C})(A + \bar{B} + C)(\bar{A} + \bar{B} + \bar{C}) = (\bar{A} + BC + \bar{B}\bar{C})(\bar{A} + \bar{B} + \bar{C}) = A\bar{B} + A\bar{C} + \bar{A}BC + \bar{A}\bar{B}\bar{C} + \bar{B}\bar{C} + \bar{B}\bar{C} = A\bar{B} + A\bar{C} + \bar{B}\bar{C} + \bar{A}BC$.

4.23 (a) $F = (B + D)(D + C)(A + D) = (BD + BC + D + DC)(A + D) = (D + BC)(A + D) = AD + D + ABC + BCD = D + ABC$. (b) $F = (B + D)(A + D)(B + C)(A + C) = (AB + D)(B + C)(A + C) = (ABC + AB + CD + BD)(A + C) = (AB + BD + CD)(A + C) = ABC + BCD + CD + AB + ABD + ACD = AB + CD$

4.24 $F = (\overline{\bar{A}\bar{B} + AB})(\overline{\bar{A} + \bar{B}})(A + B) = (\bar{\bar{A}}\bar{\bar{B}})(\overline{AB})(\bar{\bar{A}} + \bar{\bar{B}} + \overline{A + B}) = (A + B)(\bar{A} + \bar{B})(AB + \bar{A}\bar{B}) = 0$.

4.25 $F = \overline{\overline{AB(B + C)} + \overline{AB}(1 + C)} = AB(B + C)(\overline{\overline{AB}(1 + C)}) = (AB + ABC)(AB + \overline{1 + C}) = AB(AB + 0\bar{C}) = AB$.

4.26 $X = \bar{A}\bar{B}\bar{C} + \bar{A}BC$. $Y = A\bar{B}\bar{C} + \bar{A}B\bar{C} + \bar{A}\bar{B}C + A\bar{B}C = A\bar{B} + \bar{A}B\bar{C} + \bar{A}\bar{B}C$. $Z = AB\bar{C} + ABC = AB$.

4.27 $\bar{A}\bar{B}\bar{C} + \bar{A}\bar{B}C + \bar{A}B\bar{C} + A\bar{B}\bar{C} + AB\bar{C} = \bar{A}\bar{B} + \bar{C}$.

4.28 $F = \bar{A}\bar{B}C + A\bar{B}C + ABC + \bar{A}BC = AC(B + \bar{B}) + \bar{A}C(B + \bar{B}) = AC + \bar{A}C = C$.

Chapter 5

5.1 $F = A\bar{B}\bar{C}\bar{D} + AB\bar{C}\bar{D} + A\bar{B}C\bar{D} + ABC\bar{D}$. The mapping is

CD \ AB	00	01	11	10
00	0	0	1	1
01	0	0	0	0
11	0	0	0	0
10	0	0	1	1

From the map, $F = A\bar{D}$.

5.2 (a) The truth table is

Table 5.7

A	0	1	0	1
B	0	0	1	1
	R	R/Y	G	Y

Hence, $R = \bar{A}\bar{B}$, $R/Y = A\bar{B}$, $G = \bar{A}B$ and $Y = AB$. Simplifying, $R = \bar{A}\bar{B} + A\bar{B} = \bar{B}$, $Y = A\bar{B} + AB = A$.

(b) The truth table is

Table 5.8

A	0	1	0	1	0	1	0	1
B	0	0	1	1	0	0	1	1
C	0	0	0	0	1	1	1	1
	R	R	R	R/Y	G	G	G	Y

From the table, $R = \bar{A}\bar{B}\bar{C} + A\bar{B}\bar{C} + A\bar{B}C + AB\bar{C} = B\bar{C} + \bar{B}\bar{C} = \bar{C}$, $Y = AB\bar{C} + ABC = AB$, $G = \bar{A}\bar{B}C + A\bar{B}C + A\bar{B}\bar{C} = A\bar{B}C + \bar{A}C = C(\bar{A} + AB) = \bar{A}C + BC$.
(c) The truth table is

Table 5.9

A	0	1	0	1	0	1	0	1	0	1
B	0	0	1	1	0	0	1	1	0	0
C	0	0	0	0	1	1	1	1	0	0
D	0	0	0	0	0	0	0	0	1	1
	R	*R*	*R*	*R/Y*	*G*	*G*	*G*	*G*	*G*	*Y*

$R = \bar{A}\bar{B}\bar{C}\bar{D} + AB\bar{C}\bar{D} + A\bar{B}\bar{C}\bar{D} + \bar{A}B\bar{C}\bar{D} = \bar{C}\bar{D}$, $Y = AB\bar{C}\bar{D} + A\bar{B}\bar{C}D$, $G = \bar{A}\bar{B}C\bar{D} + A\bar{B}C\bar{D} + \bar{A}BC\bar{D} + ABC\bar{D} + \bar{A}\bar{B}\bar{C}D = C\bar{D} + \bar{A}\bar{B}\bar{C}D$.

5.3 The truth table is

Table 5.10

A	0	1	0	1	0	1	0	1	0	1	0	1	0	1	0	1
B	0	0	1	1	0	0	1	1	0	0	1	1	0	0	1	1
C	0	0	0	0	1	1	1	1	0	0	0	0	1	1	1	1
D	0	0	0	0	0	0	0	0	1	1	1	1	1	1	1	1
F	0	0	0	0	0	0	0	1	0	1	1	1	1	1	1	1

Hence, $F = ABC\bar{D} + A\bar{B}\bar{C}D + \bar{A}B\bar{C}D + AB\bar{C}D + \bar{A}\bar{B}CD + A\bar{B}CD + \bar{A}BCD + ABCD = AD + BD + CD + ABC$.

5.4 $0 = \bar{A}\bar{B}\bar{C}\bar{D}$, $1 = A\bar{B}\bar{C}\bar{D}$, $2 = \bar{A}B\bar{C}\bar{D}$ etc. through to $9 = A\bar{B}\bar{C}D$.

5.5 $A = 1 + 3 + 5 + 7 + 9$, $B = 2 + 3 + 6 + 7$, $C = 4 + 5 + 6 + 7$, and $D = 8 + 9$.

5.6 The truth table is

Table 5.11

A	0	1	0	1	0	1	0	1	0	1	0	1	0	1	0	1
B	0	0	1	1	0	0	1	1	0	0	1	1	0	0	1	1
C	0	0	0	0	1	1	1	1	0	0	0	0	1	1	1	1
D	0	0	0	0	0	0	0	0	1	1	1	1	1	1	1	1
F	0	1	0	1	0	1	0	1	0	1	0	1	0	1	0	1

$F = A\bar{B}\bar{C}\bar{D} + AB\bar{C}\bar{D} + A\bar{B}C\bar{D} + ABC\bar{D} + A\bar{B}\bar{C}D + AB\bar{C}D + A\bar{B}CD + ABCD = A$.

5.7 From the same truth table as 5.3, $F = ABC\bar{D} + A\bar{B}\bar{C}D + AB\bar{C}D = ABD + A\bar{C}D$.

5.8 $B = ID + (H + S)D = D(I + H + S)$.

5.9

CD \ AB	00	01	11	10
00	0	0	0	0
01	0	1	0	0
11	0	0	1	1
10	0	0	1	0

Looping the 0 cells, $F' = \bar{C}\bar{D} + A\bar{C} + \bar{A}C + \bar{A}\bar{B} + \bar{B}\bar{D}$. Changing the signs and complementing, $F = (C + D)(A + \bar{C})(\bar{A} + C)(B + D)$.

5.10 (a) Mapping, $F = ABC + A\bar{B}CD + \bar{A}BCD + ABC\bar{D} + ABD + \bar{B}C\bar{D} + \bar{A}B\bar{C}D$ gives

CD \ AB	00	01	11	10
00	0	0	0	0
01	0	1	1	0
11	0	1	1	1
10	1	0	1	1

Looping the 0 cells, $\bar{F} = \bar{C}\bar{D} + \bar{A}\bar{B}D + \bar{A}B\bar{D} + \bar{B}\bar{C}$.

(b) Mapping, $F = ABC + A\bar{B}CD + \bar{A}B\bar{C}D + ACD$

CD \ AB	00	01	11	10
00	0	0	0	0
01	0	1	0	0
11	0	0	1	1
10	0	0	1	0

$\bar{F} = \bar{A}C + \bar{B}\bar{D} + \bar{A}\bar{B} + \bar{C}\bar{D} + A\bar{C}$.

5.11 (a) The map is

CD \ AB	00	01	11	10
00	0	0	0	0
01	1	1	1	1
11	1	1	1	1
10	1	1	1	1

Looping the 1 cells, $F = C + D$.
(b) Looping the 0 cells, $\bar{F} = \bar{C}\bar{D}$. Complementing gives $F = \overline{\bar{C}\bar{D}} = C + D$.

5.12 The mapping is

CD \ AB	00	01	11	10
00	1	1	0	0
01	0	0	0	0
11	0	0	1	1
10	1	1	1	1

Looping the 0 cells, $F' = A\bar{C} + \bar{C}D + \bar{A}D$, so $F = (\bar{A} + C)(C + \bar{D})(A + \bar{D})$.

5.13 The map is

CD \ AB	00	01	11	10
00	0	0	1	1
01	1	0	1	1
11	1	1	0	0
10	0	1	1	1

From the 0 cells, $F' = \bar{A}\bar{B}\bar{D} + ACD + \bar{A}BC$. Looping the 1 cells, $F = A\bar{C} + \bar{A}\bar{B}D + \bar{A}BC + A\bar{D}$.

5.14 (a) $\bar{F} = \overline{A\bar{B}C + \bar{A}BD} = (\overline{A\bar{B}C})(\overline{\bar{A}BD}) = (\bar{A} + B + \bar{C})(A + \bar{B} + \bar{D}) = \bar{A}\bar{B} + \bar{A}\bar{D} + AB + B\bar{D} + A\bar{C} + \bar{B}\bar{C} + \bar{C}\bar{D} = \bar{A}\bar{B} + AB + \bar{B}\bar{C} + \bar{C}\bar{D} + B\bar{D}$.
(b) Mapping and looping the 0 cells

CD \ AB	00	01	11	10
00	0	0	0	0
01	0	1	0	0
11	0	1	0	1
10	0	0	0	1

$\bar{F} = \bar{C}\bar{D} + \bar{A}\bar{B} + AB + \bar{B}\bar{C} + B\bar{D}$

5.15 The map is

C \ AB	00	01	11	10
0	1	0	0	1
1	1	0	1	1

$F = \bar{B} + AC$, $\bar{F} = B\bar{C} + \bar{A}B$, $F = \overline{\overline{B\bar{C} + \bar{A}B}} = \overline{B\bar{C}}\,\overline{\bar{A}B} = (\bar{B} + C)(A + \bar{B}) = A\bar{B} + \bar{B} + AC + \bar{B}C = \bar{B} + AC$.

5.16 From the map

C \ AB	00	01	11	10
0	1	1	1	0
1	0	0	1	0

$F = \bar{A}\bar{C} + AB$, $\bar{F} = \bar{A}C + A\bar{B}$, $F = \overline{\overline{\bar{A}C + A\bar{B}}} = \overline{\bar{A}C}\,\overline{A\bar{B}} = (A + \bar{C})(\bar{A} + B) = AB + B\bar{C} + \bar{A}\bar{C}$ $= \bar{A}\bar{C} + AB$.

5.17 The map is

CD \ AB	00	01	11	10
00	1	0	0	1
01	1	0	0	0
11	0	0	1	1
10	0	0	0	1

$F = \bar{A}\bar{B}\bar{C} + A\bar{B}C + BCD$.

5.18

C \ AB	00	01	11	10
0	1	0	0	0
1	0	0	1	0

$F = \bar{A}\bar{B}\bar{C} + ABC$.

5.19

CD \ AB	00	01	11	10
00	0	0	0	1
01	1	1	1	1
11	1	1	1	1
10	1	1	1	1

$F = C + D + A\bar{B}$.

5.20

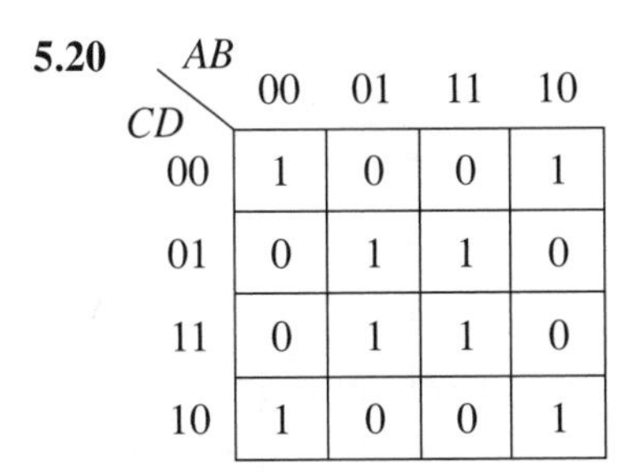

CD \ AB	00	01	11	10
00	1	0	0	1
01	0	1	1	0
11	0	1	1	0
10	1	0	0	1

$F = \bar{B}\bar{D} + BD.$

Chapter 6

6.1 (a) OR. (b) AND. (c) $\overline{ABC}$.

6.2 Double inverting, $F = A + \overline{\overline{\bar{B}\bar{C}}} = A + \overline{B + C}$. The circuit is shown in Fig. 6.18.

6.3 Outputs: gate 1, $F_1 = \overline{AB}$; gate 2, $F_2 = \overline{A\overline{AB}}$; gate 3, $F_3 = \overline{B\overline{AB}}$; gate 4, $F_4 = \overline{(\overline{A\overline{AB}})(\overline{B\overline{AB}})} = (A\overline{AB}) + (B\overline{AB}) = A(\bar{A} + \bar{B}) + B(\bar{A} + \bar{B}) = A\bar{B} + \bar{A}B =$ exclusive OR.

6.4 (a) Double inverting: $\bar{\bar{F}} = \overline{A + BD + CD} = \overline{\bar{A}\overline{BD}\,\overline{CD}}$. The NAND circuit is shown in Fig. 6.19(a). (b) Converting to POS form: $\bar{F} = \bar{A}\overline{BD}\,\overline{CD} = \bar{A}(\bar{B} + \bar{D})(\bar{C} + \bar{D}) = \bar{A}\bar{B}\bar{C} + \bar{A}\bar{D}$ $F = \overline{\bar{A}\bar{B}\bar{C} + \bar{A}\bar{D}} = (\overline{\bar{A}\bar{B}\bar{C}})(\overline{\bar{A}\bar{D}}) = (A + B + C)(A + D)$. Double inverting, $F = \overline{\overline{(A + B + C)(A + D)}} = \overline{(\overline{A + B + C}) + (\overline{A + D})}$. The NOR gate circuit is shown in Fig. 6.19(b).

Fig. 6.18

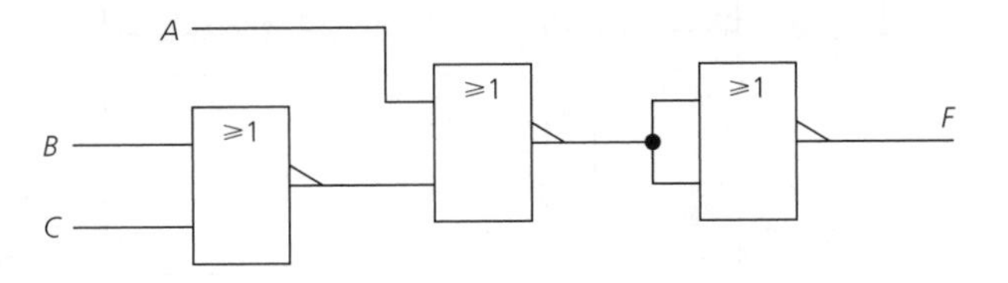

Fig. 6.19

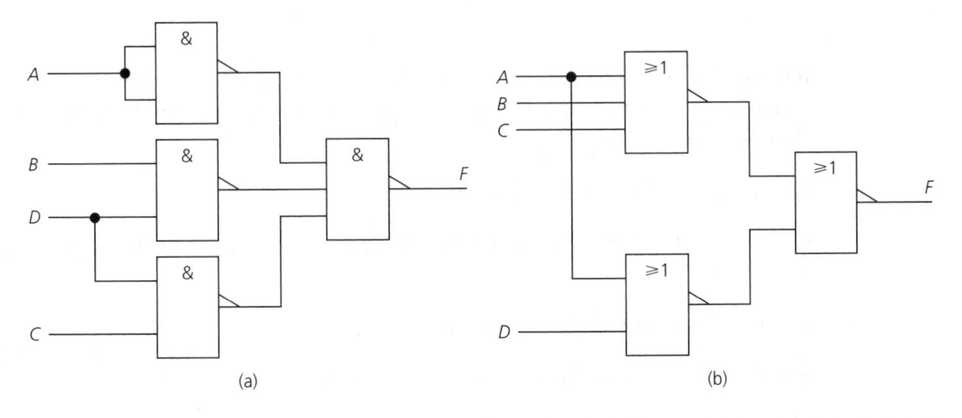

Fig. 6.20

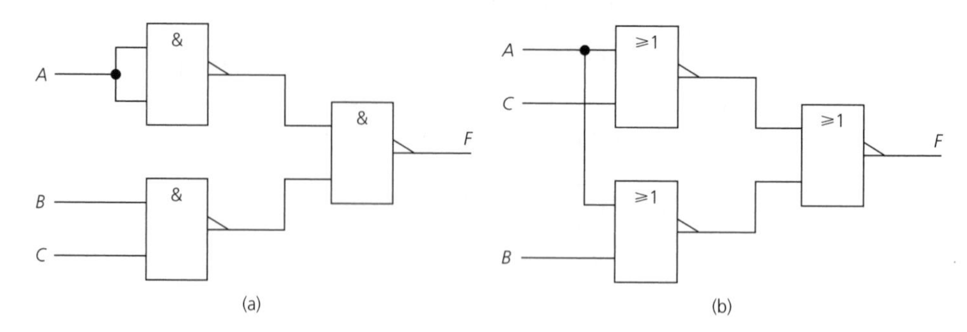

Fig. 6.21

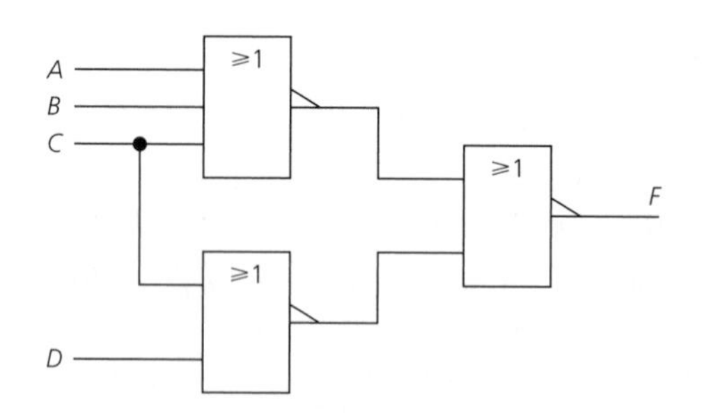

Fig. 6.22

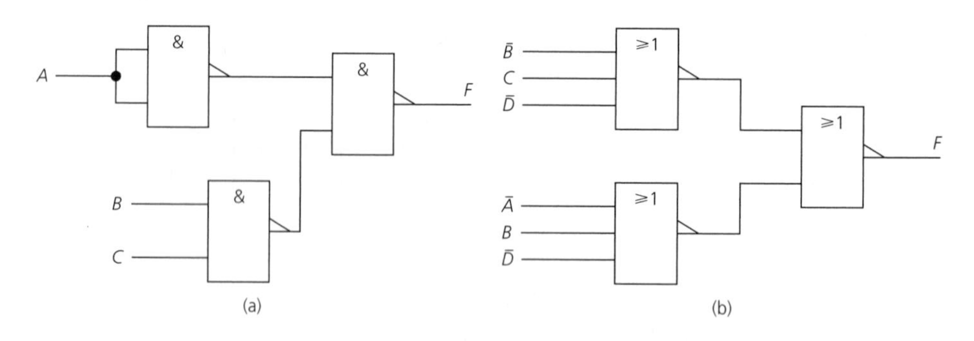

6.5 (a) $F = A + BC$. Using the OR/AND rule the circuit is shown in Fig. 6.20(a). (b) Converting to POS form: $F = (A + C)(A + B)$. Using the AND/OR rule the circuit is shown in Fig. 6.20(b).

6.6 See Fig. 6.21. (b) $F = \bar{A}\bar{B}\bar{C}\bar{D}$.

6.7 (a) $F = A + ABC + BC = A + BC$. See Fig. 6.22(a). (b) $F = (\bar{B} + C + \bar{D})(\bar{A} + B + \bar{D})$. See Fig. 6.22(b).

6.8 $F = A\bar{B} + \bar{A}B\bar{C} + A\bar{B}C$. See Fig. 6.23.

6.9 $F = (A + \bar{C})(A + \bar{B})(\bar{A} + C)(\bar{A} + B)$. See Fig. 6.24.

Fig. 6.23

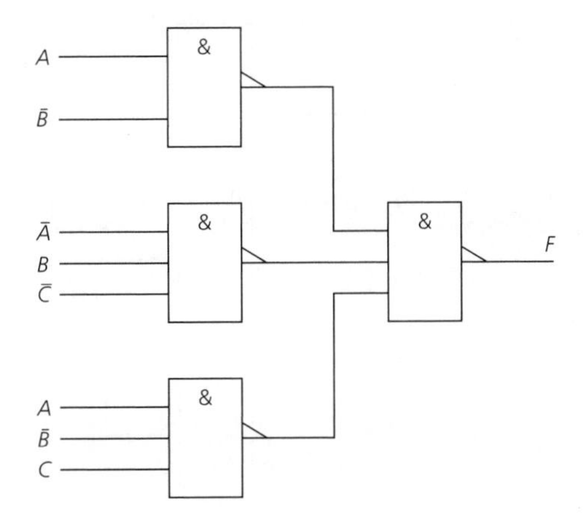

Fig. 6.24

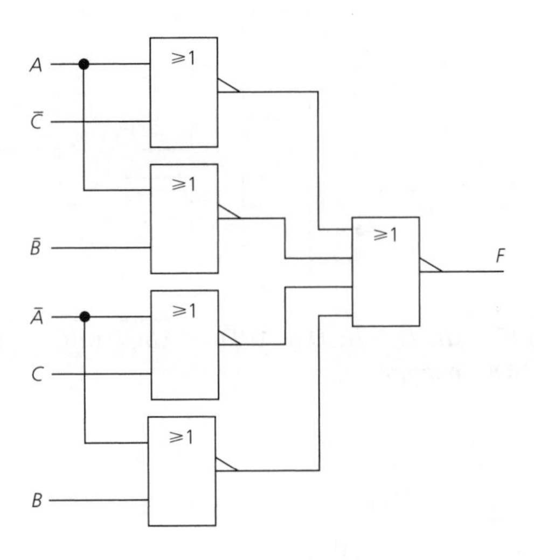

6.10 The mapping is

CD \ AB	00	01	11	10
00	1	1	0	1
01	1	1	0	1
11	0	0	0	0
10	1	0	0	1

From the 0 cells, $F' = AB + BC + CD$ and, hence, $F = (\bar{A} + \bar{B})(\bar{B} + \bar{C})(\bar{C} + \bar{D})$. See Fig. 6.25.

Fig. 6.25

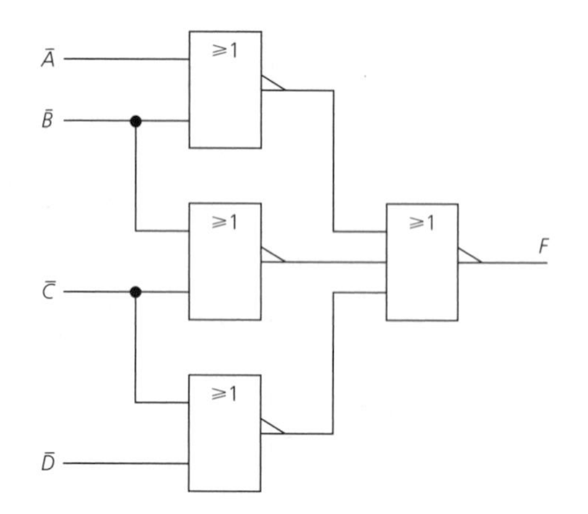

Fig. 6.26

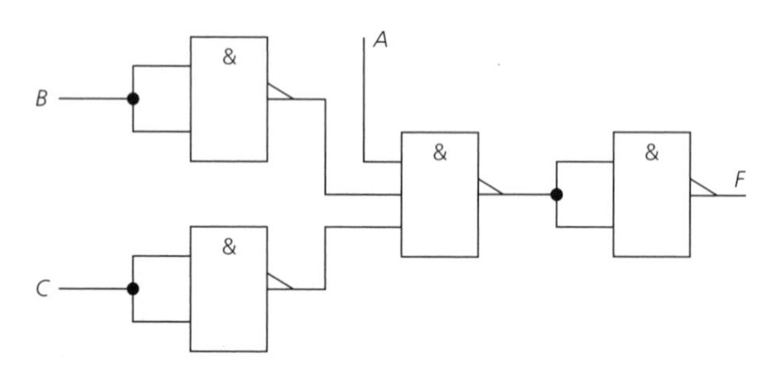

6.11 (a) $F = ABC\bar{D} + B\bar{C}D + \bar{A}\bar{B}\bar{D} = \overline{\overline{(ABC\bar{D})(B\bar{C}D)(\bar{A}\bar{B}\bar{D})}} = \overline{(\overline{ABC\bar{D}})(\overline{B\bar{C}D})(\overline{\bar{A}\bar{B}\bar{D}})}$.
(b) The SOP mapping is:

CD \ AB	00	01	11	10
00	1	0	0	0
01	0	1	1	0
11	0	0	0	0
10	1	0	1	0

From the 0 cells, $F' = A\bar{B} + CD + B\bar{C}\bar{D} + \bar{A}B\bar{D}$, so $F = (\bar{A} + B)(\bar{C} + \bar{D})(\bar{B} + C + D)(A + \bar{B} + D)$.

6.12 $F = (A + B)(A + \bar{B}\bar{C})(\bar{B}\bar{D}) = (A + A\bar{B}\bar{C} + AB)(\bar{B}\bar{D}) = A\bar{B}\bar{D}$. See Fig. 6.26.

6.13 No simplification possible. See Fig. 6.27. One triple 3-input NAND gate IC and one quad 2-input NAND gate IC.

6.14 (a) Gates 1, 4 and 5 act as an OR gate, and gates 2 and 3 act as AND gates. Therefore, $F = (B + C)\bar{A} + (A + C)\bar{B}$. Inputs at odd levels appear complemented at the output and, hence, $F = (B + C)\bar{A} + (A + C)\bar{B}$.

Fig. 6.27

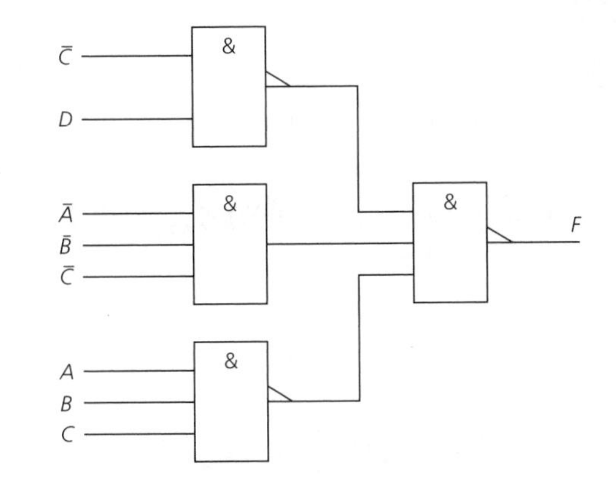

Fig. 6.28

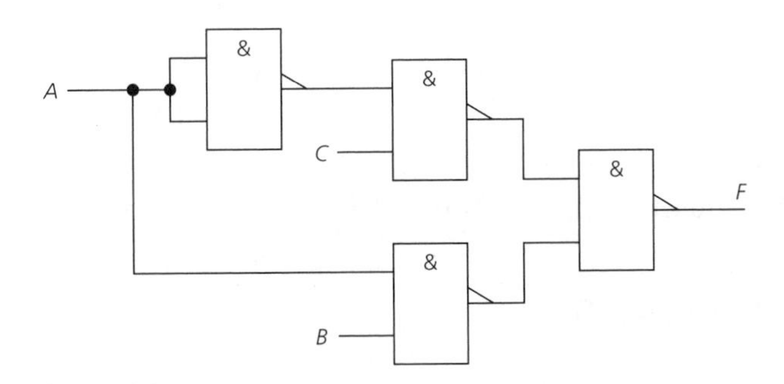

(b) Output of gate 4 = $\overline{BC}$, output of gate 5 = $\overline{AC}$, output of gate 2 = $\overline{A\overline{BC}}$, output of gate 3 = $\overline{B\overline{AC}}$, output of gate 1 = $F = \overline{\overline{(A\overline{BC})}\,\overline{(B\overline{AC})}} = A(\overline{BC}) + B(\overline{AC}) = A(\bar{B} + \bar{C}) + B(\bar{A} + \bar{C})$.

6.15 $F = \bar{A}\bar{B}\bar{C}\bar{D} + A\bar{B}\bar{C}\bar{D} + \bar{A}B\bar{C}D + AB\bar{C}D + \bar{A}BCD + ABCD$. $\bar{F} = \bar{A}B\bar{C}\bar{D} + AB\bar{C}\bar{D} + \bar{A}\bar{B}CD + A\bar{B}CD$. All other combinations are don't cares.

The mapping is

CD \ AB	00	01	11	10
00	1	0	0	1
01	×	1	1	×
11	0	1	1	0
10	×	×	×	×

From the looped 1 and × cells, $F = BD + \bar{B}\bar{D}$. From the looped 0 and × cells, $F' = B\bar{D} + \bar{B}D$, hence, $F = (\bar{B} + D)(B + \bar{D})$.

6.16 (a) $F = (\bar{A}C\overline{AB} + \overline{\bar{A}C}AB)(A + C) = [\bar{A}C(\bar{A} + \bar{B}) + AB(A + \bar{C})](A + C) = (\bar{A}C + \bar{A}\bar{B}C + AB + AB\bar{C})(A + C) = (\bar{A}C + AB)(A + C) = \bar{A}C + AB$. See Fig. 6.28.

Chapter 7

7.1 $4 = 5R_{min}/(1 + R_{min})$, $4 + 4R_{min} = 5R_{min}$. Therefore, $R_{min} = 4\ k\Omega$.

7.3 (a) $V_{CE(SAT)} \approx 0.2$ V. (b) $t_f \approx 150$ ns. (c) $t_r \approx 300$ ns.

7.4 (a) $V_{IL} = 3.5\ V = 5 - (n \times 1.6 \times 10^{-3})$, or $n = 7.2 = 7$. (b) Total load resistance $= 4000/4 = 1000\ \Omega$. $V_{IL} = (5 \times 1000)/(130 + 1000) = 4.425$ V. (c) Worst case noise margin $= 2.4 - 2 = 0.4$ V.

7.5 (a) (i) *A*, 6; *B*, 2; *C*, 2; D, 3; *E*, 3. (ii) *A*, 3; *B*, 2; *C*, 1; *D*, 4; *E*, 1.

7.6 (a) Connected to a bus. (b) Has three output states; namely, HIGH, LOW, and high impedance.

7.7 (a) $I_{C(SAT)} = (5 - 0.2)/2470 = 1.943$ mA. (b) $(2 - V_{BE(SAT)})/(47 \times 10^3) = (1.943 \times 10^{-3})/200$. $V_{BE(SAT)} = 1.54$ V.

7.8 3 V; $4.5 = 3.7(1 - e^{t/(8 \times 10^{-9})})$; $t = 4.16$ ns.

7.9 $I_{C(SAT)} = (5 - 0.2)/2200 = 2.18$ mA; $V_{CB(SAT)} = 0.75 - 0.2 = 0.55$ V.

7.11 (a) 400 mV. (b) $5/3 = 1.67$ V. (c) $12/3 = 4$ V, destroyed!

7.12 (a) $F = \bar{A}\bar{B}\bar{C}\bar{D}\bar{E}\bar{F} = \overline{A + B + C + D + E + F}$, i.e. wired-AND of inverters gives the NOR function. (b) $F = \bar{A}\bar{B}(\bar{C} + \bar{D})$.

Chapter 8

8.1 Comparing with equation (8.2), $F = \bar{A}\bar{B}\bar{C} + A\bar{B}\bar{C} + AB\bar{C} + \bar{A}\bar{B}C + A\bar{B}C$.

8.2 Connect inputs D_0 and D_3 to logic 1 and inputs D_1 and D_2 to logic 0.

8.3 $F = \bar{A}\bar{B}\bar{C}\bar{D} + \bar{A}B\bar{C}\bar{D} + \bar{A}\bar{B}C\bar{D} + A\bar{B}C\bar{D} + \bar{A}BC\bar{D} + \bar{A}\bar{B}\bar{C}D + \bar{A}B\bar{C}D + \bar{A}B\bar{C}\bar{D}$.

8.4 See Fig. 8.32.

Fig. 8.32

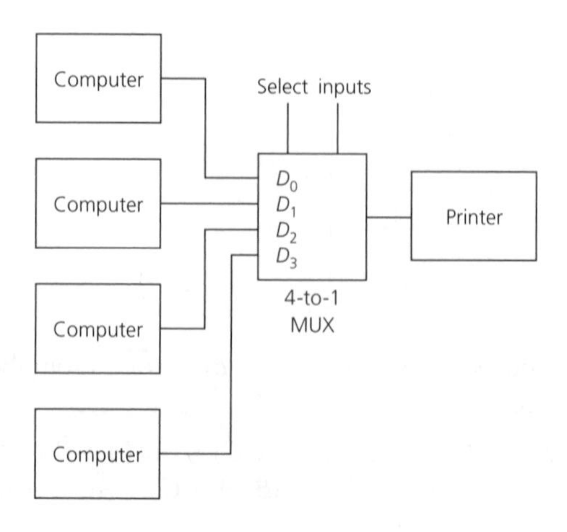

8.5 There are three input variables and three select inputs. Each input combination, e.g. $\bar{A}\bar{B}\bar{C}$, is represented by 1 in the truth table and, hence, must be connected to logic 1. All other terms are connected to logic 0. Thus, $D_0 = D_2 = D_3 = D_5 = 1$. $D_1 = D_4 = D_6 = D_7 = 0$.

8.6 There are only two select inputs, so one of the input variables must be connected to a data input. The truth table is

A	0	1	0	1	0	1	0	1
B	0	0	1	1	0	0	1	1
C	0	0	0	0	1	1	1	1
F	1	0	1	1	0	1	0	0
	D_0	D_1	D_2	D_3				

When $AB = 00$, $F = 1$ for $C = 0$ and 0 for $C = 1$; in both cases $F = \bar{C}$ and, hence, D_0 is connected to $\bar{C}$.
When $AB = 10$, $F = 0$ for $C = 0$ and 1 for $C = 1$; in both cases $F = C$ and so D_1 is connected to C.
When $AB = 01$, $F = 1$ for $C = 0$ and 0 for $C = 1$; D_2 is connected to $\bar{C}$.
When $AB = 11$, $F = 1$ for $C = 0$ and 0 for $C = 1$; D_3 is connected to $\bar{C}$.

8.7 See Fig. 8.33.

8.8 Connect input variables A, B and C to the select inputs of an 8-to-1 multiplexer. Connect data inputs D_0, D_2 and D_4 to logic 0 and all other data inputs to logic 1.

8.9 Using an 8-to-1 multiplexer, connect variables A, B and C to the select inputs. Connect data inputs D_0, D_5 and D_6 to logic 1, data inputs D_1 and D_4 to logic 0, data inputs D_3 and D_7 to variable D, and data input D_2 to the complement of input variable D.

8.10 Connect input variables A, B and C to the select inputs. Connect data inputs D_0, D_3, D_5 and D_7 to logic 1, and data inputs D_1, D_2, D_4 and D_6 to logic 0.

Fig. 8.33

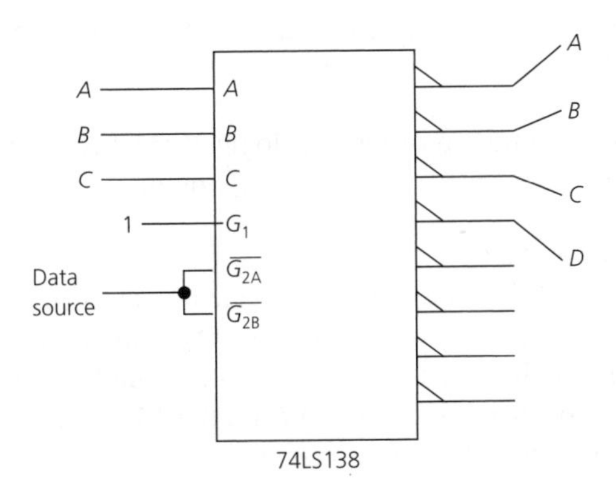

Fig. 8.34

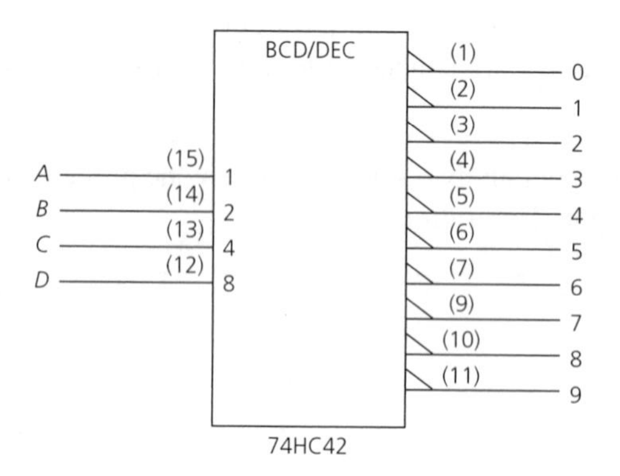

8.11 The truth table is

BCD in				*Decimal out*									
D	*C*	*B*	*A*	0	1	2	3	4	5	6	7	8	9
0	0	0	0	0	1	1	1	1	1	1	1	1	1
0	0	0	1	1	0	1	1	1	1	1	1	1	1
0	0	1	0	1	1	0	1	1	1	1	1	1	1
	etc.							etc.					

See Fig. 8.34.

8.12 See Fig. 8.35.

8.13 (a) $C(\overline{A\bar{B} + \bar{A}B}) + \bar{C}(A\bar{B} + \bar{A}B) = C(\overline{A\bar{B}})(\overline{\bar{A}B}) + A\bar{B}\bar{C} + \bar{A}B\bar{C} = C(\bar{A} + B)(A + \bar{B}) + A\bar{B}\bar{C} + \bar{A}B\bar{C} = C(\bar{A}\bar{B} + AB) + A\bar{B}\bar{C} + \bar{A}B\bar{C} = \bar{A}\bar{B}C + ABC + A\bar{B}\bar{C} + \bar{A}B\bar{C} = C(\bar{A}\bar{B} + AB) + \bar{C}(A\bar{B} + \bar{A}B) = C \oplus A \oplus B$. (b) $S = \bar{A}\bar{B}C + ABC + A\overline{BC} + \bar{A}B\bar{C}$. These are the same terms as in part (a). $C = AB\bar{C} + ABC + A\bar{B}\bar{C} + \bar{A}BC = C(A\bar{B} + \bar{A}B) + AB(C + \bar{C}) = C(A \oplus B) + AB$.

8.14 (a) $\Sigma_1 = 1$, $\Sigma_2 = 1$, $\Sigma_3 = 1$, $\Sigma_4 = 1$, $C_{out} = 0$. (b) $\Sigma_1 = 0$, $\Sigma_2 = 1$, $\Sigma_3 = 0$, $\Sigma_4 = 0$, $C_{out} = 1$. (c) $\Sigma_1 = \Sigma_2 = \Sigma_3 = \Sigma_4 = 0$, $C_{out} = 1$.

8.15 (a) In subtract position, i.e. logic 1. (b) 1. (c) 1011 and 0110, 11 and 6. (d) 0101, 5. (e) There is a carry-out, but it is ignored.

8.16 See Fig. 8.36.
Since 1 out of 32 lines is being selected 5 select lines are needed. Input select lines *A*, *B*, and *C* are applied to all four ICs. The outputs of the ICs are connected to separate inputs on one-half of the 153 dual 4-to-1 multiplexer. Input select lines *D* and *E* are connected to select inputs *A* and *B* on the 153 to determine which of its four inputs are connected to the output *1Y*.

Fig. 8.35

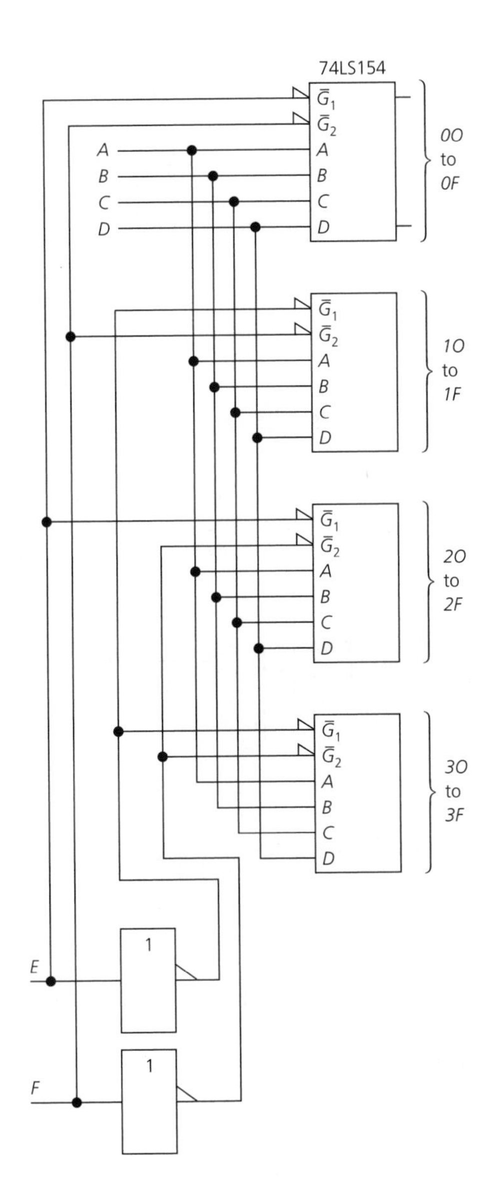

Chapter 9

9.1 See Fig. 9.31.

9.2 See Fig. 9.32.

9.3 See Fig. 9.33.

9.4 See Fig. 9.34.

9.5 (a) (i) $Q = 1$, $Q = 0$. (ii) None. (iii) None. (b) T flip-flop.

Fig. 8.36

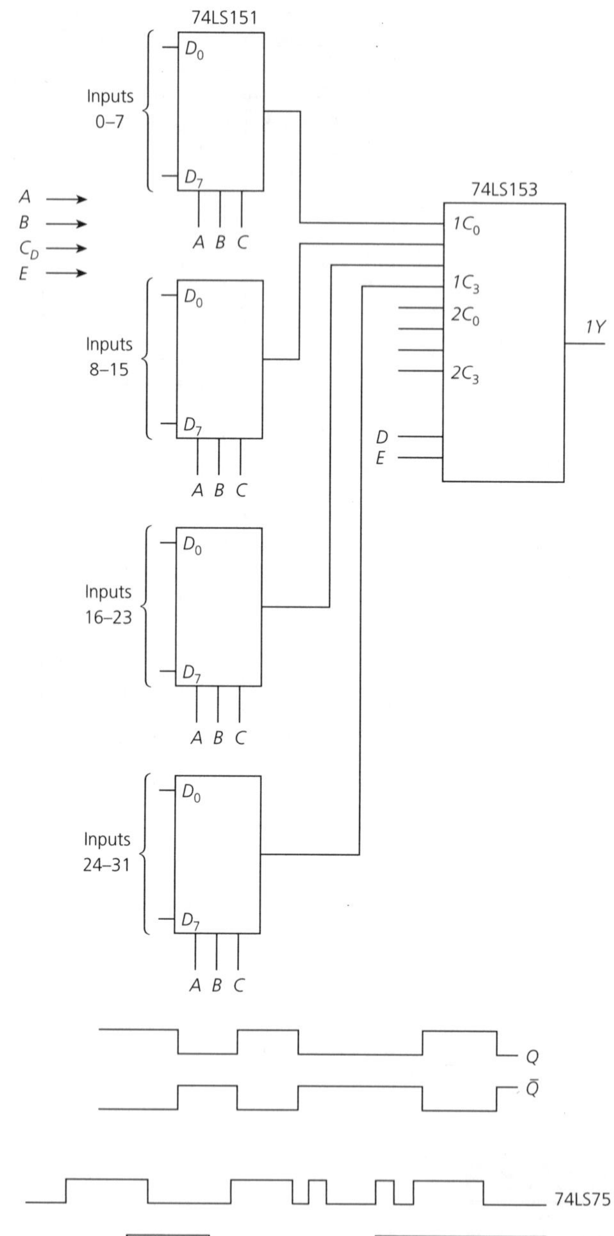

Fig. 9.31

Fig. 9.32

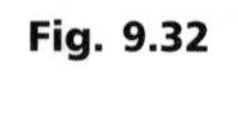

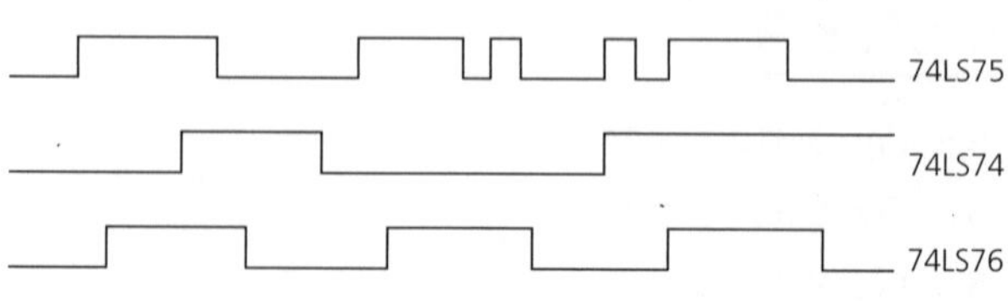

Fig. 9.33

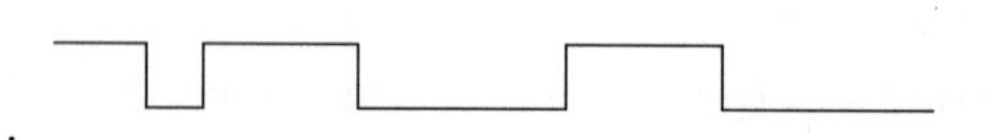

Fig. 9.34

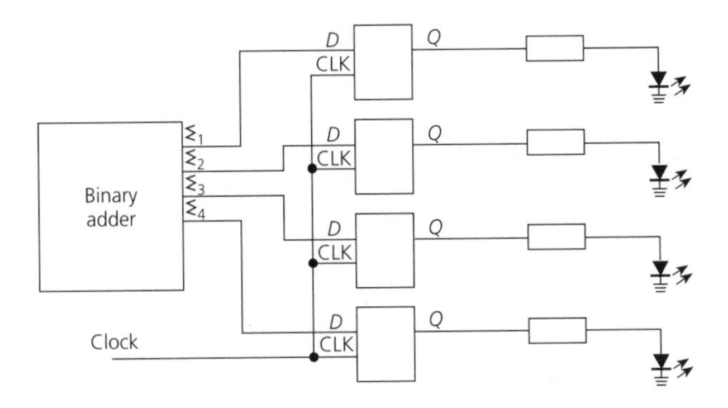

9.6 (a) (i) Set $Q = 0$. (ii) Set $Q = 1$. $Q = 0$. (b) (i) Yes. (ii) +5 V. (iii) Set.

9.8 (a) 1 (provided D was at 1 when the leading edge of the clock pulse occurred). (b) $Q = 0$. (d) (i) 20 ns. (ii) 5 ns.

9.9 Connect the input data to the J terminal and also via an inverter to the K terminal.

9.10 The truth table is

C1k	J_A	K_A	Q_A	Q'_A	J_B	K_B	Q_B	Q'_B	*Count*
0	1	1	0		0	1	0		0
1	1	1	0	1	0	1	0	0	1
2	1	1	1	0	1	1	0	1	2
3	0	1	1	0	0	1	1	0	0

The count is 3.

9.11 The output waveform has a frequency of 1 MHz and a mark–space ratio of 1/4.

Chapter 10

10.1 (a) Counts down from 15 to 0. Hence, the count is 15. (b) Only FFA is driven by the clock so the circuit is non-synchronous. (c) Each stage is required to toggle so $J = K = 1$ for all flip-flops.

10.2 The circuit must count from 000 up to 110. When, at the next clock pulse, FFA sets, the circuit must reset. Hence, connect Q_A, Q_B and Q_C to a NAND gate and apply its output to the reset line.

10.4 Connect the 74LS90 as a divide-by-5 counter with its input at CLKB. Output Q_D must be connected to input CKA of the 74LS93. The '93 has pin 12 connected to pin 1 so that it acts as a divide-by-16 circuit. The output is taken from Q_D. Then, count $= 5 \times 16 = 80$.

10.5 The counter will reset when $Q_A = 0$, $Q_B = Q_C = 1$. Hence, count is from decimal 0 to decimal 5, a count of 6.

10.6 The operation is shown in Table 10.4.

Table 10.4

Clock pulse	J_A	K_A	Q'_A	J'_B	K'_B	Q'_B	J_C	K_C	Q'_C	J_D	K_D	Q'_D	Count
0	1	1	–	0	1	–	0	1	–	0	1	–	0
1	1	1	1	1	1	0	0	1	0	0	1	0	1
2	0	1	0	1	1	1	1	1	0	0	1	0	2
3	0	1	0	0	1	0	1	1	1	0	1	0	4
4	1	1	1	0	1	0	0	1	0	1	1	1	9
5	1	1	0	0	1	0	0	1	0	0	1	0	0

The counter has an irregular count of 5.

10.7 (a) Start the count at any desired point by loading the appropriate data. After 9 steps from the starting point take ENP LOW. Suppose that the start is 0000. Then, on the next clock pulse after the count is 1000 (decimal 8), the counter should reset to 0000. For this, connect Q_A and Q_D to a NAND gate and the output of the gate to the ENP terminal. (b) Maximum count = $2^{14} - 1$ = 16 383. Only 12 Q output pins are available because of package limitations.

10.8 (b) (i) 000, 001, 010, 011, 100, 101, 110, 111. (ii) 111, 110, 101, 100, 011, 010, 001, 000.

10.9

Table 10.5

Clock pulse		D_A	Q_A	Q'_A	D_B	Q_B	Q'_B	Reset	Count	Decoded output
0	1	0	–	0	0	–	0	0	1	
1	1	0	1	0	0	0	0	1	0	
2	0	1	0	1	0	1	1	2	0	3

10.10 7_{10} to 0_{10}.

10.11 (b) The first counter IC is connected as a decade counter and the second IC is connected as a divide-by-5 circuit, therefore, overall count = 10 × 5 = 50.

10.12 (b) (i) 8. (ii) 7. (c) (i) 001. (ii) 111.

10.13 (b) Input frequency = 1000 Hz, so Q_A = 500 Hz, Q_B = 250 Hz, Q_C = 125 Hz, and Q_D = 62.5 Hz.

10.14 (a) $\bar{Q}_D\bar{Q}_C\bar{Q}_B Q_A$. (b) $Q_D\bar{Q}_C\bar{Q}_B Q_A$. (c) $Q_D Q_C\bar{Q}_B Q_A$.

10.15 Connect $\bar{Q}_A$, Q_B, $\bar{Q}_C$ and Q_D to a 4-input NAND gate. Connect the gate output to the reset terminal.

10.16 Connect Q_A to CLKB to get a count of 10. Then reset the circuit when the count is 0110 by connecting $\bar{Q}_D Q_C Q_B \bar{Q}_A$ to an AND gate and the gate output to the linked R_{01} and R_{02} terminals. Connect R_{91} and R_{92} to earth.

10.17 (a) Q_A connected to J_B, $\bar{Q}_B$ connected to J_A, $K_A = K_B = 1$. (b) Connect Q_A to CLB, Q_B to both CLC and CLD, $\bar{Q}_D$ to J_C. All K terminals to 1. (c) Two divide-by-3 counters connected in cascade. (d) One flip-flop connected as a toggle connected to a divide-by-3 circuit.

10.18 (a) $4096 = 2^{12}$, so there are 12 stages. (b) Clock frequency $= 4096 \times 66 = 270.336$ kHz.

10.19 The count starts from 1111 = decimal 15. Each clock pulse toggles FFA, FFB toggles every other clock pulse, FFC toggles every fourth clock pulse, and FFD toggles every eighth clock pulse. Hence, the counter counts down from 15 to 0.

10.20 See Fig. 10.39.

10.21 See Fig. 10.40.

Fig.10.39

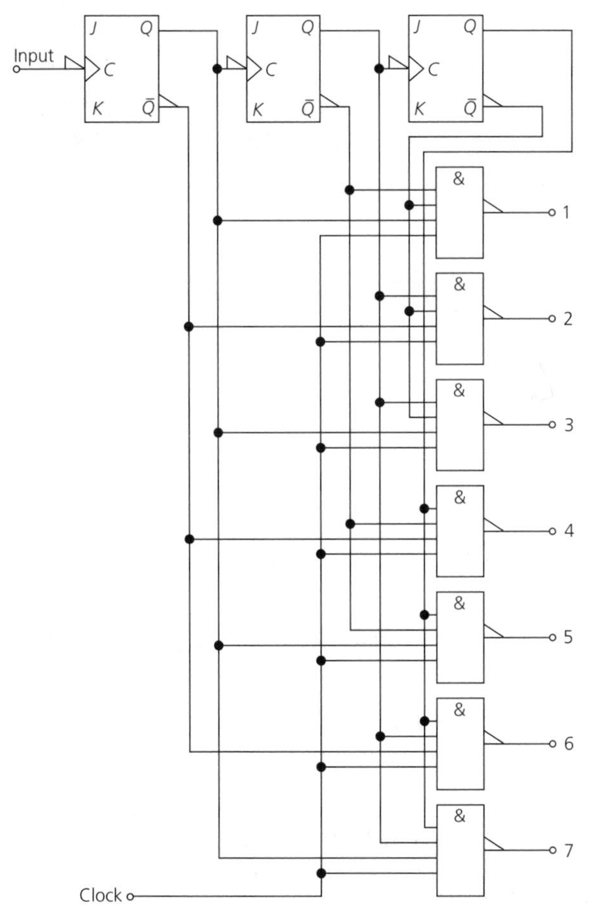

Fig. 10.40

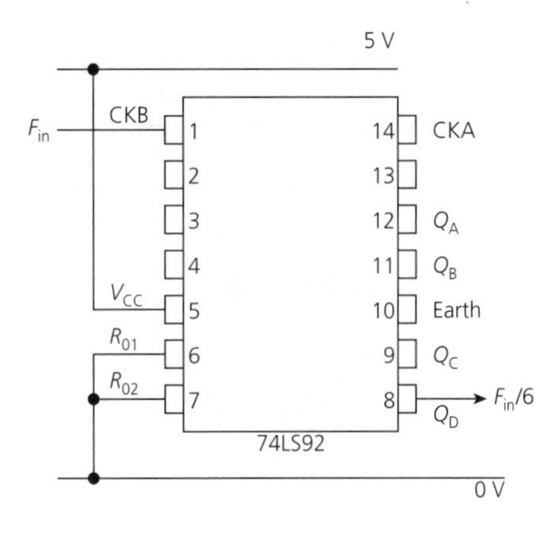

Fig. 10.41

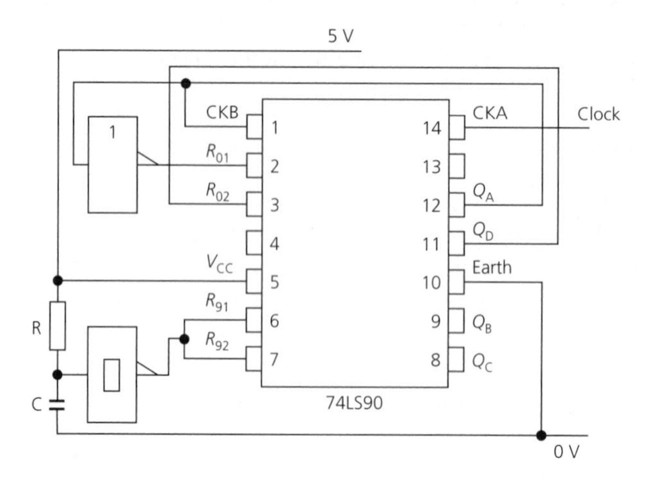

10.22 See Fig. 10.41.

Chapter 11

11.1 (a) (i) Clock pulse periodic time = $1/(2 \times 10^6) = 0.5$ μs. Time to clear = $8 \times 0.5 =$ 4 μs. (ii) All stages load simultaneously in 0.5 μs. (b) (i) 0010. (ii) 0001.

11.2 (a) See Table 11.3.

Table 11.3

Clock pulse	0	1	2	3	4	5	6	7	8
Data on D_A	1	0	1	1	0	0	0	0	0
Q_A	0	1	0	1	1	0	0	0	0
Q_B	0	0	1	0	1	1	0	0	0
Q_C	0	0	0	1	0	1	1	0	0
Q_D	0	0	0	0	1	0	1	1	0

(b) Input data = 0110 or decimal 6. (i) New data = 1100 or decimal 12. (ii) New data = 0011 or decimal 3. Therefore, (i) right shifting = multiplication by 2. (ii) Left-shifting = division by 2.

11.3 Use a PISO shift register. Reset all stages to 0 and then load the parallel number. Right shift the data out of the register to obtain serial output.

11.4 SRG8: shift register 8-bit; CLK: clock input. $\overline{\text{CLR}}$: active-LOW clear and: both *A* and *B* must be HIGH for data input. ID: controlled by CI, i.e. by the clock: right shifting.

11.5 Use, first, right-shifting and then left-shifting.

Chapter 12

12.1 (a) Number of locations = 2^{11} = 2048. There are eight bits per location, so bits stored = 8 × 2048 = 16384. (b) Organization = 2048 × 8 = 2k × 8.

12.2 Four RAMs are required. Parallel the 10 address lines. Use a 2-to-4 line decoder whose active-LOW outputs are connected to the $\overline{CS}$ inputs of the four RAMs. The inputs to the decoder will be A_{10} and A_{11}.

12.3 (a) Locations = 2^{12} = 4096. (b) 8k memory locations require 13 address pins, since 2^{13} = 8192. There are eight bits per location, so eight data pins, plus one R/$\overline{W}$ pin, one $\overline{CS}$ pin and earth and V_{CC} pins. Total pins = 25. (c) The memory matrix is 8 × 8, so the row and column addresses are 000 to 111. Hence, address 110100 is for row 4 and column 6.

12.4 (a) 4096 = 2^n. Hence, n = 12. Organization = 4k × 4. Also required are R/$\overline{W}$, $\overline{CS}$, V_{CC} and earth pins. (b) Select chip and select read or write. Because it cannot be written to. To reduce the number of package pins needed.

12.5 The ROM is shown in Fig. 12.26.

12.6 $F = \bar{A}\bar{B}\overline{C} + AB\overline{C} + \bar{A}BC + ABC.$

12.7

Fig. 12.26

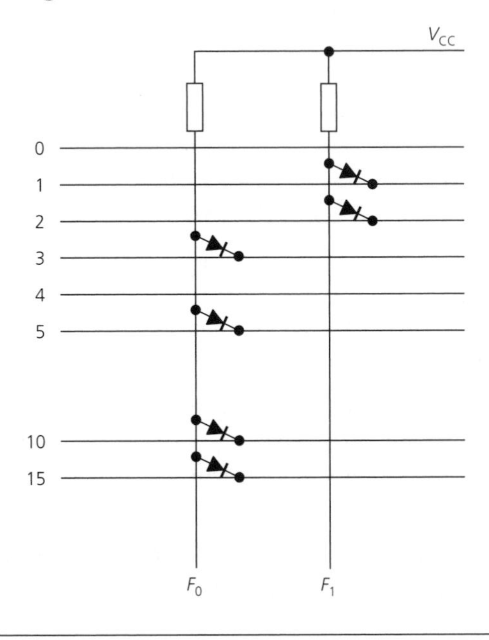

Fig. 12.27

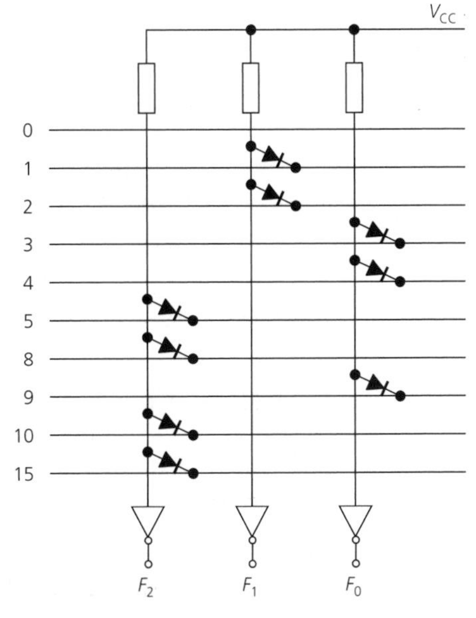

Table 12.1

CS	*OE*	*WE*	*Mode*	*Input/output*
H	×	×	not selected	high impedance
L	H	H	output disabled	high impedance
L	L	H	read	data out
L	×	L	write	data in

12.9 (a) Connect the address inputs A_0 to A_9 to the A_0 to A_9 pins of each IC, i.e. connect them in parallel. Connect the $\overline{CS}$, $\overline{RAS}$ and $\overline{CAS}$ pins together. The data outputs of the four ICs now each provide one bit of the 4-bit output word. (b) Parallel the input address lines A_0 to A_9. Address inputs A_{10} and A_{11} are applied to the inputs of a 2-to-4 line decoder with active-LOW outputs which are connected to the four $\overline{CS}$ pins. Parallel the eight data output pins of the four ICs. Connect the $\overline{RAS}$ and $\overline{CAS}$ pins together.

12.10 (a) Maximum stored words = 2^{15} = 32 768. (b) 1k = 1024, 128k = 128 × 1024 = 131 072; 131 072 = 2^n, hence, $n = 17$.

12.11 (a) Organization = 4 × 4.

(b)

Table 12.2

Row address	D_3	D_2	D_1	D_0
0	1	1	0	0
1	1	1	0	1
2	0	1	0	0
3	1	0	1	1

(c) $D_0 = A\bar{B} + AB$, $D_1 = AB$, $D_2 = \bar{A}\bar{B} + A\bar{B} + \bar{A}B$, $D_3 = \bar{A}\bar{B} + A\bar{B} + AB$.

12.12 (a) 512/32 = 16, (512/32) × 4 = 64. (b) 1M/32 = 32.8 × 32 = 256.

12.13 (a) (i) 1024 × 1024 = 1 048 576. (ii) 1 048 576 × 4 = 4 194 304. (b) (i) 64 × 1024 = 65 536. (ii) 65 535 × 16 = 1 048 576.

12.14 (a) (i) 32k requires 15 address pins. Therefore, lowest address = 0000H and highest address = 7FFFH. (ii) 1M requires 20 address pins. Therefore, lowest address = 0000H and highest address = FFFFFH. (b) Size of each address = 8000H and 1000000H, respectively.

12.15 Four ICs must be selected to give a 16-bit output word. Connect the address lines CS and WE in parallel with the corresponding pins on the other ICs.

12.16 (a) 16M × 1. (b) Three-state. (c) 4096 × 4096. (d) 1.

12.17 The required ROM is shown in Fig. 12.25.

Chapter 13

13.1 (a) (i) 16 inputs. (ii) 8 outputs. (b) Use De Morgan's rules to convert into $\bar{F} = \overline{ABCDEFGH}$ and then invert the output.

13.2 See Fig. 13.20.

13.3 $F_0 = ABC + \bar{A}\bar{B}\bar{C}$, $F_1 = \bar{A}B\bar{C} + \bar{A}\bar{B}C + A\bar{B}\bar{C}$.

13.4 (a) 2. (b) 6 and 4. (c) *I* and *I*. (d) 18. (e) 18L4.

13.5 See Fig. 13.21.

13.6 See Fig. 13.22.

13.7 (a) Reduced chip count, simplified PCB board, increased functional density, higher reliability, lower system cost. (b) lower power dissipation, bipolar fuse-link PLDs cannot be tested, bipolar can be programmed only once.

13.8 (a) See Fig. 13.23(a). (b) See Fig. 13.23(b). (c) $F_0 = A\bar{A}B\bar{B} = 0$. $F_1 = 1$ (with no input terms connected, the product term will float to 1).

Fig. 13.20

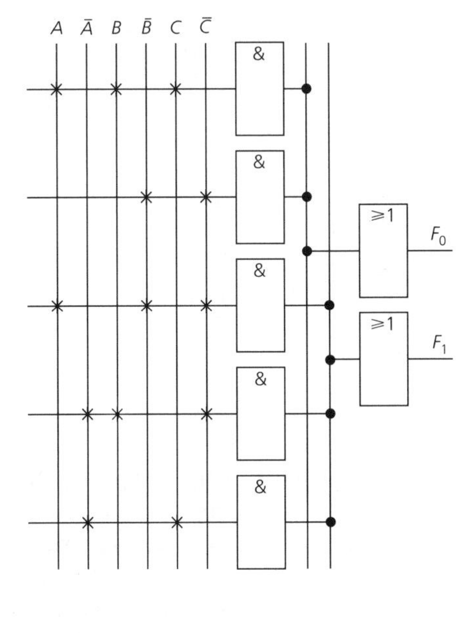

Fig. 13.21

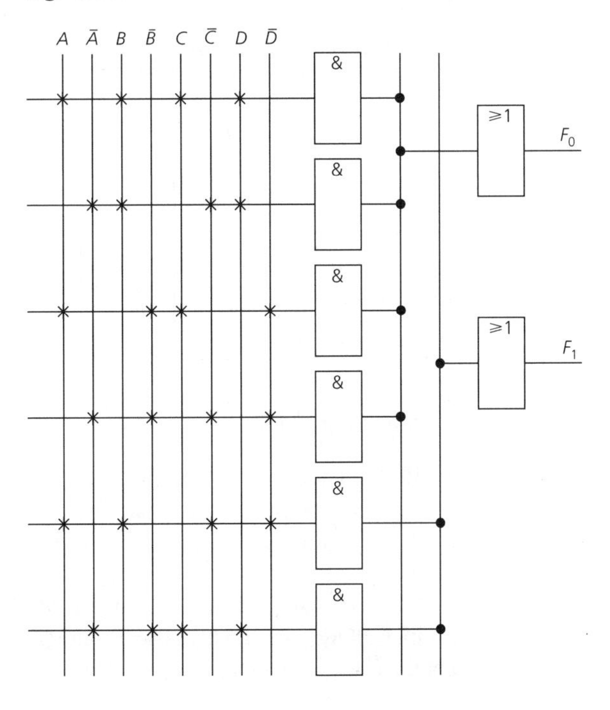

Fig. 13.22

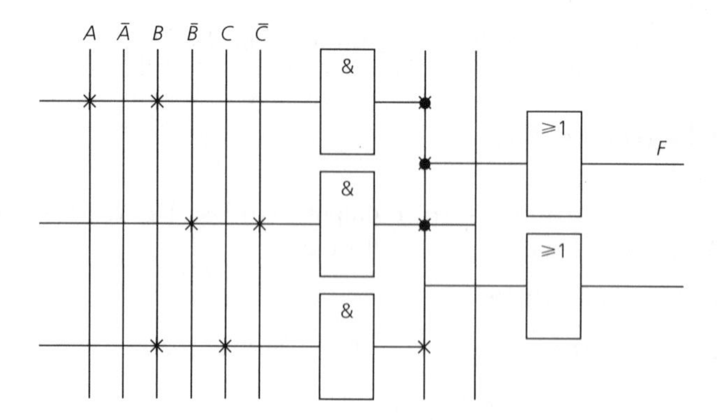

Fig. 13.23

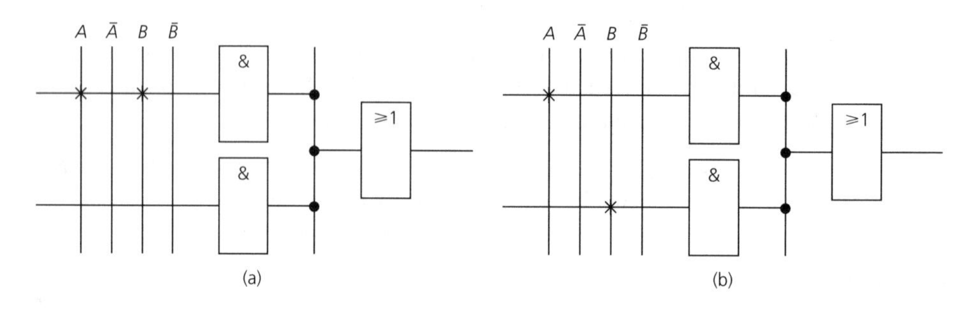

Fig. 13.24

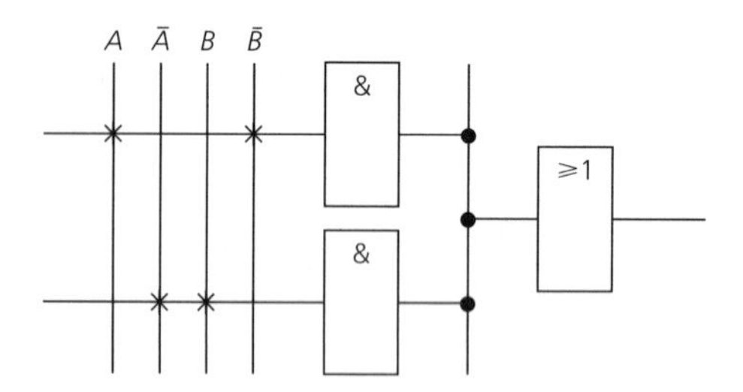

13.9 See Fig. 13.24.

13.10 Implementation of $\overline{F}$, $\overline{G}$, and $\overline{H}$ requires the use of three of the 2-input NOR gates. Implementation of $\overline{I}$ requires the use of one of the two 4-input NOR gates. Implementation of $\overline{J}$ requires the use of a 5-input NOR gate which does not exist. Hence, group I into two parts, $(AD + BC + \overline{A}B)$ and $(\overline{A}C\overline{D} + A\overline{B}\overline{C})$. Implement $P = AD + BC + \overline{A}B$ using the other 4-input NOR gate, and $Q = \overline{A}C\overline{D} + A\overline{B}\overline{C}$ using another 2-input NOR gate. Lastly, use P and Q as the inputs to another 2-input NOR gate to obtain $\overline{J}$.

Chapter 14

14.1 The truth table is as shown in Table 14.1 except for rows 6 and 9. The expressions for these are: $a = \bar{A}\bar{B}\bar{C}\bar{D} + \bar{A}B\bar{C}\bar{D} + AB\bar{C}\bar{D} + A\bar{B}C\bar{D} + ABC\bar{D} + \bar{A}\bar{B}\bar{C}D + A\bar{B}\bar{C}D$, and $\bar{A}BC\bar{D}$; d $= \bar{A}\bar{B}\bar{C}\bar{D} + \bar{A}B\bar{C}\bar{D} + AB\bar{C}\bar{D} + A\bar{B}C\bar{D} + \bar{A}BC\bar{D} + \bar{A}\bar{B}\bar{C}D + \bar{A}\bar{B}\bar{C}\bar{D}$. Mapping gives

CD \ AB	00	01	11	10
00	1	1	1	0
01	1	×	×	1
11	×	×	×	×
10	0	1	1	1

CD \ AB	00	01	11	10
00	1	1	1	0
01	1	×	×	1
11	×	×	×	×
10	0	1	0	1

From the maps, $a = D + \bar{A}\bar{C} + B + AC$ and $d = D + \bar{A}B + \bar{A}\bar{C} + A\bar{B}C$.

14.2 The required connections are shown in Fig. 14.22.

14.4 See Table 14.2 on p. 320. 5 = *abcdefg*, 9 = *abcdefg*.

14.5 (a) No. (b) No. (c) Yes.

14.7 R = (5 – 1.6 – 0.2)/(20 × 10^{-3}) = 160 Ω. Power dissipated = (20 × $10^{-3})^2$ × 160 = 64 mW.

14.8 (a) *c, d, j, k*. (b) *a, b, c, d, g, h, k, o*. (c) *a, b, c, g, h, k, l, o*. (d) *a, b, i, m*.

Fig. 14.22

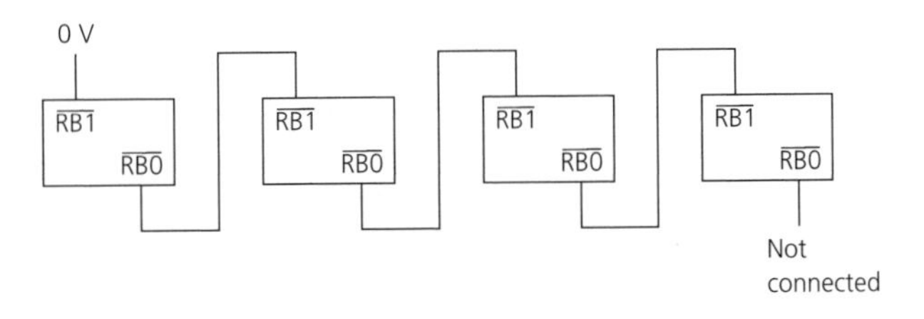

Chapter 15

15.1 (a) 12/2 = 6 ≅ 6.2 kΩ; 6/2 = 3 kΩ; 3/2 = 1.5 kΩ. (b) Largest output voltage is when all switches are at 1. Then, I = (5/12 + 5/6.2 + 5/3 + 5/1.5) = 6.223 mA. Output voltage = 6.233 × 2 = 12.446 V. (c) Resolution = 12.446/16 = 0.778 V.

15.2 (a) Steps = 2^4 = 16. Maximum output voltage = (2^4 – 1) × 0.2 = 3.0 V. 2 V = 1010 and 2.2 V = 1011. The output may be either digital word or its value may change between them. (b) 100_2 = 4. Analogue voltage = 4 × 100 = 400 mV.

15.3 (b) 2^5 = 32; 16/32 = 0.5 V. MSB = 0.5 × maximum voltage = 8 V.

15.4 (a) Number of levels = 2^8 = 256. Hence, 255 steps and a resolution of 10. 255 = 39.216 mV. 0 V = level 127, so + 2.5 V = level (255 + 127)/2 = 191 = 101111111. (b) −2.5 V = level (127 + 0)/2 = 64 = 00100000.

15.5 (a) Maximum analogue frequency = 10/2 = 5 kHz. (b) Maximum conversion time = 1/5000 = 200 μs.

15.6 V_0 = 5/2 = 2.5 V.

15.7 (a) 2^{16} = 65 536. So, 1 part in 65 536. (b) 12/65 536 = 183 μV.

15.11 (a) 2^8 = 256; 5/256 = 19.53 mV. (b) 2^{16} = 65 536; 5/65 536 = 76.3 μV.

15.12 Each resistor is double the value of the preceding one. Hence, the values are 1, 2, 4, 8, 16, 32, 64, and 128 kΩ.

15.15 For the LSB input, the input resistance = $8R$. Hence, $I = 5/8R$. Output voltage = $0.05 = (5/8R) \times 1000$; $R = 5000/0.4$ = 12.5 kΩ.

Index

Index of Integrated Circuits